SURYAKANT PATEL

Biosynthesis and the Integration of Cell Metabolism

BOOKS IN THE BIOTOL SERIES

The Molecular Fabric of Cells
Infrastructure and Activities of Cells

Techniques used in Bioproduct Analysis
Analysis of Amino Acids, Proteins and Nucleic Acids
Analysis of Carbohydrates and Lipids

Principles of Cell Energetics
Energy Sources for Cells
Biosynthesis and the Integration of Cell Metabolism

Genome Management in Prokaryotes
Genome Management in Eukaryotes

Crop Physiology
Crop Productivity

Functional Physiology
Cellular Interactions and Immunobiology
Defence Mechanisms

Bioprocess Technology: Modelling and Transport Phenomena
Operational Modes of Bioreactors

In vitro Cultivation of Micro-organisms
In vitro Cultivation of Plant Cells
In vitro Cultivation of Animal Cells

Bioreactor Design and Product Yield
Product Recovery in Bioprocess Technology

Techniques for Engineering Genes
Strategies for Engineering Organisms

Technological Applications of Biocatalysts
Technological Applications of Immunochemicals

Biotechnological Innovations in Health Care

Biotechnological Innovations in Crop Improvement
Biotechnological Innovations in Animal Productivity

Biotechnological Innovations in Energy and Environmental Management

Biotechnological Innovations in Chemical Synthesis

Biotechnological Innovations in Food Processing

Biotechnology Source Book: Safety, Good Practice and Regulatory Affairs

BIOTECHNOLOGY BY OPEN LEARNING

Biosynthesis and the Integration of Cell Metabolism

PUBLISHED ON BEHALF OF :

Open universiteit
Valkenburgerweg 167
6401 DL Heerlen
Nederland

and

Thames Polytechnic
Avery Hill Road
Eltham, London SE9 2HB
United Kingdom

Butterworth-Heinemann Ltd
Halley Court, Jordan Hill, Oxford OX2 8DP

 PART OF REED INTERNATIONAL BOOKS

OXFORD LONDON BOSTON
MUNICH NEW DELHI SINGAPORE SYDNEY
TOKYO TORONTO WELLINGTON

First published 1992

British Library Cataloguing in Publication Data

A catalogue record for this book is available from the British Library

Library of Congress Cataloguing in Publication Data

A catalogue record for this book is available from the Library of Congress

ISBN 0 7506 1506 0

Composition by Thames Polytechnic
Printed and Bound in Great Britain by Thomson Litho Ltd, East Kilbride, Scotland

The Biotol Project

The BIOTOL team

OPEN UNIVERSITEIT, THE NETHERLANDS
Dr M. C. E. van Dam-Mieras
Professor W. H. de Jeu
Professor J. de Vries

THAMES POLYTECHNIC, UK
Professor B. R. Currell
Dr J. W. James
Dr C. K. Leach
Mr R. A. Patmore

This series of books has been developed through a collaboration between the Open universiteit of the Netherlands and Thames Polytechnic to provide a whole library of advanced level flexible learning materials including books, computer and video programmes. The series will be of particular value to those working in the chemical, pharmaceutical, health care, food and drinks, agriculture, and environmental, manufacturing and service industries. These industries will be increasingly faced with training problems as the use of biologically based techniques replaces or enhances chemical ones or indeed allows the development of products previously impossible.

The BIOTOL books may be studied privately, but specifically they provide a cost-effective major resource for in-house company training and are the basis for a wider range of courses (open, distance or traditional) from universities which, with practical and tutorial support, lead to recognised qualifications. There is a developing network of institutions throughout Europe to offer tutorial and practical support and courses based on BIOTOL both for those newly entering the field of biotechnology and for graduates looking for more advanced training. BIOTOL is for any one wishing to know about and use the principles and techniques of modern biotechnology whether they are technicians needing further education, new graduates wishing to extend their knowledge, mature staff faced with changing work or a new career, managers unfamiliar with the new technology or those returning to work after a career break.

Our learning texts, written in an informal and friendly style, embody the best characteristics of both open and distance learning to provide a flexible resource for individuals, training organisations, polytechnics and universities, and professional bodies. The content of each book has been carefully worked out between teachers and industry to lead students through a programme of work so that they may achieve clearly stated learning objectives. There are activities and exercises throughout the books, and self assessment questions that allow students to check their own progress and receive any necessary remedial help.

The books, within the series, are modular allowing students to select their own entry point depending on their knowledge and previous experience. These texts therefore remove the necessity for students to attend institution based lectures at specific times and places, bringing a new freedom to study their chosen subject at the time they need and a pace and place to suit them. This same freedom is highly beneficial to industry since staff can receive training without spending significant periods away from the workplace attending lectures and courses, and without altering work patterns.

Contributors

AUTHORS

Dr T.G. Cartledge, Nottingham Polytechnic, Nottingham, UK

Dr R.O. Jenkins, Leicester Polytechnic, Leicester, UK

Dr C.K. Leach, Leicester Polytechnic, Leicester, UK

EDITOR

Dr G.D. Weston, Leicester Polytechnic, Leicester, UK

SCIENTIFIC AND COURSE ADVISORS

Dr M. C. E. van Dam-Mieras, Open universiteit, Heerlen, The Netherlands

Dr C. K. Leach, Leicester Polytechnic, Leicester, UK

ACKNOWLEDGEMENTS

Grateful thanks are extended to all those who have contributed to the development and production of this book. In addition to the authors, editor and course advisors, special thanks to Miss K. Brown, Mrs N.A. Cartledge, Miss H. Leather, Mr K. Sharpe, Miss J. Skelton, and Professor R. Spier. The development of the BIOTOL texts has been funded by COMETT, The European Community Action programme for Education and Training for Technolgy, by the Open universiteit of The Netherlands and by Thames Polytechnic. Thanks are also due to the authors and editors of Open universiteit materials upon which some of the text was based.

Project Manager Dr J.W. James

Contents

	How to use an open learning text	viii
	Preface	ix
1	**Introduction,** C.K. Leach	1
2	**Uptake of nutrients,** Dr R.O. Jenkins	11
3	**Nitrogen and sulphur assimilation,** Dr R.O. Jenkins	41
4	**Amino acid and nucleotide biosynthesis,** Dr T.G. Cartledge	65
5	**The biosynthesis of lipids,** Dr T.G. Cartledge	109
6	**The biosynthesis of carbohydrates,** Dr T.G. Cartledge	143
7	**The integration and regulation of metabolism,** Dr R.O.Jenkins	171
8	**Control of metabolic pathway flux,** Dr R.O.Jenkins	191
	Responses to SAQs	225
	Appendices 1,2,3.	249

How to use an open learning text

An open learning text presents to you a very carefully thought out programme of study to achieve stated learning objectives, just as a lecturer does. Rather than just listening to a lecture once, and trying to make notes at the same time, you can with a BIOTOL text study it at your own pace, go back over bits you are unsure about and study wherever you choose. Of great importance are the self assessment questions (SAQs) which challenge your understanding and progress and the responses which provide some help if you have had difficulty. These SAQs are carefully thought out to check that you are indeed achieving the set objectives and therefore are a very important part of your study. Every so often in the text you will find the symbol Π, our open door to learning, which indicates an activity for you to do. You will probably find that this participation is a great help to learning so it is important not to skip it.

Whilst you can, as a open learner, study where and when you want, do try to find a place where you can work without disturbance. Most students aim to study a certain number of hours each day or each weekend. If you decide to study for several hours at once, take short breaks of five to ten minutes regularly as it helps to maintain a higher level of overall concentration.

Before you begin a detailed reading of the text, familiarise yourself with the general layout of the material. Have a look at the contents of the various chapters and flip through the pages to get a general impression of the way the subject is dealt with. Forget the old taboo of not writing in books. There is room for your comments, notes and answers; use it and make the book your own personal study record for future revision and reference.

At intervals you will find a summary and list of objectives. The summary will emphasise the important points covered by the material that you have read and the objectives will give you a check list of the things you should then be able to achieve. There are notes in the left hand margin, to help orientate you and emphasise new and important messages.

BIOTOL will be used by universities, polytechnics and colleges as well as industrial training organisations and professional bodies. The texts will form a basis for flexible courses of all types leading to certificates, diplomas and degrees often through credit accumulation and transfer arrangements. In future there will be additional resources available including videos and computer based training programmes.

Preface

A knowledge of the structure and activities of cells, the basic functional units of biological systems, is an essential foundation to all branches of biotechnology and applied biology. Cells are the framework within which a wide variety of chemical and physical processes takes place.

We can conveniently divide these processes into two groups: those involving relatively small, low molecular weight components and those involving the production and use of macromolecules. The former of these we may call intermediary metabolism, it eventually involves the processes concerned with the conversion of nutrients into the precursors of the cells' macromolecules. The study of the latter group which predominantly incorporates the production of DNA, RNA and protein is frequently described as molecular biology or molecular genetics. The division is rather artificial, however, and in fact the former and latter groups merge in the continuum of cellular activity.

The BIOTOL series of texts offers opportunities to gain a comprehensive knowledge of cells and to orientate this understanding into areas of special interest. Two BIOTOL texts ('The Molecular Fabric of Cells,' and 'The Infrastructure and Activities of Cells,') use the theme of cells as biological 'factories' to explore the composition, structure and general properties of cells. Within these 'factories' many activities take place. Nutrients are imported into cells and are modified to provide useable energy and building blocks (precursors) for cell synthesis or for the elaboration of secondary products. We can also identify processes of information storage, and retrieval.

The present text is the final one of a series of three BIOTOL texts devoted to developing a firm knowledge of intermediary metabolism. The title of the first, 'Principles of Cell Energetics', reflects the importance that must be placed on energetic considerations concerning cell metabolism. This approach is extended in the second text, 'Energy Sources for Cells,' to include autotrophic modes of energy generation and some of the more specialist organic energy sources used by several heterotrophs. These first two texts provide an in-depth understanding of the nature and diversity of the processes (catabolism and autotrophy) by which cells may generate both a usable source of energy and the materials needed to make the precursors required for the synthesis of cell constituents. In the present text, we show how the products of catabolism, autotrophic energy generation and carbon fixation are used to make the essential building blocks for the manufacture of cellular components.

The text has been written on the assumption that the reader is familiar with the major molecular species found in biological systems and with the central catabolic pathways which operate under aerobic and anaerobic conditions. It is also assumed that the reader is familiar with the principle features of autotrophic physiology and metabolism. Although these assumptions have been made, extensive use is made of molecular and other 'reminders' within the text.

The text begins with a brief orientating chapter which leads into a discussion of the sources and mechanisms of uptake of nutrients and the assimilation of sulphur and nitrogen. This, together with the knowledge assumed above, is used to discuss the biosynthesis of the major groups of biochemicals (amino acids, nucleotides, lipids and carbohydrates). No attempt is made to cover *all* of the biochemicals within these groups nor to include all *groups* of biochemicals. Such an approach would be a daunting one. Instead, emphasis is placed on developing a knowledge of the pattern of biosynthesis and on exploring how biosynthesis is interconnected both physically and chemically with catabolism. Strategies for investigating the biochemical problems posed by biosynthesis are developed within the text.

The integration and regulation of catabolic and biosynthetic processes are essential to cells if they are to maintain a balanced metabolism in which energy generation and precursor production are harmonised. This aspect of metabolism is a recurrent theme throughout but it especially is the focus of the final two chapters. The text is, therefore, much more than a collection of metabolic pathways. Although important, the pathways themselves are merely the foundations upon which an understanding of intermediary metabolism can be developed.

Intermediary metabolism underpins so many aspects of applied biology and biotechnology that its importance cannot be overstressed. We urge readers to make full use of the in-text activities (ΠI) and self-assessed questions (SAQs) to maximise the benefit to be gained from the text.

One final point we wish to make is that, as you will learn from the text, biosynthesis shows remarkable similarities between species including animals, plants and micro-organisms. Nevertheless many differences do exist and the authors have taken opportunities to illustrate these differences especially in the biotechnologically-important groups - micro-organisms and plants. This text has, therefore, a different flavour from many biochemistry texts which carry a human/medical orientation. This, together with its reader-orientated activities makes this and its partner BIOTOL texts, a learning package of distinction.

Scientific and Course Advisors: Dr MCE Van Dam-Mieras
Dr C K Leach

Introduction

1.1 Three Biotol texts on cells' metabolism 2

1.2 The fuelling reactions are diverse 2

1.3 Biosynthetic processes are similar in all systems 5

1.4 Precursor molecules for biosynthesis 6

1.5 The arrangements of chapters 8

Introduction

1.1 Three Biotol texts on cells' metabolism

major areas of metabolism

Metabolism includes a wide range of processes which can be divided into three major groups. These are:

- the generation of a usable form of energy and reducing power;
- the generation of simple molecules to act as precursors of cell constituents;
- the synthesis of new cell constituents (biosynthesis).

The first two of these groups of processes may be regarded as 'fuelling' reactions since they are the processes which provide the fuel for new cell synthesis. The final group is concerned with the use of the products of the first two to make new cell constituents.

To cover the necessary breadth of chemical and biological knowledge to understand metabolism, three Biotol texts have been produced. The first two texts, ('Principles of Cell Energetics' and 'Energy Sources for Cells') deal with the underlying thermodynamics and enzymology and examine the processes which lead to the production of cellular energy (ATP), reducing power (NADH/NADPH) and simple organic molecules. In other words, these two texts deal with the cellular fuelling reactions. In this, the third text, the biosynthetic processes of cells are examined. It has been written on the assumption that the reader has knowledge of the fuelling reactions including the mechanisms for generating ATP, reducing power and simple organic metabolites by heterotrophic and autotrophic systems.

The purpose of this introductory section is to explain, in outline, the layout of this text and its relationship to the other two Biotol texts committed to cell metabolism. It will also enable readers to check whether they have the requisite knowledge to benefit from this text.

1.2 The fuelling reactions are diverse

1.2.1 The generation of cellular energy and reducing power

Central to our understanding of metabolism is the knowledge that the three main products of the fuelling reactions (energy, reducing power, simple organic molecules), are the starting materials for biosynthetic reactions. We can represent this relationship by a simple diagram:

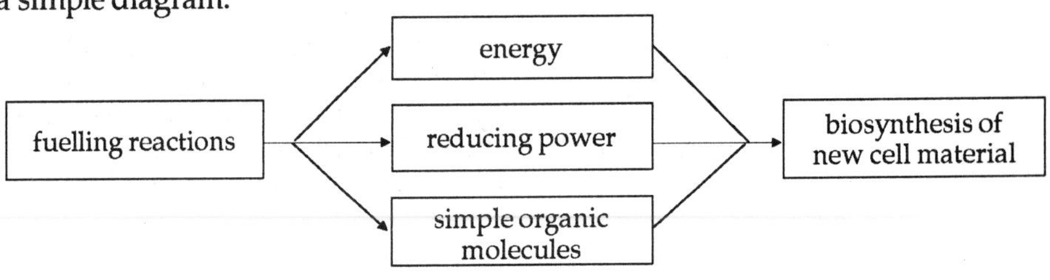

diverse fuelling reactions

The symmetry of this diagram is, however, misleading. Although the biosynthetic processes are similar but not identical in all systems, there is a much greater diversity in the fuelling reactions. Let us see if we can work out the reasons for this.

Π Make a list of the sources of energy which may be used by cells.

In your list you probably had a wide range of organic substrates such as carbohydrates, proteins and fats. You may also have included some aromatic and aliphatic hydrocarbons. Careful thought may have enabled you to include light as a source of energy. With such a diversity in energy sources, it is not surprising cellular energy-generating mechanisms are also diverse.

Π Write down a list of agents or systems that can produce reducing power in the form of NADH or NADPH.

heterotrophic systems

This one is not quite so straightforward. Heterotrophic systems (those which use organic nutrients) use the oxidation of organic substrates to produce reducing power mainly, but not exclusively, in the form of NADH or NADPH. Typically we can draw such reactions as:

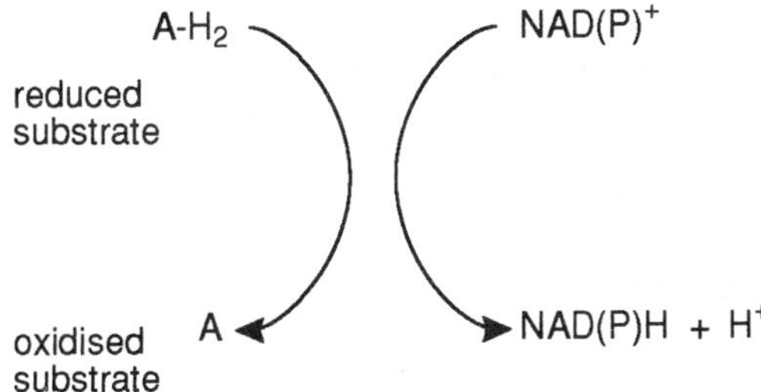

where A represents an organic molecule.

photo-autotrophs & chemo-autotrophs

In photoautotrophs reducing power is generated by the energy of light in activating electrons to a lower redox potential. In chemoautotrophs, reduced inorganic substrates such as NH_4^+ and H_2S are used to produce reducing power.

Π Write down as many processes as you can by which ATP can be produced from ADP + Pi.

Three types of phosphorylation

We anticipate that you would include in your list oxidative phosphorylation, photophosphorylation and substrate level phosphorylation. The mechanism(s) used by cells for the production of ATP is, of course, related to the energy source that is being used. ATP synthesis and the generation of reducing power are generally linked. Thus in photoautotrophs, reducing power is generated by the energy of light, harvested by chlorophyll, activating electrons to a lower redox potential (ie they become more reducing). ATP is generated by passing some of these strongly reducing (activated) electrons down an electron transport chain. This results in protons (H^+) being pumped across membranes the subsequent pH gradient established across the membrane being used to drive ATP synthesis in a process known as photophosphorylation.

pumping of protons

reduced inorganic substrates

In chemoautotrophs, reduced inorganic substrates (such as H_2S), are used to produce reducing power and to provide electrons for passage down an electron transport chain accompanied by the production of ATP.

reduced organic substrates

Heterotrophs use reduced organic substrates to produce reducing power mainly in the form of NAD(P)H, whilst ATP is generated by substrate level phosphorylation (anaerobes) or by substrate level phosphorylation and oxidative phosphorylation (aerobes). Substrate level phosphorylation is, of course, phosphorylation which takes place when a substrate is converted to a product - for example when phosphoenol pyruvate is converted to pyruvate by the enzyme pyruvate kinase. We remind you that oxidative phosphorylation is the process in which ATP is generated when reduced substrates (eg NADH and $FADH_2$) are oxidised via an electron transport chain. Some anaerobic heterotrophs can, however, produce ATP by using an electron transport mechanism in which an inorganic substrate (eg NO_3^- or $SO_4^{=}$) acts as an oxidant in place of O_2.

There are many different varieties of organic substrates that may be used by heterotrophs. These include biological products (eg carbohydrates, proteins, lipids, nucleic acids) and aromatic and aliphatic hydrocarbons and their derivatives. Almost all organic molecules, with the possible exception of some man-made polymers such as plastics and halogenated derivatives, can be catabolised (broken down and oxidised) to yield ATP and reducing power.

With these points in mind, it is not surprising that there is a very extensive range of metabolic routes and processes involved in the generation of reducing power and cellular energy.

1.2.2 Generation of metabolic precursors

carbon cycle

breakdown of organic substrates

Before we turn our attention to biosynthesis, let us examine the generation of the other products of the fuelling reactions, namely the production of simple organic molecules. In heterotrophs these products arise mainly through the breakdown of the organic substrates being used as an energy source.

Π How are these simple organic molecules produced in chemoautotrophs and photoautotrophs?

carbon dioxide fixation

The simple answer is from carbon dioxide (CO_2) by a process known as carbon dioxide fixation. The most common, but not ubiquitous, of the metabolic pathways is known as the Calvin (or Calvin-Benson) cycle.

The material discussed so far is summarised in Figure 1.1. The two Biotol texts, 'Principles of Cell Energetics' and 'Energy Sources for Cells', examine in depth the processes represented by the arrows in Figure 1.1. If in reading this section you have realised that you are not sufficiently familiar with the fuelling reactions of heterotrophs and autotrophs, we would suggest that you take the opportunity to read the Biotol texts mentioned above before proceeding.

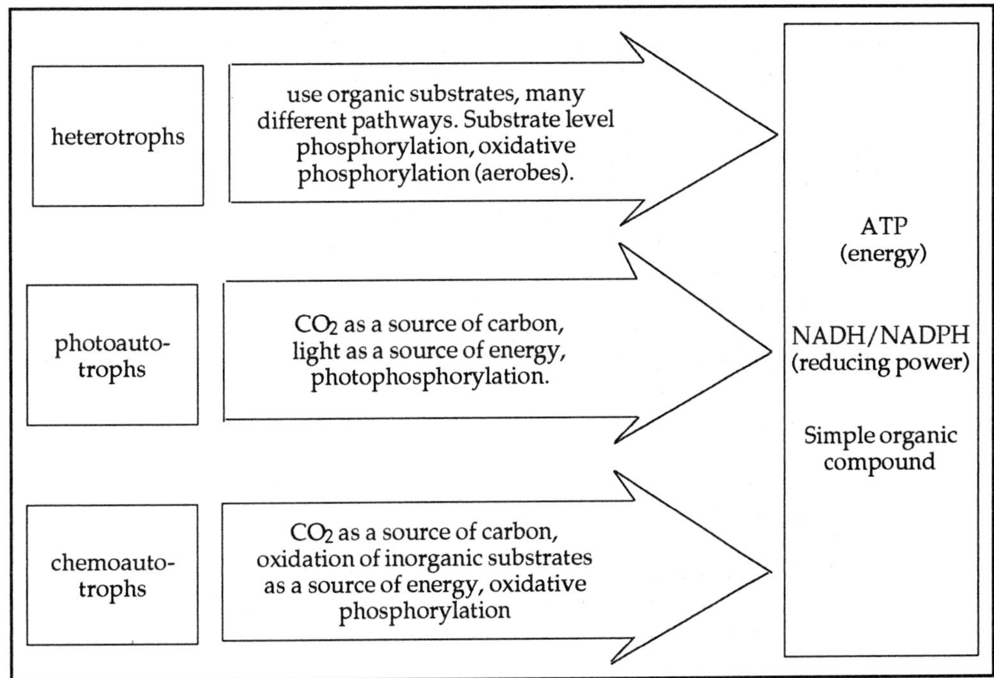

Figure 1.1 The fuelling reactions of cell metabolism.

1.3 Biosynthetic processes are similar in all systems

We begin this section by asking you to respond to a question:

Π Write down the names of the nucleotide bases that are found in nucleic acids. Are different nucleotides used by plants, animals and micro-organisms?

nucleic acids

We would anticipate that you would have written down adenine, guanine, cytidine, thymine and probably uracil. The first four are found in DNA from all sources. There are, of course, differences in the quantities of each nucleotide and in their sequence along the DNA strands. Nevertheless, the point we are attempting to make is that all organisms will need to be able to synthesise the same four bases to make DNA. They will also need to make uracil in order to make RNA.

amino acids

We can apply the same argument to the majority of proteins. We know that about 20 different amino acids are found in proteins irrespective of the organisms from which the proteins are derived. These amino acids are, of course, arranged in different orders and in different ratios within these proteins. Nevertheless all organisms will need to produce the same 20 or so amino acids. It is not surprising, therefore, to find that the biosynthetic pathways are similar (but not necessarily identical) in most organisms.

Variations are shown particularly by cells that:

- live in extreme environments;
- are components of multicellular systems.

primary metabolism

secondary metabolism

Cells that live in extreme environments often need to produce specialised structures to protect themselves from environmental damage. In multicellular systems there is also greater opportunity for cellular specialisation bringing with it a greater opportunity for a diversity of biosynthetic products. These include such products as pigments (eg in flowers), hormones and alkaloids. This diversity is mainly in what we might regard as secondary metabolism. Primary metabolism (ie metabolism directly related to cell growth), especially the biosynthesis of the core compounds from which cells are made, remains remarkably similar.

1.4 Precursor molecules for biosynthesis

12 precursor molecules

Despite the vast array of molecules needed to make cellular structures, biosynthesis begins with 12 quite simple compounds. The names and structures of these are shown in Figure 1.2.

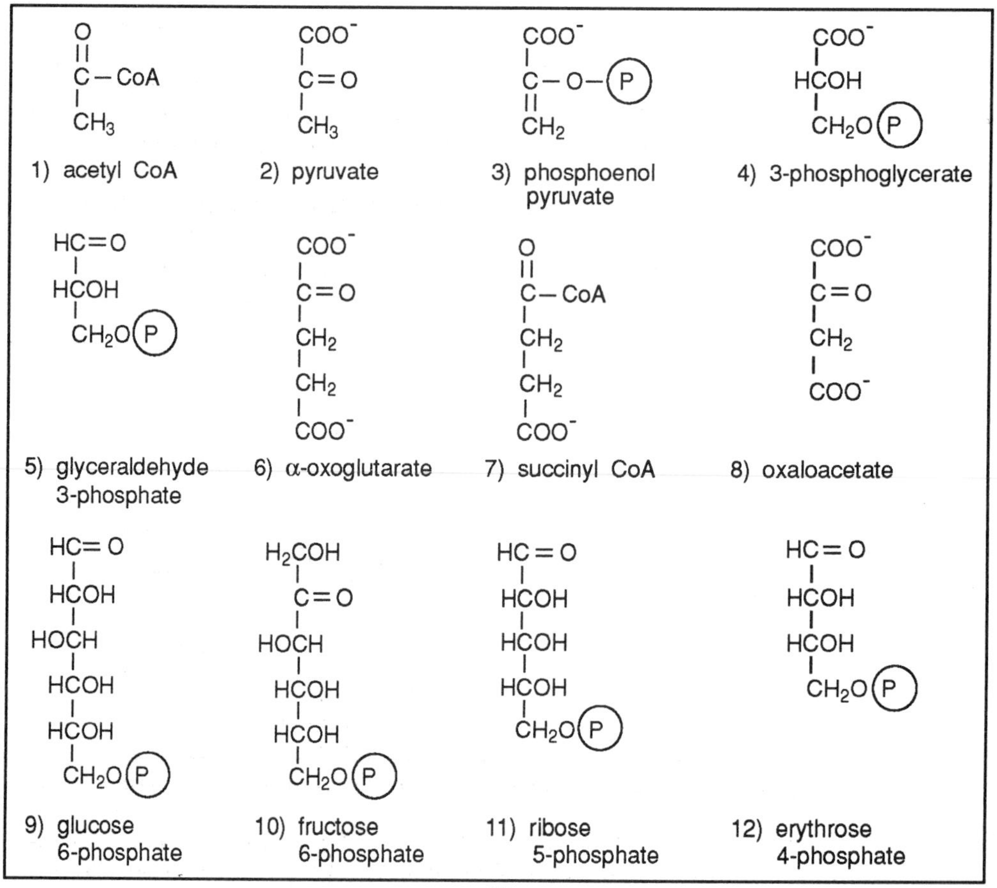

O ‖ C–CoA | CH_3 1) acetyl CoA	COO^- | C=O | CH_3 2) pyruvate	COO^- | C–O–(P) ‖ CH_2 3) phosphoenol pyruvate	COO^- | HCOH | CH_2O(P) 4) 3-phosphoglycerate
HC=O | HCOH | CH_2O(P) 5) glyceraldehyde 3-phosphate	COO^- | C=O | CH_2 | CH_2 | COO^- 6) α-oxoglutarate	O ‖ C–CoA | CH_2 | CH_2 | COO^- 7) succinyl CoA	COO^- | C=O | CH_2 | COO^- 8) oxaloacetate
HC=O | HCOH | HOCH | HCOH | HCOH | CH_2O(P) 9) glucose 6-phosphate	H_2COH | C=O | HOCH | HCOH | HCOH | CH_2O(P) 10) fructose 6-phosphate	HC=O | HCOH | HCOH | HCOH | CH_2O(P) 11) ribose 5-phosphate	HC=O | HCOH | HCOH | CH_2O(P) 12) erythrose 4-phosphate

Figure 1.2 The 12 compounds used for biosynthesis. (P) = phosphate.

You will recognise many of the compounds as products of the central fuelling process.

Π You may like to see if you can identify which catabolic processes generate the compounds listed in Figure 1.2. If you can identify at least one pathway for each compound, you clearly have a good knowledge of the catabolic fuelling reactions.

Compounds 1-5, 9-10 occur in the Embden Meyerhof pathway. Compounds 6-8 occur in the TCA cycle and compounds 11 and 12 in the pentose phosphate pathway.

All the amino acids, carbohydrates, lipids and nucleic acids needed for cell synthesis are made from the compounds shown in Figure 1.2. We will learn in later chapters that we can group biosynthetic reactions into clusters or families according to the precursor molecules used and the nature of the products made.

Π In addition to these simple precursors and a source of energy and reducing power to drive biosynthesis, other components are needed. Can you list what they are? (Examine the compounds listed in Figure 1.2 and think about the products that will be synthesised - what is missing?)

source of nitrogen and sulphur

The answer we want includes a source of nitrogen which is needed to make amino acids and nucleotides. We would also anticipate the need for a source of sulphur in order to make the sulphur-containing amino acids (cysteine, cystine and methionine) and the sulphonated carbohydrates. There will also be a requirement for the appropriate enzymes to catalyse the biosynthetic reactions. We could also add to this list, elements that are needed to maintain the structure and activity of cell components, such as magnesium, iron, calcium, potassium and the trace elements.

This text is largely concerned therefore with the processes by which the products of the fuelling reactions, together with utilisable nitrogen (usually NH_4^+) and sulphur (usually $SO_4^=$) sources, are used to make the building blocks for cell synthesis. We can represent these processes diagrammatically, as shown in Figure 1.3.

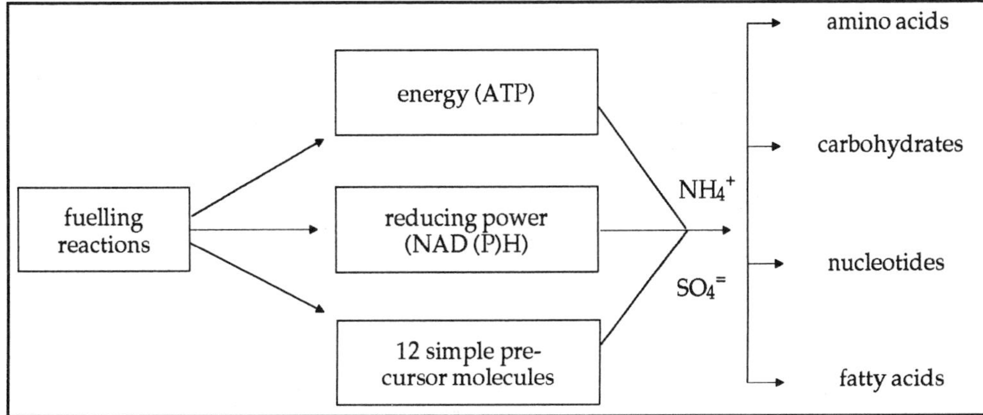

Figure 1.3 The relationship between the products of the fuelling reactions and the cellular building blocks.

You should, therefore, anticipate finding chapters devoted to the pathways and processes leading to each of these major groups of biochemicals. We have, for simplicity, omitted the sequences leading to many of the minor, although important, cell components (eg some cofactors, porphyrins). Three other important aspects of metabolism are, however, included. These are the uptake of nutrients by cells, the integration of the fuelling and biosynthetic reactions and the regulation of metabolism.

1.5 The arrangements of chapters

nutrient uptake

It is perhaps obvious that before any biosynthesis can take place, nutrients have to be absorbed from the cell's environment. We examine the uptake of nutrients in chapter 2. This is followed by the examination of the special status of nitrogen and sulphur uptake and metabolism. Armed with this knowledge and that of the fuelling reactions gained from the Biotol texts, 'Principles of Cell Energetics' and 'Energy Sources for Cells', you will be ready to study the metabolic pathways which lead to the production of the building blocks of cellular materials. Thus, in subsequent chapters the biosynthesis of fatty acids, sugars, amino acids and nucleotides is examined. These monomers provide the building blocks for the cellular macromolecules. In the chapters relating to fatty acid and sugar biosynthesis we also consider the incorporation of these monomers into lipids and polysaccharides. The polymerisation of nucleotides and amino acids to form nucleic acids and proteins is, however, a rather special case. The polymerisation of these is specified by the sequence of nucleotides in the genome and the consideration of the processes of nucleotide polymerisation (gene replication and transcription) and amino acid polymerisation (translation) should more correctly be discussed in the context of molecular genetics. These processes are described briefly in the Biotol text , 'Infrastructure and Activities In Cells' and more fully in the Biotol texts, 'Genome Management in Prokaryotes' and 'Genome Management in Eukaryotes'.

biosynthesis of building blocks

polymerisation

Thus we can represent the subject matter covered by this text within the context of the total metabolic activities of cells by the shaded boxes shown in Figure 1.4.

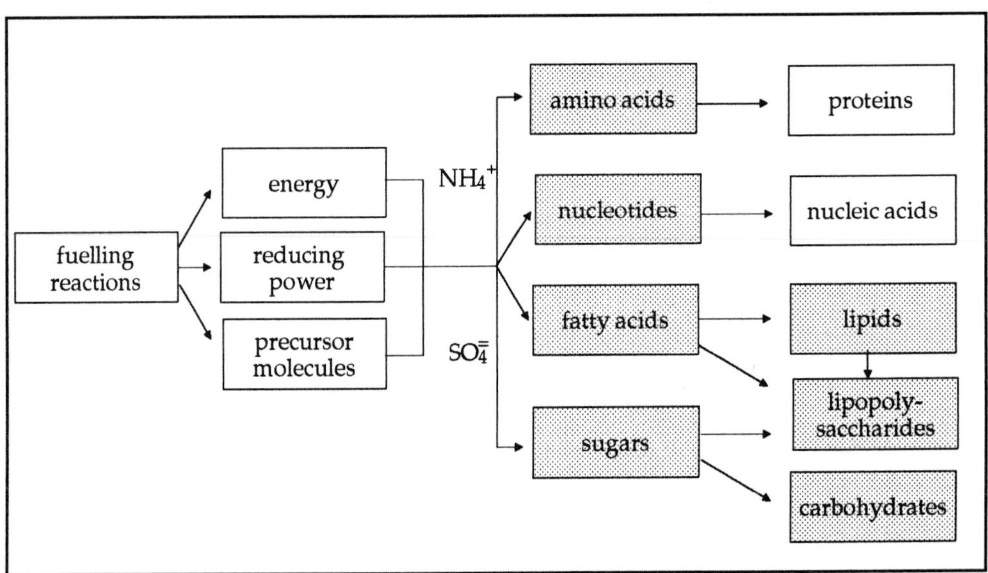

Figure 1.4 The area of metabolism covered by this text (shaded boxes).

The assembly of macromolecules into cellular structures such as the membranes, ribosomes, Golgi apparatus etc, are dealt with in the context of cell biology (Biotol text 'Infrastructure and Activity of Cells').

regulation

It should be self-evident that, if cells are not to be wasteful, the various precursors for biosynthesis must be produced in the proportions in which they are required. A key question is, therefore, how do cells regulate the various fuelling reactions to provide precursor molecules in the ratios in which they are needed? Likewise it is important for cells to produce the building blocks in sufficient but not excessive quantities. The final two chapters of this text explore the ways in which the fuelling and biosynthetic reactions are integrated and regulated to produce the building blocks for cell synthesis in a balanced manner that can be sustained.

Uptake of Nutrients

Introduction 12

2.1 The nutritional requirements of cells 12

2.2 Membranes as permeability barriers 15

2.3 The distinction between passive diffusion and carrier-mediated transport 18

2.4 Passive diffusion through protein channels. 21

2.5 Carrier-mediated transport 23

2.6 Passive transport systems. 25

2.7 Active transport systems 26

2.8 Group translocation across membranes 28

2.9 Binding proteins 30

2.10 The utilisation of substrates that cannot pass through the membrane 33

2.11 Experimental approaches to studying transport systems 35

Summary and objectives 40

Uptake of Nutrients

Introduction

The interior of the cell consists of an aqueous solution of salts, amino acids, vitamins, coenzymes, and a wide variety of other soluble materials; this is the cell pool. The cell pool is retained by the plasma membrane and when this is destroyed most of these materials are able to leak out, only substances too large to pass through the cell wall (if present) are retained. Components enter the pool either as nutrients taken up from the environment or as material synthesised from other constituents in the cell. Most nutrients enter the cell by specific transport systems which require their binding to membrane proteins.

In this chapter we will examine the processes by which cells take up nutrients from their environment. Firstly, the nutritional requirements of cells and the role of the plasma membrane as a permeability barrier will be examined. Much of the chapter is then devoted to describing the characteristics of the various classes of uptake processes associated with biological membranes. The chapter is not intended to be a comprehensive catalogue of specific uptake processes, since they are extraordinarily diverse with regards to the individual components involved. You will find that much of our knowledge of uptake processes comes from studying bacteria and this has potential biotechnological applications. In the final section of the chapter we will examine some of the experimental approaches which have led to major advances in our understanding of transport processes.

At the end of this chapter there is a flow diagram (Figure 2.16) which summarises the relationship between the different classes of transport systems. As you work through the chapter you might find it useful to refer to this diagram to see where each transport system fits into the overall scheme of membrane transport.

2.1 The nutritional requirements of cells

To grow, organisms must draw from the environment all the substances they require for the synthesis of their cell material and for the generation of energy. These substances are termed nutrients.

macro and micronutrients and trace elements

The chemical composition of the cell indicates the major nutritional requirements of cells. The bulk of the organic mass of animals and prokaryotes (95%) is made up of carbon, oxygen, hydrogen, nitrogen, phosphorus and sulphur; when supplied in a culture medium these are called macronutrients. Nutritional studies show that potassium, magnesium, calcium and iron are required in relatively small amounts; these are called micronutrients. Other micronutrients, such as manganese, cobalt, copper, molybdenum and zinc, are required in very small amounts and are referred to as trace elements. Indeed, it is often difficult to demonstrate that the trace elements are essential for growth since they are often present as contaminants of the major inorganic constituents and this alone may be sufficient to satisfy the nutritional requirements of the organism. Plant cells have the same requirements as animal cells but in quite different proportions. Table 2.1 shows the general physiological functions of the principal elements.

Element	Physiological Functions
hydrogen	constituent of cellular water, organic cell materials
oxygen	constituent of cellular water, organic cell materials; as O_2, electron acceptor in respiration of aerobes
carbon	constituent of organic cell materials
nitrogen	constituent of proteins, nucleic acids, coenzymes
sulphur	constituent of proteins (as amino acids cysteine and methionine), of some coenzymes (eg CoA, cocarboxylase)
phosphorus	constituent of nucleic acids, phospholipids, coenzymes
potassium	one of the principal inorganic cations in cells; cofactor for some enzymes
magnesium	important cellular cation; inorganic cofactor for very many enzymatic reactions, including those involving ATP; functions in binding enzymes to substrates; constituent of chlorophylls
manganese	inorganic cofactor for some enzymes, sometimes replacing Mg
calcium	important cellular cation; cofactor for some enzymes (for example proteinases); important in membrane stability
iron	constituent of cytochromes and other heme or nonheme proteins; cofactor for a number of enzymes
cobalt	constituent of vitamin B_{12} and its coenzyme derivatives
copper, zinc, nickel, molybdenum	inorganic constituents of special enzymes

Table 2.1 General physiological properties of the principal elements.

Some cells have additional, specific mineral requirements. For example, sodium is required at relatively high concentration by certain marine bacteria and is also present in nearly all animal cells.

Π Looking at Table 2.1, which elements are important micronutrients as: 1) constituents of enzymes, 2) constituents of cytochromes, 3) cofactors of enzymes?

Elements listed below phosphorus in Table 2.1 are classed as micronutrients.

1) Cu, Zn, Ni, Mo are constituents of enzymes.

2) Fe is a constituent of cytochromes.

3) K, Mg, Mn, Ca, Fe, Co, are cofactors of certain enzymes.

metallic elements absorbed as salts

We will now consider the form in which nutrients enter the cell. In culture media, all the required metallic elements can be supplied in the form of cations of inorganic salts. Phosphorus can also be used as a nutrient when provided in inorganic form, as phosphate salts.

carbon sources

All naturally-produced organic compounds and many synthetic ones can be used as a source of carbon and energy for at least one type of organism. It is thus impossible to describe concisely the chemical nature of the carbon sources which can be used by cells. Since most organic substrates are at the same general oxidation level as organic cell constituents, they do not have to undergo a primary reduction to serve as sources of cell carbon. Most organic nutrients enter the cell in a simple form, for example: as carbohydrate monomers rather than polymers: as fatty acids instead of lipids: as amino acids and not as proteins.

growth factors

In addition to the primary carbon sources, cells may require growth factors. These are organic compounds which serve as a precursor or a constituent of the cell's organic material, but which it cannot synthesise from simpler carbon sources. A wide array of compounds function as growth factors in one organism or another but they fall generally into one of three categories:

- amino acids - required as constituents of proteins;
- purines and pyrimidines - required as constituents of nucleic acids;
- vitamins - a diverse collection of organic compounds that are required for the functioning of certain enzymes.

nitrogen sulphur

Nitrogen and sulphur occur in the organic compounds of the cell principally in reduced form as in amino ($-NH_2$) and sulphydryl groups (-SH), respectively. Some cells use the oxidised inorganic states, such as nitrates and sulphates, and reduce them in the cytoplasm. This will be examined in detail in a later chapter. Other cells are unable to bring about a reduction of one or both of these ions and must be supplied with the element in a reduced form, as ammonium salts and sulphide. The nitrogen and sulphur requirements can often be met by organic nutrients that contain these two elements in reduced organic combination eg the amino acids methionine and cysteine.

oxygen carbon dioxide

Gases such as oxygen and carbon dioxide are required by most cells. Nitrogen, hydrogen or methane may be utilised by some.

SAQ 2.1

Ingredients of a culture medium designed for the growth of certain bacteria are shown below:

Water	1l
K_2HPO_4	1g
Glucose	5g
NH_4Cl	1g
$MgSO_4$ $7H_2O$	200mg
Fe_2SO_4 $7H_2O$	10mg
$CaCl_2$	10mg
Methionine	10mg
Trace elements	(Mn, Mo, Cn, Co, Zn) as inorganic salts, 0.02-0.5mg of each

1) Explain why glucose is provided in relatively large quantities but methionine in only small quantities.

2) In what chemical forms will nitrogen and sulphur enter the cell?

3) What ingredient could possibly provide both nitrogen and sulphur to the cells? Why is it unlikely to do so in this medium formulation?

4) If the methionine was replaced with a low level of yeast extract, which is a complex ingredient containing many amino acids, vitamins etc, would you expect the medium to support a more diverse or less diverse population of heterotrophic organisms? Explain your answer.

2.2 Membranes as permeability barriers

phospholipid in the permeability barrier

You need to be familiar with the composition, organisation and function of biological membranes. You will recall that membranes are composed of a phospholipid bilayer which has proteins embedded in it. The membrane has a hydrophobic matrix, formed by the nonpolar fatty acid tails of the lipids with proteins dispersed within it. Proteins may also make contact with the interior and exterior of the cell (span the membrane). Figure 2.1 will remind you of the structural organisation of biological membranes.

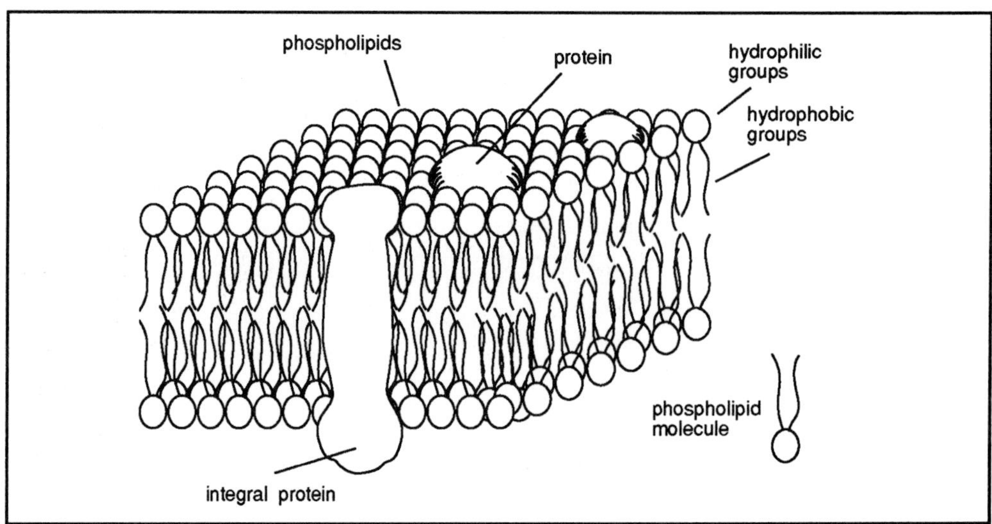

Figure 2.1 Structure of a biological membrane. Although there are some chemical differences, the overall structure shown is similar in prokaryotes and eukaryotes.

Despite its thinness (7nm), the plasma membrane functions as a tight permeability barrier, so that passive movement of most solute molecules does not readily occur. The permeability of lipid bilayers has been measured in two well-defined synthetic systems: lipid vesicles and planar bilayer membranes.

lipid vesicles used to study permeability

Lipid vesicles (or liposomes) are aqueous compartments enclosed by a lipid bilayer (see Figure 2.2). They can be formed by suspending a suitable lipid, such as phosphatidyl choline, in an aqueous solute. The mixture is then sonicated (agitated by high-frequency sound waves) to give dispersion of closed vesicles containing the solute that was used in their preparation. By suspending the vesicles in another aqueous medium the rate of diffusion of the solute from the inner compartment to the surrounding solution can be determined.

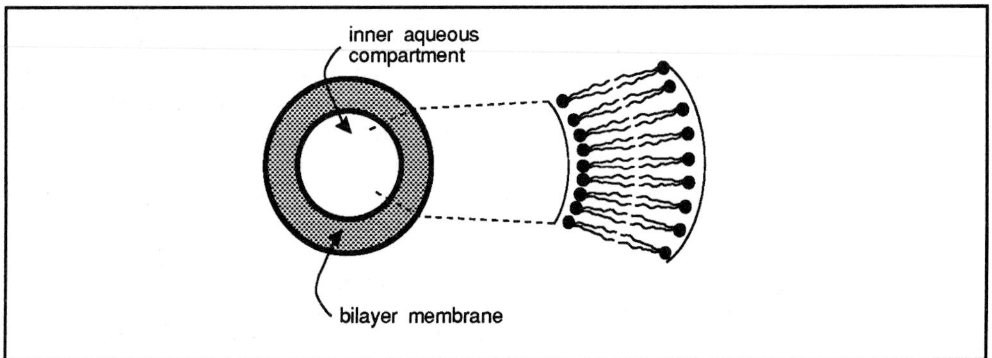

Figure 2.2 Diagram of a lipid vesicle.

Planar bilayer membranes can be formed across a 1mm hole in a partition between two aqueous compartments using a fine paint brush and a membrane-forming solution, such as phosphatidyl choline in decane.

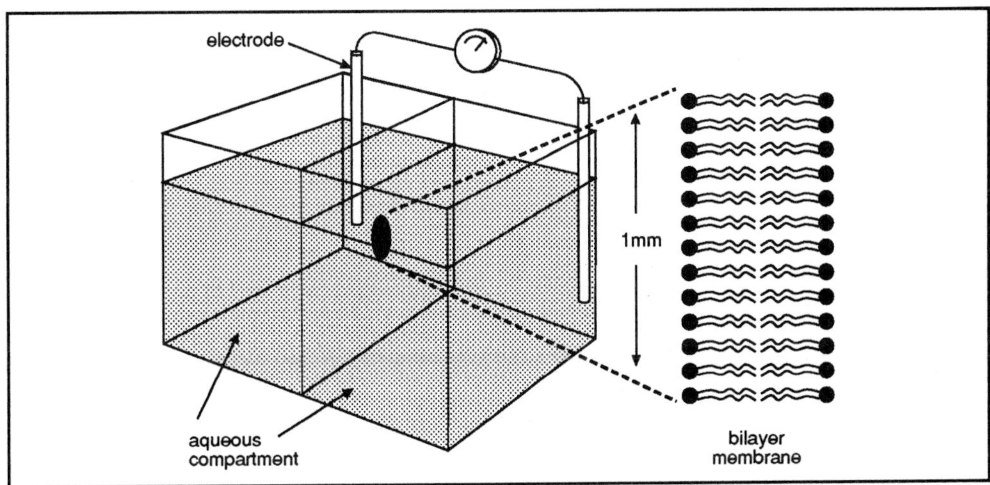

Figure 2.3 Experimental arrangement for the study of planar bilayer membranes. A bilayer membrane is formed across a 1mm hole in a septum that separates two aqueous compartments.

The electrical conductance properties of the lipid bilayer can be readily measured by inserting electrodes into each aqueous compartment. For example, its permeability to ions is determined by measuring the current across the membrane as a function of the applied voltage.

current across the membrane

Π Will the current across a membrane be higher for a solute of low permeability compared with one of high permeability?

The correct answer is no, it will be lower. The relationship between voltage (V), current (C) and resistance (R) is $V/R = C$. So, if the permeability of an ion is low (high resistance) then the current must also be comparatively low if the voltage remains constant.

size,fat solubility and charge govern permeability

Studies using these synthetic systems have shown that the rate at which molecules can penetrate the membrane is determined by factors such as size, charge, solubility in lipid, and the concentration of those particles on either side of the membrane. Small non-polar molecules quickly diffuse across the membrane. This is also true for small uncharged polar molecules. The greater their molecular weight, the slower polar substances will pass the membrane. If, in addition, a polar molecule is charged, it will be virtually impossible for it to pass through the membrane. Polarity and charge attract water to the molecule and create a hydration coat which stops the diffusion of a particle across the hydrophobic matrix of the membrane.

Π Explain what is meant when molecules are described as being polar but uncharged.

Polar molecules have an uneven distribution of electrons because some groups on the molecules attract electrons more strongly than other groups. This generates partial charges on parts of the molecule - a partial charge has a value of less than one unit of charge. A charged molecule has one or more whole charges.

SAQ 2.2

List the following substances in order of permeation (at pH 7) across a lipid bilayer: glucose, glucose 6-phosphate, oxygen, hydrogen, Na^+ ions, glycerol and water.

2.3 The distinction between passive diffusion and carrier-mediated transport

Solute may undergo a net movement or flux across a membrane in either of two ways, passive diffusion or carrier-mediated transport. Passive diffusion is often referred to as simple diffusion or nonmediated uptake.

The two processes can be distinguished experimentally by measuring the rate of transport under different conditions. Passive diffusion is at all times directly dependant on the difference in concentration of the solute. In other words, if the concentration across the membrane increases or decreases the rate of passive diffusion is also increased or decreased.

passive diffusion defined

Passive diffusion is also non-specific and the compound usually distributes itself in such a way that its concentration inside the cell is the same as outside implying a bidirectional movement of solute. The solute molecule is neither chemically modified nor bound to another molecular species in its passage through the membrane. Water and gases such as O_2 and CO_2 are the main nutrients that enter cells by passive diffusion.

carrier-mediated transport defined

Now let us compare the characteristics of simple diffusion with those of carrier-mediated transport, the kinetics of which have three identifying characteristics:

- saturability;
- specificity;
- specific inhibition.

Saturability means that the transport system can become saturated with the substance transported, just as enzymes become saturated with their substrates. This is shown in Figure 2.4, which also shows how carrier-mediated transport may be distinguished experimentally from passive diffusion.

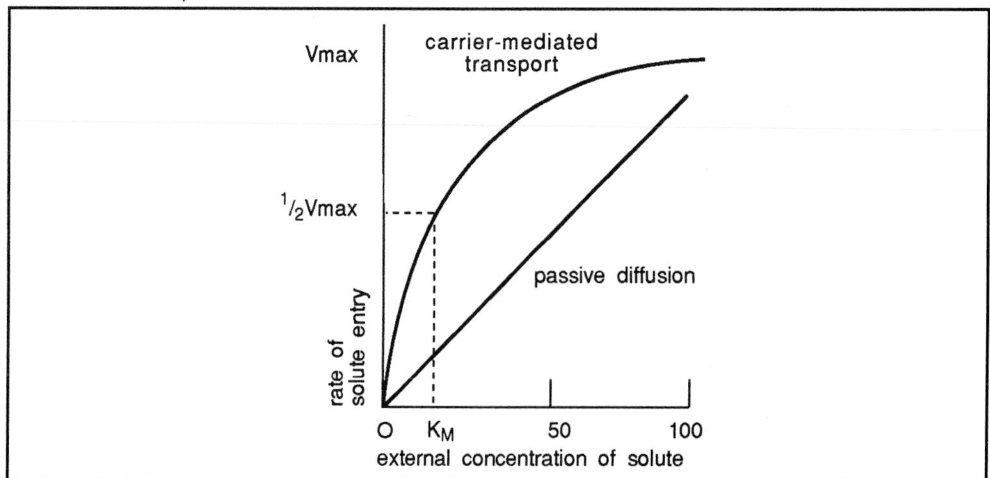

Figure 2.4 The relationship between uptake and external concentration in passive diffusion and carrier-mediated transport. Note that in carrier transport the uptake rate shows saturation at high external concentration.

Saturability suggests that each membrane transport system contains a specific site to which the substrate must bind reversibly in order to be transported across the membrane. It also suggests that the rate of binding of the substrate on one side of the membrane, its transport across the membrane, or its release on the other side of the membrane can set an upper limit to the rate of transport, just as the rate of formation (or breakdown) of an enzyme-substrate complex can limit the rate of an enzyme reaction.

For the transport of substrate by means of a carrier, Michaelis-Menten kinetics hold that in steady state the kinetics of the process are described by the equation

$$V = \frac{Vmax \, . \, [S]}{K_M + [S]}$$

V = transport rate

Vmax = maximum transport rate

[S] = concentration of substance to be transported

K_M = Michaelis-Menten constant.

Π If Vmax is 50nmol min^{-1} and K_M is 12mmol l^{-1}, calculate the change in [S] needed to increase the velocity of the process from 5nmol min^{-1} (=0.1 Vmax) to 45nmol min^{-1} (=0.9Vmax). (Do this on a piece of paper and then check your calculation with that of ours printed below).

The way we have calculated this is:

$$V_1 = 0.1 \, Vmax = \frac{Vmax.[S_1]}{K_M + [S_1]}$$

$$\text{Thus } 0.1 = \frac{[S_1]}{K_M + [S_1]}$$

$$10[S_1] = K_M + [S_1]$$

$$[S_1] = \frac{K_M}{9}$$

$$= 1.33 \text{ mmol } l^{-1}$$

$$\text{Similarly } V_2 = 0.9 \, Vmax = \frac{Vmax.[S_2]}{K_M + [S_2]}$$

$$\text{Thus } 0.9 = \frac{[S_2]}{K_M + [S_2]}$$

$$[S_2] = 9.K_M$$

$$= 108 \text{mmol } l^{-1}$$

Thus to achieve the increase in the velocity from 5nmol min^{-1} to 45nmol min^{-1}, we have to increase the concentration by a factor of 81 (ie $[S_2]/[S_1] = 81$).

Π Is carrier-mediated transport more effective at low external nutrient concentrations than passive diffusion?

The answer is usually yes. It is indicated by the steepness of the initial slope of the line for carrier-mediated transport shown in Figure 2.4. The lower the K_M value the more efficient the transport system is at low nutrient concentrations.

carrier-mediated transport shows specificity

Specificity of transport means that in a group of very closely related substances some may be transported rapidly, some slowly and others not at all. For example, some transport systems promote the influx (uptake) of D-glucose and a number of closely related monosaccharides but have little or no activity towards D-fructose or disaccharides such as lactose. Carrier-mediated transport systems may also show stereospecificity, for example, the amino acid transport system of animal cell membranes transports the L-amino acids more actively than the D-isomers.

inhibition can be competitive or non-competitive

Specific inhibition of some carrier-mediated transport systems occurs competitively with substances closely related to the substrate. For instance, the closely related amino acids L-alanine, L-serine and glycine compete with each other for uptake. This is one reason why amino acid imbalance in a culture medium leads to poor growth of organisms requiring them - an excess of one amino acid prevents the uptake of the other. A similar kind of competition for uptake is also shown among closely related cations, such as Na^+, K^+ and Rb^+. The amount of inhibition depends on the affinity of the substances for the carrier and on the ratio of their concentrations. Transport systems may be inhibited noncompetitively and this form of inhibition is not related to the concentration of substance which is to be transported.

Π The effect of an inhibitor on transport system (A) and transport system (B) is shown below. For each, identify the type of inhibition as either competitive or non competitive. Give your reason.

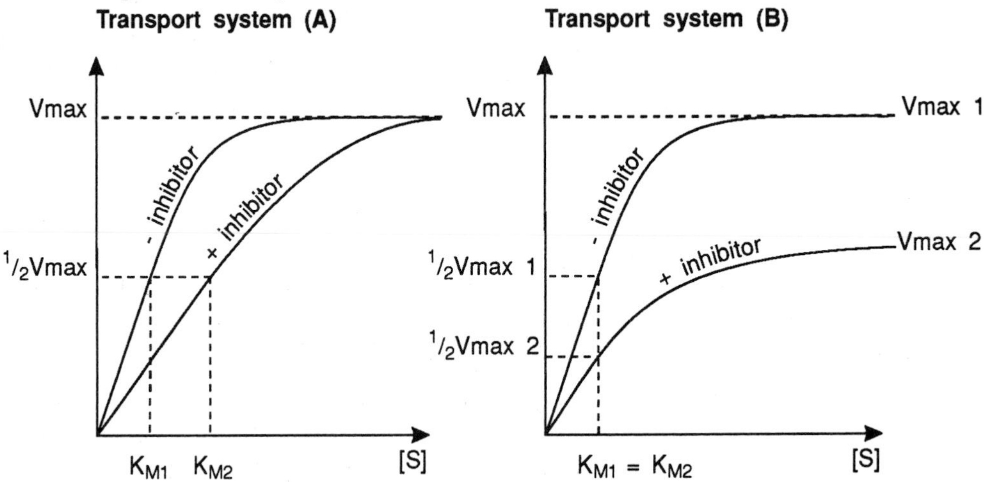

(A) This is a case of competitive inhibition because the effect of the inhibitor decreases as the concentration of the substance increases. Note that the inhibitor has not reduced Vmax.

(B) This is a case of non competitive inhibition because inhibition is not reduced by high [S]. Note that Vmax is lower in the presence of the inhibitor.

Note that in (A) K_M is much higher in the presence of the inhibitor but in (B) it is the same in the presence and absence of inhibitor.

An example of the characteristics of the carrier-mediated transport mentioned above can be found in the transport of glucose across the erythrocyte membrane. Kinetic experiments have led to the conclusion that there are about 300,000 functional glucose carriers for each erythrocyte. The rate of glucose transport depends on the glucose concentration in the extracellular liquid. The glucose carrier shows broad specificity, which means that, apart from D-glucose, other sugars can be transported too, like D-mannose, D-galactose and D-ribose. The affinity of the carrier for each of these sugars is expressed by the K_M.

example: glucose carrier

Π The glucose carrier in erythrocytes has K_M values for the transport of D-glucose, D-mannose and D-fructose of 2.6, 18.5 and 2000 mmol l^{-1} respectively. How would the uptake of glucose from an equimolar mixture of 1) D-glucose and D-mannose and 2) D-glucose and D-fructose compare with that from glucose alone?

The K_M values tell us that the carrier has the highest affinity for D-glucose, then D-mannose and is physiologically insignificant for the transport of fructose. Therefore we would expect D-mannose to competitively inhibit the rate of D-glucose transport but D-fructose to have no measurable effect.

carriers act as enzymes

These properties of carrier-mediated transport strongly suggest that enzyme-like molecules are involved which are capable of binding specific substrates and transporting them across the membrane. Such enzymes are called transport systems or carriers. You might find that some other texts also use the terms translocases or permease enzymes to describe these molecules.

Π Write down what effect you would expect temperature to have on the rate of passive diffusion and on carrier-mediated transport.

For passive diffusion across a membrane the temperature coefficient is usually that of physical diffusion, which means that the rate is increased 1.4 fold for each 10°C rise in temperature. For carrier-mediated transport there will be an optimum temperature above which the transport rate will begin to decline. This reflects the involvement of proteins which become denatured at higher temperatures.

2.4 Passive diffusion through protein channels.

protein channels are also used in passive diffusion

Apart from the proteins involved in carrier-mediated transport other types of proteins are involved in transport phenomena. These proteins form hydrophilic protein channels in membranes thereby allowing the passage of small molecules and ions. In these processes no prior binding is involved.

Michaelis-Menten kinetics are not obeyed and the passage is regarded as a form of passive diffusion. Whereas passive diffusion not involving proteins is generally slower than carrier-mediated transport, passive diffusion through protein channels is generally faster.

receptor and voltage switches

There are two types of protein channels for passive diffusion:

1) Gated channels. These protein channels are opened only temporarily. Regulation takes place through changes in membrane potential (voltage-operated channels) or by binding of a ligand to a receptor on the membrane (receptor-operated channels). Examples of both types of gated channels can be found in the excitation

of skeletal muscle, involving channels for Ca^{2+}, Na^+ and K^+. Gated channels are illustrated in Figure 2.5.

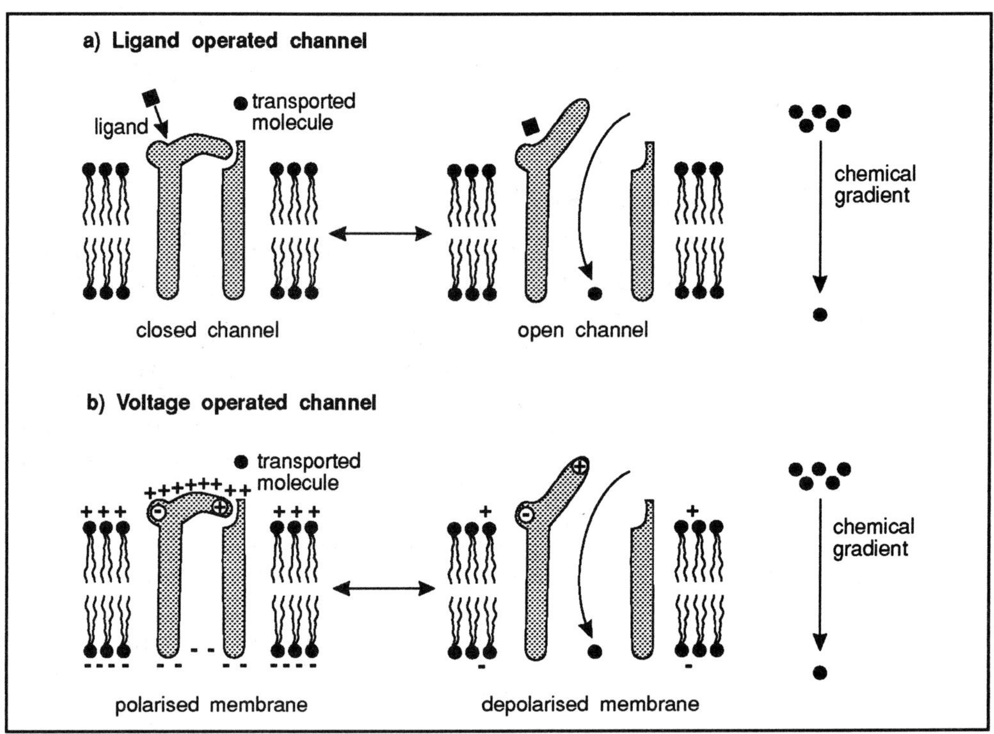

Figure 2.5 Passive diffusion through gated channels. a) Receptor-operated channel: channel is regulated by the binding of a ligand, such as an ion or small molecule. b) Voltage-operated channel: here the opening of the channel is determined by the polarisation or depolarisation of the membrane.

2) Open channels. These are channels which are permanently open and often have fairly large pores (up to 0.4nm) allowing not only the passage of ions but also that of other relatively small molecules. These hydrophilic channels can be found in prokaryotes as well as in eukaryotes.

porin is an open channel protein

mitochondria and chloroplasts have open channels

A well known example of an open channel is the protein porin. This protein has a relative molecular mass of 37,000 and occurs in the cell wall of Gram-negative bacteria. The internal diameter of the pore is approximately 1nm and is large enough to admit molecules with a molecular mass less than 600 to the periplasmic space, the space between the cell wall and the plasma membrane. The various transport systems in the plasma membrane are responsible for selected transport into the cytoplasm. Passive diffusion channel proteins in the outer membranes of mitochondria and chloroplasts show many similarities to porin.

Π Why do the identifying characteristics of carrier-mediated transport (saturability, specificity and specific inhibition) not apply to passive diffusion through protein channels?

Passive diffusion through protein channels does not involve the binding of substrate molecules to the transport protein and, therefore, does not show saturability, specificity or specific inhibition.

Π Before you go on to the next section, look at the overall scheme (Figure 2.16) and highlight the processes that have been considered so far.

All headings under the main heading of passive diffusion should have been highlighted.

2.5 Carrier-mediated transport

passive diffusion is different from passive transport

In the previous section we discriminated carrier-mediated transport from diffusion. We can further distinguish two main types of transport, called active and passive transport. The characteristic difference between the two is that passive transport does not require energy, whereas active transport does. Passive transport, also called facilitated diffusion, should not be confused with passive diffusion. They are similar in the sense that the substrate moves down a concentration gradient and the processes do not require the expenditure of energy, but remember that the three identifying characteristics of carrier-mediated transport discussed earlier do not apply to passive diffusion.

active and passive transport distinguished

We shall now consider how to distinguish between active and passive transport. Active transport has three very important characteristics that distinguish it from passive transport:

- it can move solute molecules up a concentration gradient;
- it is dependent upon metabolic energy;
- it is normally unidirectional.

If energy generation stops or energy coupling to the carrier is blocked, some active transport systems may also engage in passive transport of their substrates across the membrane. This will occur in either direction, depending on the relative concentration of the substrates across the membrane and at equilibrium entry and exit occur at the same rate. This is illustrated in Figure 2.6.

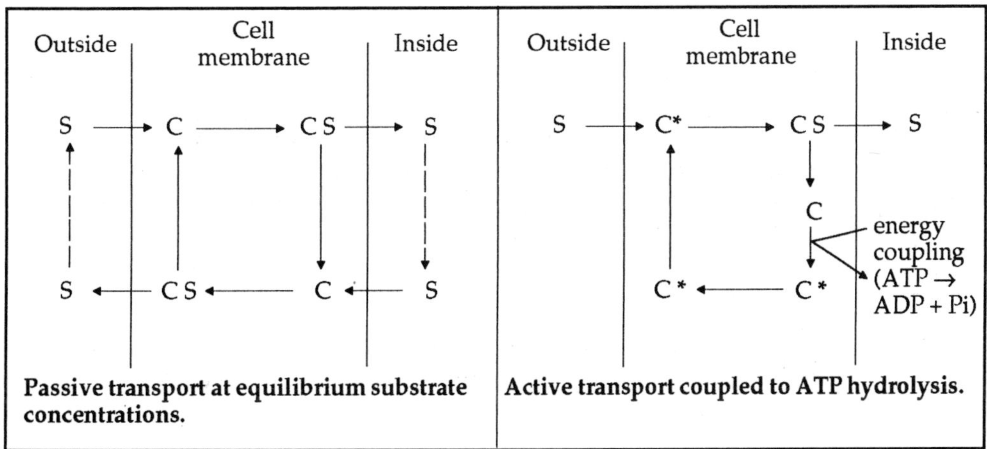

Figure 2.6 Models of passive and active transport.
Key: C = carrier; C* = energised carrier; S = substrate; SC = carrier/substrate complex.

conformational change occurs

The models propose that a specific protein containing a binding site complementary to the substrate transported serves as a carrier for the substrate. It allows movement of the substrate molecules across the membrane by changing the orientation of the binding site so that it now faces the other compartment, into which the bound substrate is discharged. In models of active transport the energised carrier molecule is thought to have reduced affinity for substrate on the cytoplasmic side of the membrane, this would account for the unidirectional transport against a concentration gradient. The reduced affinity is probably caused by a conformational change in the protein brought about by energisation of the carrier; for example, by hydrolysis of ATP. This is illustrated in Figure 2.7.

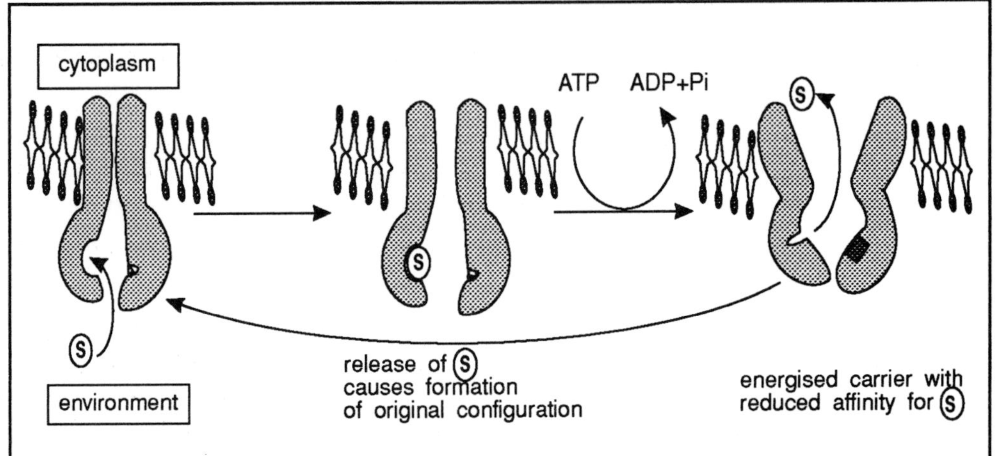

Figure 2.7 Transport proteins act as channels. In active transport a conformational change brought about by coupling of ATP hydrolysis accounts for a unidirectional transport of substrate against a concentration gradient. The energised carrier has a reduced affinity for substrate on the cytoplasmic side of the membrane.

Π Would you expect a reduced affinity for substrate on the cytoplasmic side of the membrane to decrease the rate of transport? Explain your answer.

The correct answer is no. This is because a reduced affinity on this side of the membrane would enhance the breakdown of the carrier-substrate complex.

carriers act as channels

All carriers transfer substrates by acting as channels across the hydrophobic part of the membrane (as shown in Figure 2.7). It was originally thought that carriers might diffuse across the membrane, but studies have revealed that membrane proteins diffuse across the membrane only very slowly or not at all. We shall see later in this chapter that some antibiotics can act as diffusional carriers across membranes.

Π Would you expect active and passive transport to be distinguishable according to thermodynamic principles?

system gains free energy in active transport

Yes they are. The distinction relies on a thermodynamic principle that an input of energy is required to transfer a solute at a given concentration into a compartment in which the concentration is higher. Conversely, there is a decrease in free energy when a solute passes into a compartment where its concentration is lower. Active transport is thus rigorously defined in thermodynamic terms as transport process in which the system gains free energy, and passive transport is defined as one in which the system loses free energy.

SAQ 2.3

Indicate which of the following statements are true:

1) Passive diffusion characteristically does not involve molecules other than the transported molecule.

2) Passive transport processes show saturation kinetics.

3) Certain carriers may transport substrate both unidirectionally and bidirectionally.

2.6 Passive transport systems.

We shall now examine different types of passive and active transport systems.

red cells use passive transport

Passive transport is a common mechanism of transport in eukaryotic cells but it is relatively rare among prokaryotes. One of the best known passive transport systems is the glucose carrier of the human erythrocyte which can transport a variety of D-sugars while comparable L-sugars are not transported.

ADP:ATP exchange

antiport

Passive transport systems are also involved in the transport of molecules within cells. For example, several passive transport systems occur in the mitochondrial inner membrane. They are specific for certain metabolites and will not transport even closely related molecules. The best studied example is the ATP-ADP carrier which transports one molecule of ADP into the mitochondrial matrix, and one molecule of ATP out of the matrix. Since they can function in either direction in response to the concentration gradient of transported metabolites these exchange systems are called passive antiport systems (Figure 2.8).

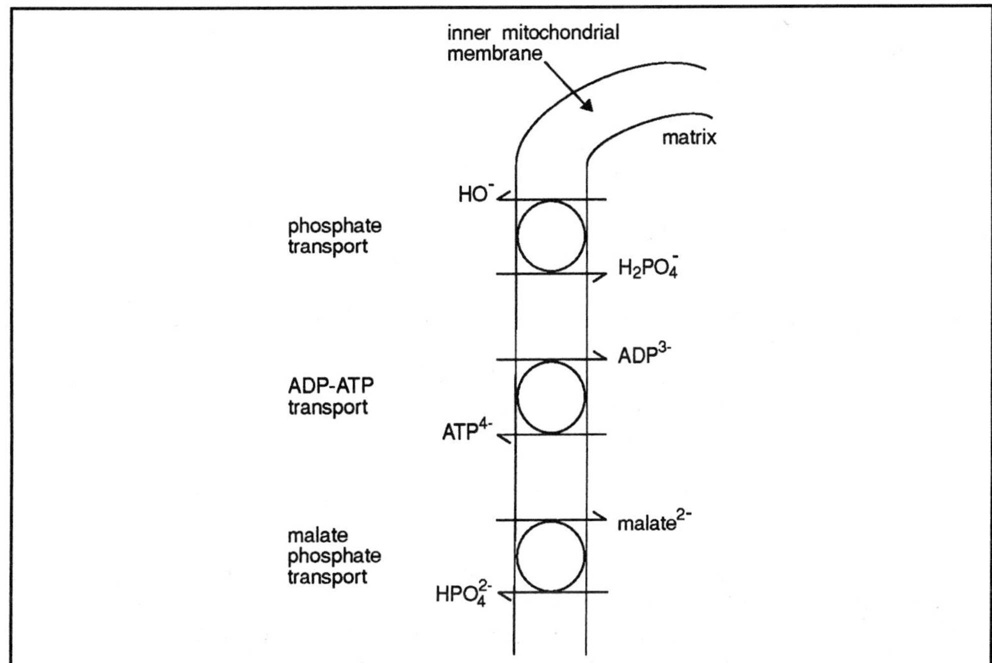

Figure 2.8 Some of the antiport transport systems of the inner mitochondrial membrane. They can function in either direction in response to the concentration gradients of the transported metabolites.

2.7 Active transport systems

The energy requirement of active transport

Active transport is one of the three major energy-requiring activities of cells - the other two being biosynthesis and mechanical movement. There are good reasons why cells invest energy in active transport:

active transport supports rapid metabolic rates

- active transport systems create concentrations of solutes within the cell that can be several hundred to a thousand times greater than those outside. Uptake of most nutrients in prokaryotes is performed by active transport. The reason for this is that they have rapid rates of metabolism but often live in environments where compounds are present at very low concentration. This can only be achieved if nutrients are concentrated in the cytoplasm;

internal concentration kept constant

- active transport systems participate in the maintenance of metabolic steady states by their ability to maintain the internal concentration of organic nutrients and metabolites relatively constant even though the concentrations outside the cell fluctuate widely. Similarly, by maintaining constant and optimum levels of inorganic ions inside the cell, particularly K^+ and Ca^{2+}, active transport has a role in the regulation of metabolic processes and helps to maintain the osmotic relationships between cells and their surrounding medium;

functions of active transport

- in animal cells active transport is also involved in many cell specific processes, including the transmission of information by the nervous system, the excitation and relaxation of muscle; the absorption of compounds from the intestine and the resorption of compounds from urine by the kidneys. It is very striking that up to 70% of the available metabolic energy in mammalian kidney cells is utilised for active-transport processes.

So, active transport effectively 'pumps' solutes across the plasma membrane, so as to accumulate them in the cytoplasm chemically unchanged but at a higher concentration. As in any other pump, this requires the expenditure of energy which means the performance of work. You should already be familiar with calculations of the amount of energy required to transport energy across biological membranes but we will remind you of the relationship $\Delta G = RT \ln \frac{C_2}{C_1}$ where C_1 and C_2 are the concentrations on either side of the membrane. (The thermodymanic principles involved are described in the Biotol text, 'Principles of Cell Energetics'). If you are in any doubt about this, try the following calculation.

Π Calculate the amount of energy required for the transport of one mole of glucose from a compartment with a concentration of 0.01 mol l^{-1} to a compartment with a concentration of 0.1 mol l^{-1} at a constant temperature of 20°C. Assume that the concentration in these compartments will not be changed by this transport.

$$(R = 8.314\ JK^{-1}\ mol^{-1})$$

Since:

$$\Delta G = RT \ln \frac{C_2}{C_1}$$

then:

$$\Delta G = 8314 \times 293 \times 2.303 \times \log \left(\frac{0.1}{0.01}\right) = 5.6 kJ$$

Π If the same solute molecule was charged would you expect the free energy increases to be higher or lower than the calculated value shown above?

In general we might anticipate that the free energy increase would be higher for a charged molecule because it is less permeable to the membrane than an uncharged molecule. This means that more energy would have to be expended to transport it. If the membrane is charged (ie it has a potential) then this will influence the amount of energy required.

Π Which phenomenon provides the energy needed for active transport?

proton motive force

The proton-motive force, generated as a result of cellular metabolism, is usually the driving force behind active transport. You will recall that cellular metabolism uses the energy released from catabolic reactions to establish gradients of ions across the membrane, accompanied by an electrical potential. Much of the evidence for the involvement of the proton-motive force in active transport comes from studies on sugar and amino acid uptake in bacteria. Bacteria predominantly couple gradients of H^+ (the coupling ion) to active transport and there are many ways by which different bacteria generate gradients of H^+ activity. You are already familiar with the two most widely used mechanisms, which are:

- hydrolysis of ATP to ADP + Pi by a membrane-bound ATPase. This brings about the extrusion of protons from the cell and occurs in anaerobic bacteria and in the plasma membrane of eukaryotes;
- operation of the respiratory chain, which occurs in aerobic organisms.

Both processes set up a H^+ gradient with the cytoplasm alkaline and electrically negative compared to the outer surface. The membrane is impermeable to both H^+ and OH^-, so that the equilibrium is not spontaneously restored. Thus the force exerted on the coupling ion, primarily the difference in electrical potential, is used by the cell to drive the movement of solute. The establishment of a proton-motive force is often referred to as primary active transport.

primary active transport

Some active transport systems in bacteria do not use the proton-motive force but are dependent upon phosphate bond energy. The mechanism by which phosphate bond energy is utilised to drive these transport systems remains unclear.

How is the proton-motive force coupled to active transport?

This is achieved by the carrier protein having specific sites for both its substrate and the coupling ion.

secondary active transport

The substrate and H^+ are carried across the membrane in a 1:1 ratio, making it possible for the carriers to 'pump' nutrients inwards. This is illustrated in Figure 2.9. The movement of solute driven by the proton motive force is called secondary active transport. Cations, such as K^+ may accumulate in the cytoplasm in response to the electrical gradient, since the interior of the cell is negative (Figure 2.9). Movement of cations in this way can set up a secondary gradient of ions which may itself serve as an energy source for the active transport of other molecules.

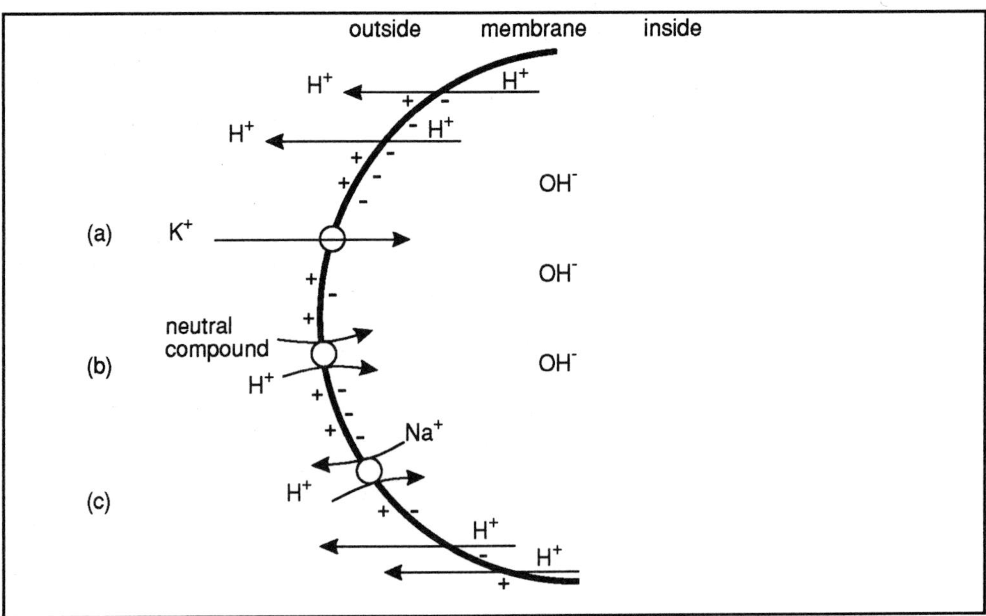

Figure 2.9 Explanation of how an ion gradient (OH^- inside, H^+ outside) could lead to transport of either cations or anions.Active transport mechanisms; a) b) and c) are referred to in the text.

three classes of secondary active transport

There are three classes of secondary active transport, these are:

symport

- symport: two molecules are transported by the same carrier in the same direction. In active symport one molecule flows down its previously established gradient and the other molecule flows with it;

antiport

- antiport: two molecules are transported on the same carrier in opposite directions. In active antiport one molecule flows down its concentration gradient, thus exchanging one gradient for another;

uniport

- uniport: only one molecule is transported per carrier. Active uniport is the flow of ions driven by the electrostatic gradient.

Π Now identify each of the three transport systems a) to c) shown in Figure 2.9 as one of the three classes of secondary transport systems defined above.

Transport system a) is a uniport.

Transport system b) is a symport.

Transport system c) is a antiport.

2.8 Group translocation across membranes

substrate altered during transport

You will recall that active transport generates a concentration gradient of transported substrate, without chemical modification of the substrate molecule. But there is another type of carrier-mediated membrane transport process across the membrane. In this a chemical on one side of the membrane is transported across the membrane and released in a chemically modified form on the other. (Examine the transport of a sugar shown in Figure 2.10). Since the product that appears inside the cell is chemically different from the external substrate no concentration gradient is produced across the membrane. This is called group translocation since it is a given chemical group that is transported and not an unaltered molecule or ion.

The most thoroughly studied group translocation system is the phosphotransferase system by which certain sugars enter bacteria at the expense of phosphoenolpyruvate (PEP). It catalyses the general reaction:

sugar (outside) + PEP (inside) $\rightarrow$ sugar-P (inside) + pyruvate (inside)

PEP donates a phosphate

Each phosphotransferase system is complex, involving the sequential action of four distinct phosphate-carrying proteins as illustrated in Figure 2.10.

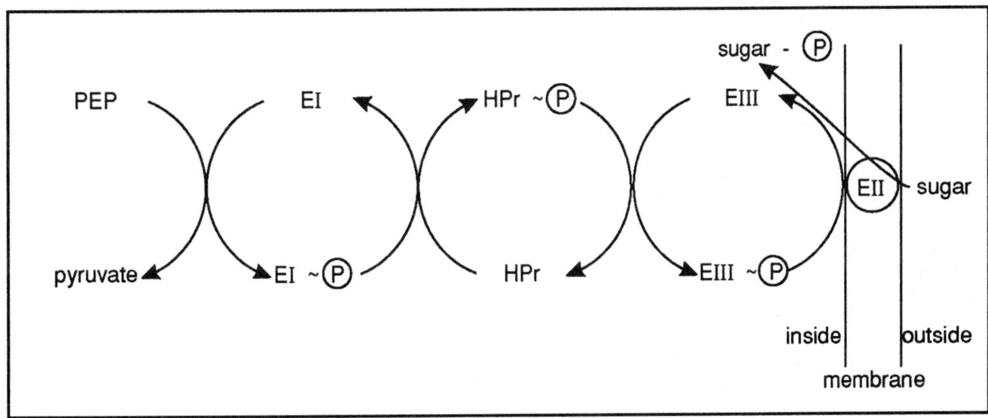

Figure 2.10 The functioning of a phosphotransferase system - an example of group translocation across a membrane. A high-energy phosphate group from phosphoenolpyruvate is transferred through a chain of four proteins to the incoming sugar molecule. Enzyme II also serves as the membrane carrier protein. (See text for details).

Enzyme II (EII), located within the membrane, serves as a carrier for the sugar substrate and accepts phosphate from Enzyme III (EIII). These two enzymes are specific for a particular sugar substrate and their synthesis is usually induced by the presence of the sugar in the cell's external environment. Enzyme I (EI) and HPr (a small temperature stable protein) are both nonspecific and participitate in all the phosphotransferase systems of a particular cell.

EI and HPr are common

The chemical change of the substrate that occurs on its entry into the cell requires the expenditure of energy in the form of a high-energy phosphate bond. However, you will recall that glucose must be phosphorylated before it is catabolised in the cell. Transport is thereby accomplished by a reaction that would also occur intracellularly even if glucose was brought in by an active transport system. Group translocation mechanisms thus conserve metabolic energy. Group translocation mechanisms involving high-energy phosphate bonds are also thought to play a role in the transport of purine and pyrimidine bases by certain bacteria.

SAQ 2.4

Which of the statements numbered 1-7 below apply to a) passive diffusion; b) active transport; c) passive transport; d) group translocation?

1) Substance modified during transport.
2) Substrate binds to carrier protein.
3) Uses but does not bind to carrier protein.
4) Uses receptor operated channels.
5) Does not use membrane protein.
6) Uses voltage operated channels
7) Consumes metabolic energy (directly or indirectly).

SAQ 2.5

What will be the consequences for a bacterium of a mutation causing a deviant non-functional HPr enzyme in that bacterium? What will be the consequences if the only non-functional enzyme is enzyme II?

SAQ 2.6

Use a simple line drawing to represent the following transport processes on the cell membrane.

1) Passive diffusion of O_2.
2) Facilitated diffusion of glycerol (GLY).
3) Shock-sensitive active transport of maltose (MAL).
4) Proton symport of lactose (LAC).
5) Sodium antiport of melibiose (MEL).
6) Group translocation of glucose (GLU).

2.9 Binding proteins

Gram-negative bacteria have binding proteins

A variety of proteins, collectively known as binding proteins occur in the periplasm of Gram-negative bacteria and have the property of binding with high affinity to specific substrates. They are not carrier proteins themselves because they are not located in the cell membrane. Nevertheless they play an essential role in the transport of certain substrates in bacteria.

cold shock changes permeability

The role of binding proteins in membrane transport can be demonstrated by damaging the cell wall of Gram-negative bacteria by cold osmotic shock treatment. This allows the binding proteins in the periplasm to leak out into the suspending medium. Cells treated in this way lose their ability to transport certain substrates but retain their ability to transport others. A direct correlation exists between the presence in the periplasm of a binding protein for a particular substance and the loss, after cold osmotic shock, of the cell's ability to transport that substance.

shock-sensitive and insensitive transport systems

Such transport systems are called shock-sensitive; those that remain in osmotically shocked cells are termed shock-insensitive. Shock-sensitive active transport systems are usually driven by the hydrolysis of ATP or some other source of high energy phosphate bond. Shock-insensitive systems, on the other hand, are usually driven by a proton-motive force.

Binding proteins have two important roles in transport:

binding proteins stimulate active transport

- by binding to the substrate they increase its effective concentration in the periplasm, providing a higher effective concentration to the carrier protein for transport into the cell;
- by interacting with the carrier protein, binding proteins stimulate active transport.

A large number of specific binding proteins have been isolated which are capable of binding certain amino acids, many sugars, and inorganic ions.

2.9.1 The exploitation of binding proteins

waste water treatment

Binding proteins, because of their high affinity and specificity, have potential for use in biotechnological processes designed to remove specific pollutants from waste water. They have the additional benefit of being very stable proteins that bind their ligands over a broad range of pH and ionic strength. The phosphate-binding protein of *Escherichia coli* has been used to develop a model system for the removal of pollutants involving the use and regeneration of the binding protein (Figure 2.11).

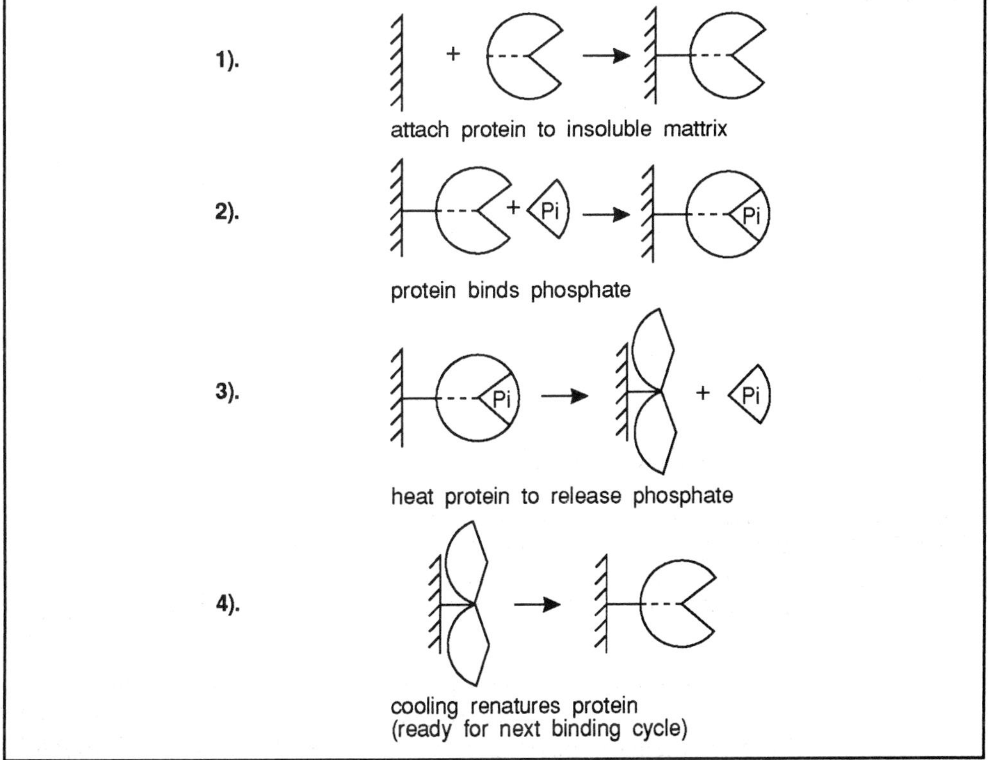

Figure 2.11 A scheme for phosphate-binding protein immobilisation and cyclic ligand binding.
1) Phosphate-binding protein of *E. coli* is immobilised to inert chemical beads by chemical binding; the beads are packed into a water-jacketed column. 2) The protein binds phosphate until all the sites are filled. 3) The bound phosphate is removed by heating the column. 4) Cooling regenerates the protein; the column can function through many repeated cycles. (Pi = inorganic phosphate).

2.9.2 The special case of iron transport

Iron transport can be regarded as a special case of mineral nutrition since it is required in fairly large amounts by most organisms (although not at the level of macronutrients), yet is normally present in the environment in a very insoluble form. Iron has two oxidation states, FeII (Fe^{2+} or ferrous iron) and FeIII (Fe^{3+} or ferrous iron). FeII iron is generally more soluble than FeIII iron and iron can, to some extent, be solubilised by reducing it to the FeII state. In solution FeII is spontaneously oxidised to FeIII. Iron solubility is modified by chelating agents. These contain atoms such as oxygen or nitrogen which can coordinately donate lone pairs of electrons to transition elements such as Fe.

ironophores increase iron transport

Many micro-organisms produce natural chelators, iron-binding organic compounds called ironophores, which solubilise Fe III iron and transport it into the cell. One major group of ironophores are derivatives of hydroxamic acid, which bind iron very strongly, as shown below.

$$\underset{\text{hydroxamate group}}{R-\underset{|}{\overset{}{N}}(OH)-\underset{\|}{C}(O)-R} + Fe^{3+} \longrightarrow R-N(O)-C(O)-R \ \ (\text{both O bound to } Fe^{3+})$$

Once the iron-hydroxamate complex has passed into the cell by active transport, the iron is reduced to the FeII state, and since the complex with FeII is less strong, the iron is released and can be used in the synthesis of iron-containing compounds (Figure 2.12).

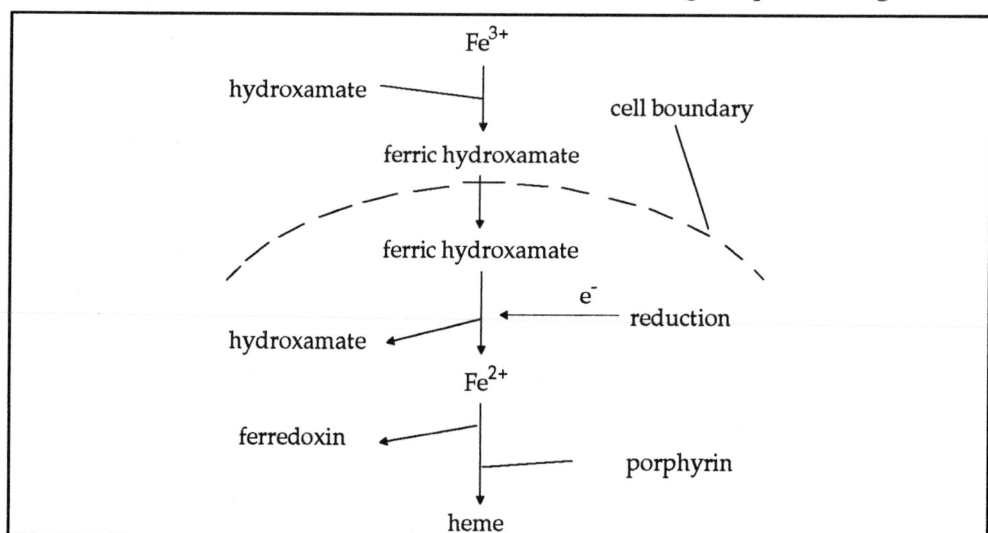

Figure 2.12 Diagram showing an example of the role of iron-binding hydroxamate in the iron nutrition of a bacterium.

natural and synthetic chelating agents

Some bacteria require but cannot synthesise hydroxamates and must have them provided in their culture medium, where they act as vitamin type growth factors. In culture media, iron can be made available by providing it in chelated form with a synthetic chelating agent, such as ethylenediamine-tetraacetic acid (EDTA). Providing iron in chelated form is especially important when organisms are being grown in media that lack natural iron chelators, such as amino acids, and where the pH is likely to rise above 6.0.

2.10 The utilisation of substrates that cannot pass through the membrane

2.10.1 The role of exoenzymes

external digestion

Some molecules (eg starch, cellulose, or RNA) and highly charged small molecules (eg nucleotides) that cannot pass through the cell membrane can nevertheless be used as substrates for growth by many organisms. These substrates are enzymatically degraded in the external medium by exoenzymes which are secreted mainly by fungi and bacteria. Examples of exoenzymes are given below:

Exoenzyme	Macromolecular substrate	Molecule that enters the cell
amylase	starch	glucose, maltose oligoglucosides
peptidase	peptides	amino acids
cellulase	cellulose	glucose, cellobiose
deoxyribonuclease	DNA	deoxyribonucleosides

The products of these degradations then enter the cell by specific transport systems. This process of extracellular digestion is in many respects similar to the ones by which complex foods are broken down in the stomach and intestines of animals.

test for extracellular digestion

The action of exoenzymes can be readily demonstrated. When microbial colonies develop on an agar medium containing particles of an insoluble macromolecule that can be digested, each colony is surrounded by an expanding clear area in which the insoluble substrate has been hydrolysed by the action of the exoenzyme.

In addition to true exoenzymes, Gram-negative bacteria produce certain incompletely excreted proteins which are also responsible for substrate degradation. These may be located in the cell membrane, the periplasm or the outer layers of the cell wall.

2.10.2 The involvement of membrane vesicles in membrane transport

endocytosis: uptake of large molecules

Macromolecules may also be transported by mechanisms which require the formation and fusion of membrane vesicles. Uptake of nutrients by this process is called endocytosis. During endocytosis the cell membrane invaginates and engulfs the particle to be transported. Pinching off results in the creation of an intracellular vesicle. Three types of endocytosis can be described:

pinocytosis

- Pinocytosis. This occurs frequently in eukaryotic cells and is a process in which small vesicles, containing extracellular fluid in which nutrients are dissolved, are formed by invagination of the plasma membrane;

receptor-bound endocytosis

- Receptor-bound endocytosis. This is a more specific process of endocytosis which takes place on specific parts of the membrane, the so called coated pits. A schematic representation of receptor-bound endocytosis is shown in Figure 2.13;

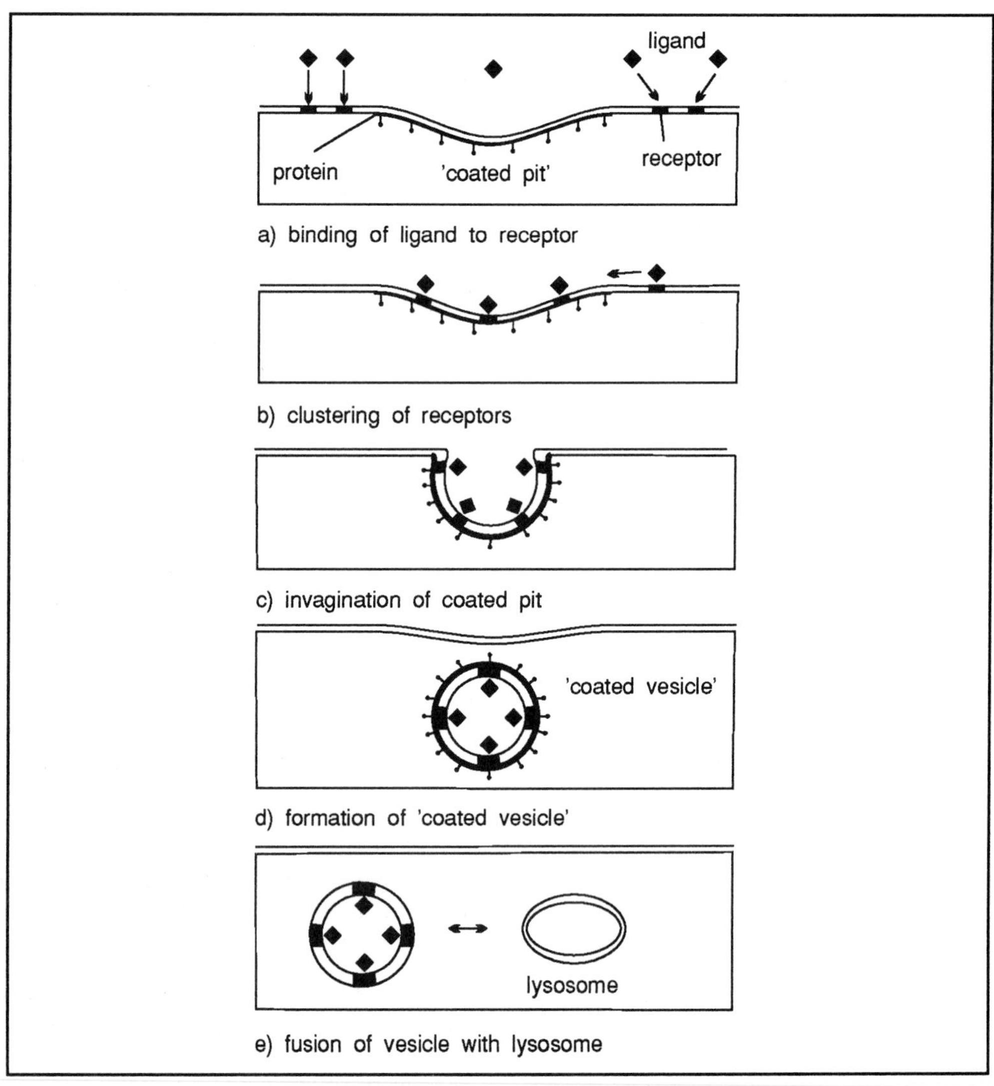

Figure 2.13 The stages of receptor-bound endocytosis. The coated pits are protein enriched regions of the plasma membrane, a) if a macromolecule binds to a specific receptor, the ligand receptor complexes in the region of the coated pit cluster together, b) after invagination, c) and pinching off, d) the vesicle is wrapped in a complex network of protein molecules derived from the coated pit. Soon after its formation (e),the vesicle loses this protein coat and is able to fuse with lysosomes so that the substances which have been accepted can be digested and degraded. Released sugars and amino acids can leave the lysosome and go into the cytoplasm.

phagocytosis

- Phagocytosis. This is a process in which certain eukaryotic cells take up extremely large particles, such as bacteria or cell debris. The intracellular vesicles that are formed, called phagosomes, fuse with lysosomes, after which the enzymes present in the lysosome degrade the engulfed particle. Phagocytosis is a form of receptor-bound endocytosis.

Π The proteins and several types of lipids in the cell membrane are asymmetrically-oriented in the two layers of the lipid bilayer. What happens to the orientation of membrane proteins when the vesicles are pinched off during endocytosis?

orientation of proteins in vesicles

In a transmembrane protein we can distinguish a cytoplasmic part and an exoplasmic part. In endocytosis the orientation of the proteins in the membrane will be maintained. This means in practice that the cytoplasmic part remains cytoplasmic after being pinched off from the membrane and the exoplasmic part is turned towards the lumen of the vesicle.

Π Now locate your position in the overall scheme shown in Figure 2.16. Identify which uptake processes have not yet been described.

You should have identified transport antibiotics as the only uptake process that has not been described yet. Before going on to them we need to look in some detail at the methods used to study membrane transport phenomena.

2.11 Experimental approaches to studying transport systems

In this section we will examine the experimental approaches which have contributed to our knowledge of membrane transport systems. You should remember that these experimental approaches invariably involve a host of analytical biochemistry techniques, particularly those of protein chemistry, which will not be described here.

2.11.1 Membrane preparations

When erythrocytes are exposed to distilled water under controlled conditions they swell and their membranes increase in permeability. As a result, the material in the cytoplasm can leak out into the surrounding medium. These preparations are called erythrocyte ghosts. Erythrocyte ghosts can be 'loaded' with different kinds of salts. For example, when isotonic NaCl solution is added to a suspension of ghosts, they shrink to normal size and their membranes return to their usual relatively-impermeable state. These preparations are called reconstituted or resealed erythrocytes. During the resealing process the ghosts entrap salts in a concentration equivalent to that in the suspending medium. Ghosts can also be loaded with ATP if it is added to the salt medium in which they are reconstituted. Studies using ghosts have provided valuable information on the operation of active transport systems across animal cell membranes.

loading of erythrocyte ghosts

For studies of transport systems in bacteria, membrane vesicles can be used. These can be prepared using simple methods involving enzymatic digestion of the cell wall and sonication of the resulting protoplasts in a suitably buffered solution.

Vesicles from *E. coli* are empty sacs, 0.5 to 1.5μm in diameter with the membrane oriented such that the cytoplasmic side is outermost and the exoplasmic side turned towards the lumen of the vesicle. They retain full activity of the major active-transport systems and also contain the components of the respiratory chain, all bound to the membrane structure. Such vesicles promote the active transport of a number of different sugars and the concentration of sugars within the vesicles may exceed by a hundredfold the external concentration of the sugar in the medium. The absence of cell wall material in vesicle preparations simplifies membrane transport studies compared with the use of whole bacterial cells.

vesicles aid the study of transport processes

2.11.2 Measurement of concentration gradients

Measurement of concentration gradients gives information on whether a given transport process occurs up or down a gradient. However, the precise magnitude of the concentration gradient of a solute across a biological membrane is not always easy to measure, since the available analytical methods may not be able to distinguish between free and bound solute. For example, Ca^{2+} is strongly bound to other intracellular solutes, particularly proteins. The bound Ca^{2+} does not contribute to the gradient of the free Ca^{2+} and an analytical method that determines total Ca^{2+} of cells or surrounding medium does not give an accurate measure of the true Ca^{2+} gradient across the plasma membrane. Special analytical procedures are often required to determine the concentration of unbound molecular species.

bound and unbound ions must be distinguished

2.11.3 Mutants of bacteria

The isolation of mutants in conjunction with physiological studies can provide valuable information on transport systems. The following example illustrates this point. When certain bacteria are grown on glucose as the sole carbon source, they are unable to utilise external citrate added to the medium. However, if such cells are removed from the glucose medium and placed in a medium where citrate is the sole carbon source, they quickly adapt and acquire the ability to utilise exogenous citrate as a carbon source. One possible explanation is that the bacteria have an inducible citrate-transport system whose biosynthesis is genetically repressed when it is not needed ie when glucose is present. This was confirmed by the isolation of mutants of such bacteria in which the ability to use citrate in response to exogenous citrate had been lost. These mutants are called transport-negative mutants and are believed to be metabolically intact in all other respects. From such experiments it has been concluded that the molecular components of membrane transport systems are genetically determined, just like the enzymes of a metabolic pathway.

transport systems can be induced

transport-negative mutants show loss of transport functions

SAQ 2.7

Having in mind what was discussed above, consider whether the bacteria will have a glucose carrier protein and/or citrate carrier protein in their cell membrane under the following conditions:

1) Wild-type bacteria with glucose and citrate in the medium.

2) Wild-type bacteria with citrate in the medium.

3) A transport-negative mutant for citrate with citrate in the medium.

4) A transport-negative mutant for glucose with glucose in the medium.

Other mutants of bacteria have been found which are capable of electron transport but unable to carry out phosphorylation of ADP. These mutants can still transport sugars which indicates that ATP formation is not required for the coupling of the energy of electron transport to the sugar-transport mechanism. This is consistent with the proton-motive force providing the energy for sugar transport in bacteria.

2.11.4 Inhibitors

Another approach is to use inhibitors in studies of membrane transport. Oligomycin, for example, is an inhibitor of oxidative phosphorylation and can again be used to demonstrate that ATP is not required for sugar transport in bacteria. Inhibitors which block or alter functional groups on proteins can be used to investigate the role of these proteins in membrane transport. A range of inhibitors is available which strongly and specifically inhibit specific transport systems.

2.11.5 Transport antibiotics

Some micro-organisms synthesise small compounds that make membranes permeable to certain ions. They are valuable experimental tools in membrane transport studies and are called ionophores. You should note that ionophores are distinct from ironophores, as described previously in Section 2.9.2.

ionophores disrupt ion gradients

Ion gradients are essential for optimal functioning of the cell and because ionophores disrupt ion gradients across membranes by causing leakage of ions, they act as antibiotics.

Because of the way in which they function, ionophores can be divided into two groups:

channel-forming ionophores

- Channel-formers, which create pores in the membrane. Their specificity is mainly determined by the internal diameter of the ion channel. Gramimycin, for example, induces an increased permeability of the membrane for Na^+ and K^+ and other monovalent cations. Its mechanism of action is shown in Figure 2.14.

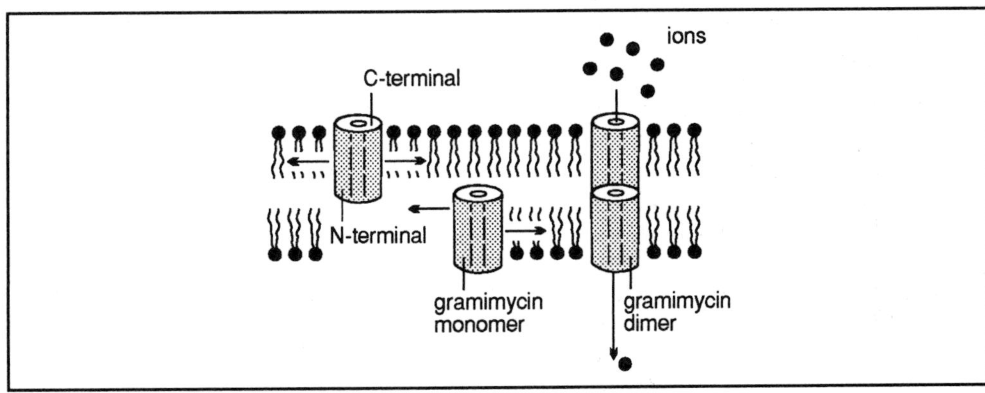

Figure 2.14 The action of a channel forming ionophore (gramimycin). Two monomers, one in each monolayer of the membrane, associate to form one complete pore in the membrane.

gramimycin monomers form channel dimers

Gramimycin is a polypeptide arranged in a helical manner with a hydrophobic periphery and a hydrophilic channel. Each polypeptide spans one half of the lipid bilayer so a channel can be opened only if a dimer is created between two gramimycin molecules. This relies on association and dissociation of the gramimycin molecules on monolayers of the membrane which lie opposite each other. In the short period of time when a channel is open (approximately one second) about 10^7 ions can cross the membrane. In Figure 2.16 channel-formers are classified under carrier-mediated transport. However, you should note that because binding of ions to the transport antibiotic molecule does not occur, they could also be classified under the open channel category of passive diffusion.

diffusional carrier ionophores

- Diffusional carriers are carriers which wrap around certain ions and thus enable transport. The arrangement of a well-known ion carrier ionophore, valinomycin, is shown in Figure 2.15. It has a hydrophilic centre which can bond six to eight oxygen atoms and take up a K^+ ion. Because of the hydrophobic exterior of the ring, the whole complex becomes soluble in the hydrophobic matrix of the membrane.

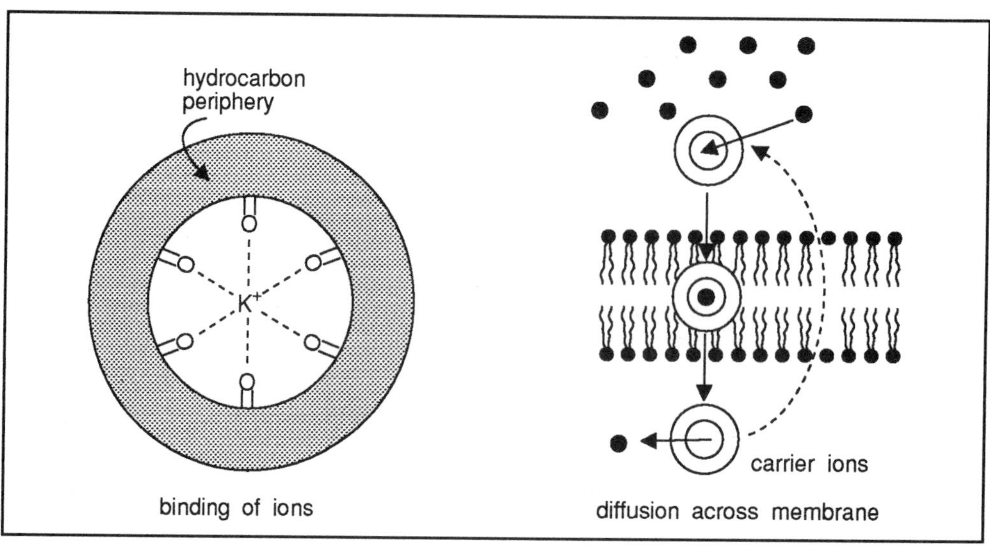

Figure 2.15 The action of valinomycin, a diffusional carrier ionophore. The hydrophobic exterior shields the ion charge and the entire complex can cross the membrane.

Π Are the transport antibiotics like valinomycin examples of passive diffusion or passive transport?

They bind to the transported molecule and, therefore, are examples of passive transport.

ion channels and carriers respond differently to temperature

Their direction of operation is determined by the electro-chemical gradient. In the case of transport through ion channels, the rate depends on the speed at which the ions diffuse while with ion carriers, transport is determined by the diffusion of the whole carrier across the membrane. In experiments it is possible to distinguish the two forms of transport by measuring the rate of transport as a function of temperature.

SAQ 2.8

Experiments employing mutant strains of bacteria that lack membrane-bound ATPase showed that certain substrates can be transported by these cells only if they are provided with an energy source, the metabolism of which is known to generate ATP by substrate-level phosphorylation. Other substrates were transported even if the metabolized substrate did not generate ATP by substrate-level phosphorylation.

Interpret these results in the light of what you know about the mechanism of active transport.

SAQ 2.9

Explain how the rate of ion transport across the membrane of a synthetic lipid vesicle depends on temperature for both channel-formers and diffusional carriers. Remember that lipids in membranes can be either in a gel phase (practically immobile fatty acid tails, arranged in orderly fashion) or in a liquid crystalline phase (interaction between fatty acids tails has decreased causing increased mobility). Will the influence of temperature cause gradual or sudden changes in the rate of transport in each case?

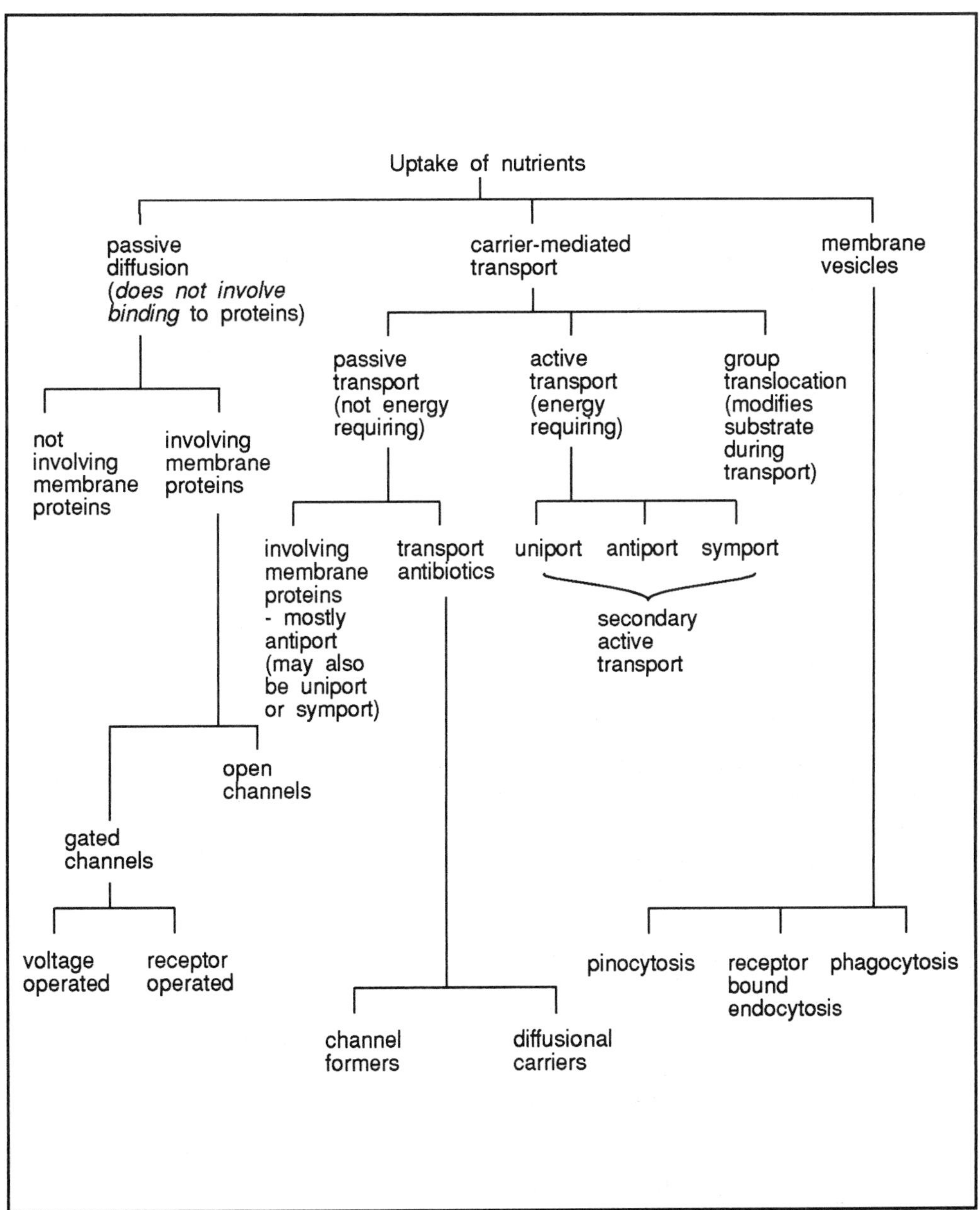

Figure 2.16 A flow diagram showing the relationship between the different classes of membrane uptake processes.

SAQ 2.10

Explain why the conductivity of an artificial lipid bilayer will not change continually but in a step-by-step manner under the influence of gramimycin molecules.

Summary and objectives

We have seen that cells have many nutritional requirements and that lipid bilayers are impermeable to all but a few nutrients. There are many different classes of transport processes in biological membranes, the relationship between them is summarised in Figure 2.16. Most nutrients enter the cell by carrier-mediated transport systems. These are quite specific and control the type of molecules entering the cell. Binding proteins and exoenzymes, although not responsible for membrane transport, aid the uptake of nutrients. There are many experimental approaches to studying uptake using both artificial and natural membranes, each with its own particular merits and limitations.

Now that you completed this chapter you should be able to:

- explain why cells need solute uptake mechanisms;
- predict the order of permeation through a lipid bilayer of given substrates;
- describe the similarities and differences between diffusion and transport phenomena and identify the class of uptake process from given characteristics;
- explain the involvement of the proton-motive force in active transport;
- draw models to describe the various classes of uptake processes;
- explain why the properties of binding proteins are important both to some cells and to biotechnologists;
- discuss experimental approaches to studying uptake processes.

Nitrogen and sulphur assimilation

Introduction	42
3.1 The requirement for nitrogen and sulphur	42
3.2 The assimilation of sulphate	43
3.3 The assimilation of ammonia	46
3.4 The assimilation of nitrate	48
3.5 The assimilation of molecular nitrogen	50
3.6 The importance of nitrogen fixation in nature	61
Summary and objectives	64

Nitrogen and sulphur assimilation

Introduction

In this chapter we will examine the biosynthetic reactions that enable nitrogen and sulphur from inorganic sources to be incorporated into organic material within cells. These are called assimilation reactions. We will not be considering the assimilation of any other elements in detail and there are three main reasons for this:

- metabolism of carbohydrates yield carbon, oxygen and hydrogen. Of the other major bioelements, (phosphorus, nitrogen and sulphur), phosphorus is often available as phosphate within the enviornment and is used within cells as phosphate. Nitrogen and sulphur occur in a variey of forms and are often chemically modified before being assimilated;
- assimilation of some inorganic forms of nitrogen and sulphur is unusual because they are reduced in the cell prior to incorporation into organic material;
- all the reactions are important in the geological cycling of elements and some have great biotechnological potential.

We will consider the cell's requirement for nitrogen and sulphur first and then go on to describe the assimilation reactions from the most common and important sources of sulphur and nitrogen in more detail.

3.1 The requirement for nitrogen and sulphur

cysteine
methionine

Both nitrogen and sulphur are fundamental requirements of living systems. Inorganic sulphur is required for the synthesis of the two sulphur-containing amino acids cysteine and methionine.

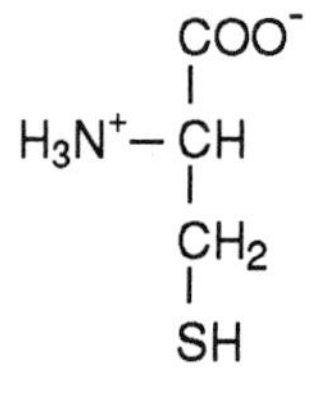

COO^- | H_3N^+—CH | CH_2 | SH

cysteine

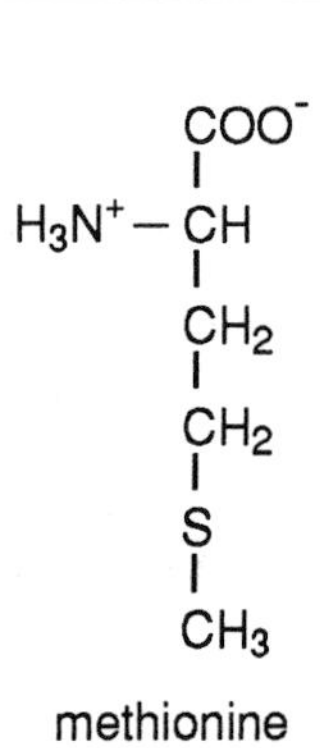

COO^- | H_3N^+—CH | CH_2 | CH_2 | S | CH_3

methionine

The main sulphur-containing compounds in the cell are proteins and some vitamins (thiamine, biotin and lipoic acid). Sulphur enters biosynthetic metabolism in a reduced state, as hydrogen sulphide (H_2S). Some organisms can utilise a more oxidised form of sulphur, such as sulphate, but these forms must all be reduced to H_2S before they can be incorporated into organic compounds.

Like sulphur, inorganic nitrogen is required for the synthesis of amino acids. These, in turn, are required for the synthesis of purines and pyrimidines.

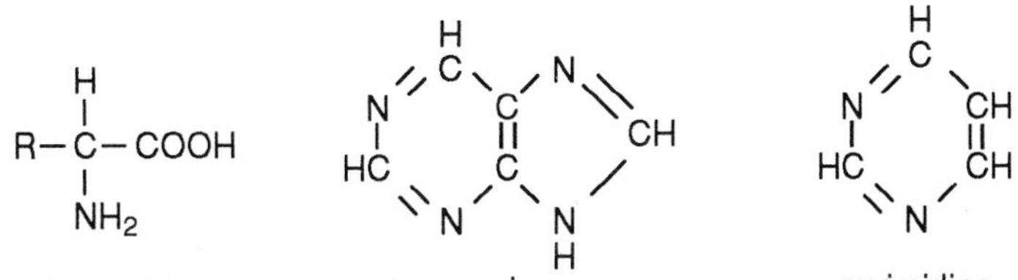

amino acid　　purine　　pyrimidine

various forms of nitrogen are used

The structures shown above are building blocks for the synthesis of many different types of nitrogen-containing compounds in the cells, such as proteins, nucleotides and nucleic acids, porphyrins and vitamins. Nitrogen enters biosynthetic metabolism in a reduced state, as ammonia (NH_3). Some organisms can only use nitrogen in this reduced state and, in culture media, the requirement for ammonia can be satisfied by the provision of nitrogen as ammonium salts eg NH_4Cl. Other organisms can also use more oxidised forms of inorganic nitrogen, the most common of which is nitrate. This can be supplied as a nitrate salt, such as KNO_3. Other inorganic sources used by certain micro-organisms include nitrite (NO_2^-), cyanide (CN^-) and hydroxylamine (NH_2OH). Free nitrogen gas (N_2), also known as molecular nitrogen or dinitrogen, can also be used by certain bacteria. Regardless of the chemical nature of the source, nitrogen actually enters into the formation of an organic compound as ammonia.

ammonia is the key reactant

Thus all of the above sources of nitrogen must be converted to ammonia, as we will see below.

nitrogen in culture medium

In culture media, the nitrogen and sulphur requirements can often be met by the provision of organic nutrients that contain these elements in reduced form. These may be as amino acids or more complex protein degradation products, such as peptides. There are a large variety of peptide mixtures that can be used. They differ in the sources of protein used (eg animal proteins, plant proteins) and in the way these proteins were hydrolysed. A common peptide mixture is peptone.

Π If peptone water was the sole source of nitrogen and sulphur for cells, would you expect the reduction of nitrogen and sulphur to occur during assimilation?

The answer is no. Peptone water is a complex mixture of proteins, polypeptides and amino acids. In this form nitrogen is already reduced to the level of ammonia and sulphur to the level of hydrogen sulphide.

SAQ 3.1

Which of the following forms of nitrogen or sulphur can be directly incorporated into organic compounds?

1) nitrate; 2) ammonia; 3) nitrite; 4) nitrogen gas; 5) hydrogen sulphide; 6) sulphate; 7) sulphite.

3.2 The assimilation of sulphate

In this section we shall consider how sulphur compounds more oxidised than hydrogen sulphide are reduced prior to their incorporation into cellular material. The great majority of micro-organisms can fulfil their sulphur requirements from sulphate. Sulphate salts, such as $MgSO_4.7H_2O$, are common ingredients of culture media.

reduction of sulphate

The process in which the sulphur atom in sulphate is reduced prior to its incorporation into cellular organic material is called assimilatory sulphate reduction and involves the transfer of eight electrons to sulphate.

The number of electrons transferred is equal to the difference in oxidation state (or oxidation number) of the sulphur atom in sulphate and sulphide. Oxidation state is defined as the number of atoms of hydrogen that will combine with (or replace) one atom of an element. For example, oxygen has an oxidation state of -2 and will combine with two hydrogens to form the neutral molecule H_2O. In an ion, the sum of the oxidation states of all atoms is equal to the charge on that ion. In simple molecules the oxidation number of an atom can be calculated by adding up the H and O atoms present and using the oxidation number of these elements, along with the charge on the ion, to calculate the oxidation state of an atom.

Π What are the oxidation states of the sulphur atom in the sulphate ion and in hydrogen sulphide (oxidation state of hydrogen is +1)?

In the sulphate ion the sum of oxidation states of the oxygen is (4) x (-2) = -8 and there is a charge on the ion of -2. The oxidation state of the sulphur atom is given by (-8) + (X) =- 2, which equals +6. In hydrogen sulphide (H_2S) each hydrogen has an oxidation state of +1. As the molecule has no charge then the oxidation state of sulphur must be -2. Since the difference in oxidation states +6 - (-2) = 8, eight electrons are required to reduce sulphate ions to hydrogen sulphide.

The pathway of assimilatory sulphate reduction to H_2S has four stages.

activated sulphate

- Synthesis of activated sulphate. The sulphate is first converted to adenosine phosphosulphate. This involves the transfer of adenosine monophosphate from ATP to sulphate and is catalysed by the enzyme sulphate adenylyl-transferase (also called ATP sulphurylase).

$$ATP + SO_4^{2-} \rightleftarrows \text{adenosine 5'-phosphosulphate} + 2PPi.$$

The enzyme adenylylsulphate kinase then further activates the sulphate by transfer of a phosphate from ATP to form phosphoadenosine phosphosulphate (adenosine 3'-phosphate 5'-phosphosulphate).

$$ATP + \text{adenosine 5'-phosphosulphate} \rightarrow \text{adenosine 3'-phosphate 5'-phosphosulphate} + ADP$$

sulphite

- Reduction of sulphate to sulphite. This is catalysed by adenosine 3'-phosphate 5'-phosphosulphate sulphate reductase which uses NADPH as electron donor.

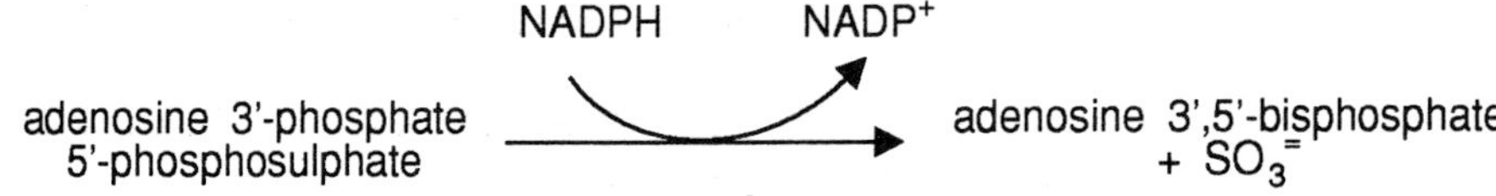

sulphide

- Reduction of sulphite to hydrogen sulphide. The final six-electron transfer is catalysed by a large enzyme called sulphite reductase, which also uses NADPH as electron donor.

$$SO_3^{=} \xrightarrow{3NADPH \rightarrow 3NADP^+} H_2S + H_2O$$

In plants the reduction of sulphate takes place in chloroplasts. The AMP part of adenosine 5′-phosphosulphate is replaced by a membrane bound carrier. Thus:

$$\text{carrier} - SH + AMP - O - S(=O)_2 - OH \longrightarrow \text{carrier} - S - S(=O)_2 - OH + AMP$$

$$\text{carrier} - S - S(=O)_2 - OH \xrightarrow{\text{reductant}} \text{carrier} - S - SH$$

The usual reductant in chloroplasts is reduced ferredoxin.

amino acid

- Formation of amino acid. The convertion of hydrogen sulphide to organic sulphur occurs by reaction with the amino acid serine to form cysteine. This is the precursor of other sulphur-containing compounds in the cell.

$$H_2S + H_2C(OH) - CH(NH_2) - COOH \rightleftharpoons H_2C(SH) - CH(NH_2) - COOH + H_2O$$

Because of the long names of some of these substrates, many shortened versions have been used in the literature. For example, adenosine 5′-phosphosulphate = APS; adenosine 3′-phosphate 5′-phosphosulphate = phosphoadenosine phosphosulphate. A common feature is to omit positional information (eg 3′, 5′).

SAQ 3.2

Decide whether each of the following statements is true or false and justify your decisions:

1) Organisms able to utilise sulphate are also likely to be able to utilise sulphite and hydrogen sulphide.

2) NADPH is required for the reaction of hydrogen sulphide to form cysteine.

3) Serine can serve as a source of carbon, nitrogen and sulphur for cells.

4) Assimilation of sulphate is more energetically demanding on cells than the assimilation of cysteine.

5) Three high energy phosphate bonds are required for the assimilation of sulphate.

3.3 The assimilation of ammonia

fixation of ammonia

The nitrogen atom of ammonia is at the same oxidation level (-3) as the nitrogen atoms in the organic constituents of the cell. The assimilation of ammonia does not, therefore, necessitate oxidation or reduction. There are three ammonia fixing reactions:

glutamate

- one forms an amino group ($-NH_2$) in glutamate;

$$\underset{\text{2-oxoglutarate}}{{}^{-}OOC-(CH_2)_2-\overset{\displaystyle O}{\overset{\|}{C}}-COO^{-}} + NH_3 + NADPH + H^+ \xrightarrow{\text{glutamate dehydrogenase}} \underset{\text{glutamate}}{{}^{-}OOC-(CH_2)_2-\overset{\displaystyle NH_2}{\overset{|}{C}}-COO^{-}} + NADP^+ + H_2O$$

asparagine

- another forms an amide group ($-CO-NH_2$) in asparagine;

$$\underset{\text{aspartate}}{{}^{-}OOC-CH_2-CHNH_2-COO^{-}} + NH_3 + ATP \xrightarrow{\text{asparagine synthetase}} \underset{\text{asparagine}}{H_2N-\overset{\displaystyle O}{\overset{\|}{C}}-CH_2-CHNH_2-COO^{-}} + AMP + PPi$$

glutamine

- another forms an amide group in glutamine.

$$\underset{\text{glutamate}}{{}^{-}OOC-(CH_2)_2-CHNH_2-COO^{-}} + ATP + NH_3 \xrightarrow{\text{glutamine synthetase}} \underset{\text{glutamine}}{H_2N-\overset{\displaystyle O}{\overset{\|}{C}}-(CH_2)_2-CHNH_2-COO^{-}} + ADP + Pi$$

transamination

These three products of NH_3 assimilation can be directly incorporated into proteins. In addition, other amino acids can be formed by transamination between glutamate and non-nitrogenous metabolites, for example, the formation of L-alanine and L-aspartate shown below:

$$\text{L-glutamate} + \text{pyruvate} \rightarrow \text{2-oxoglutarate} + \text{L-alanine}$$

$$\text{L-glutamate} + \text{oxaloacetate} \rightarrow \text{2-oxoglutarate} + \text{L-aspartate}$$

Similarly, the amide group of glutamine is the source of the amino group in many metabolites, such as carbamyl phosphate, NAD^+ and guanosine triphosphate. The glutamine is converted back to glutamate in these reactions. An example is shown below:

$$\text{uridine triphosphate} + \text{glutamine} + \text{ATP} \rightarrow \text{cytidine triphosphate} + \text{glutamate} + \text{ADP} + \text{Pi}$$

We have seen that there are three ammonia assimilation reactions and that nitrogen from ammonia can be converted to all the nitrogen-containing compounds of cells. Since there are three ammonia assimilation reactions, you may wonder if they are used equally or whether there is a one major route of NH_3 assimilation.

high ammonia concentration

There is not a simple answer to this since it concerns the regulation of the three assimilation reactions. The route of NH_3 assimilation depends upon the concentration of NH_3 in the cell. At high concentration two sequential reactions, catalysed by glutamate dehydrogenase and glutamine synthetase, lead to the formation of glutamine:

This pathway functions only when the concentration of available NH_3 is high because the substrate affinity of glutamate dehydrogenase is relatively low and the enzyme does not function effectively at low concentration of NH_3.

low ammonia concentration

At low NH_3 concentrations the glutamine synthetase reaction becomes the major route of NH_3 assimilation. Under these conditions a new enzyme, glutamate synthetase, is induced which catalyses the reaction:

glutamine + 2-oxoglutarate → 2 glutamate

The enzyme is sometimes called GOGAT, which describes its catalytic function (glutamine-oxoglutarate amino transferase); it transfers an amino group from glutamine to 2-oxoglutarate.

Π Write the sequence for reactions which occur at low NH_3 concentration. Remember that two enzymes are involved, glutamine synthetase and GOGAT. You should first write the reactions catalysed by each of these enzymes and then the overall reaction underneath. Attempt this without looking at our solution below

glutamate + NH_3 + ATP —(glutamine synthetase)→ glutamine + ADP + Pi

glutamine + 2-oxoglutarate —(glutamate synthase)→ 2 x glutamate

overall reaction: 2-oxoglutarate + NH_3 + ATP → glutamate + ADP + Pi

The low ammonia route is only used when it is needed. This means that it does not operate when the concentration of ammonia is high. In most bacteria for example, there are regulatory mechanisms that 'switch off' GOGAT by converting it to an inactive form when ammonia concentrations are high.

Π Can you think of a reason why 'switching off' the low ammonia route is desirable for the cells when the ammonia concentration is high? Re-examine the equations if you need to.

You will recall or can see that the low ammonia route uses ATP while the high ammonia route does not. The cell saves energy by switching off the low ammonia route.

SAQ 3.3

Match the enzymes labelled 1) to 7) with the statements labelled a) to j). Each statement may be used once, more than once or not at all.

1) GOGAT;

2) glutamate dehydrogenase;

3) sulphite reductase;

4) asparagine synthase;

5) glutamine synthetase;

6) sulphate adenylyl transferase;

7) glutamate/pyruvate transaminase.

a) requires ATP;

b) requires NADPH;

c) requires NADH;

d) requires NH_3;

e) requires glutamate;

f) functions only when the concentration of NH_3 is high;

g) can be converted to an inactive form;

h) ADP is formed;

i) glutamine is formed;

j) glutamic acid (glutamate) is formed.

3.4 The assimilation of nitrate

The nitrate ion can be used by many micro-organisms and plants as a source of nitrogen.

Π Complete the following statement by entering the missing numbers. The oxidation state of the nitrogen atom in NO_3^- is____ and to be assimilated to NH_3

must be reduced to an oxidation number of _____. This requires the transfer of _____ electrons to NO_3^-.

The missing numbers are +5, -3 and 8, in that order. Nitrate is reduced to ammonia before it is incorporated into organic material.

The process of assimilatory nitrate reduction is catalysed by two enzyme complexes called assimilatory nitrate reductase and assimilatory nitrite reductase. The reaction sequence is shown below:

$$NO_3^- \rightarrow NO_2^- \rightarrow NH_3$$

nitrate reduction

The nitrate reductase belongs to a small group of enzymes that contain molybdenum as cofactor (Mo-co) which serves as an electron carrier within the enzyme. Electrons are taken indirectly from NADH or NADPH.

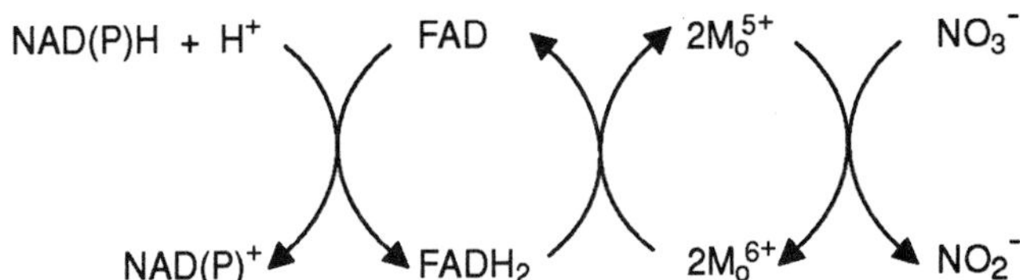

nitrite reduction

Nitrite reductase is a single enzyme that carries out a complex six-electron transfer reaction. The mechanism of transfer is unclear although hydroxylamine (NH_2OH) is known to be formed as an intermediate of the process.

We can represent this as:

reductant reductant

$$NO_2^- \longrightarrow [NH_2OH] \longrightarrow NH_3$$

The reduction of nitrate to ammonia is thermodynamically capable of generating sufficient energy to phosphorylate ADP to ATP. However, no ATP is produced and the reduction process can be viewed as actually costing ATP.

Π Why do you think that the reactions should be viewed as costing the cell ATP?

The electrons utilised to reduce nitrate could have been directed through an ATP-generating electron transport chain.

In eukaryotes, NADH and NADPH serve as sources of electrons in nitrite reduction and the electron transfer sequence is shown below.

$$NAD(P)H \rightarrow Fe\text{-}S \rightarrow FAD \rightarrow cytochrome\ b_{557} \rightarrow NO_3^-$$

Fe-S is an iron-haem protein rich in sulphydryl groups.

The assimilatory nitrate reductase in anaerobic bacteria does not use NADH or NADPH as electron donor but the actual donor is as yet unknown.

Π List the components of the electron-transport chain shown above in order of descending reduction potential. What is the function of Fe-S, FAD and cytochrome b_{557} in the sequence?

The order of descending reduction potential is the reverse order to that shown; NAD(P)H has the lowest reduction potential because it most readily transfers its electrons (it is the strongest reductant). Fe-S, FAD and cytochrome b_{557} are electron carriers and they 'channel' the electrons to NO_3^-, releasing free energy in small amounts.

Π NADH or NADPH are regarded as sources of electrons but from what are the electrons ultimately derived?

The electrons are derived from the compound acting as the energy source. This may be an organic or inorganic compound, depending on the metabolic category to which the organism belongs.

Π Distinguish between the terms assimilatory nitrate reduction and dissimilatory nitrate reduction.

The distinction between these processes is most readily made by considering what the nitrate is being used for by the cell. In assimilatory nitrate reduction, nitrate is used as a source of nitrogen for biosynthetic purposes. In dissimilatory nitrate reduction, the nitrate is used as a terminal electron acceptor in respiration. This process, called anaerobic respiration, is considered in detail in the Biotol text, 'Principles of Cell Energetics'.

3.5 The assimilation of molecular nitrogen

nitrogen fixation

Nitrogen in the N_2 molecule (dinitrogen) has an oxidation state of zero and must be reduced to ammonia (oxidation state -3) prior to incorporation into nitrogenous components of the cell. The process is called nitrogen fixation.

Reduction of N_2 to NH_3 requires some tough chemistry. The N_2 molecule is very stable and considerable energy is required to split the N-N bond.

Π How many electrons are required to reduce N_2 to $2NH_3$?

The change in oxidation state of the nitrogen atom (0 to -3) indicates that six electrons are required for each N_2. However, we shall see that when the reduction of N_2 to NH_3 is catalysed by biological systems eight electrons are required. This is because molecular hydrogen is also formed in the reaction. The role of hydrogen in this reaction will be considered later in this section.

3.5.1 Nitrogen-fixing organisms

The ability to fix nitrogen is the property of certain bacteria; no eukaryotic organisms fix nitrogen.

The nitrogen-fixing organisms can be divided into a number of ecological and physiological groups, as shown in Figure 3.1.

	Phototrophic	**Chemotrophic**
Free-living, aerobic	Cyanobacteria	*Azotobacter* group *Mycobacterium* Methane oxidisers *Thiobacillus*
Free-living, anaerobic	Cyanobacteria Purple bacteria Green bacteria	*Clostridium* *Klebsiella* *Bacillus* *Desulfovibrio* *Desulfotomaculum* Methanogenic bacteria
Symbiotic, aerobic	Cyanobacteria (+ fungi, ferns)	*Rhizobium* (+legumes, grass) *Azospirillum* (+ grass) *Frankia* (+ alder, hawthorn, etc)
Symbiotic, anaerobic	None known	*Citrobacter* (+ termites)

Figure 3.1 Types of nitrogen-fixing bacteria.

symbiotic nitrogen fixers

The symbiotic nitrogen fixers do so in close association with a plant or a lichen. You will recall that a lichen is itself a symbiotic relationship between a fungus and a lower alga. In the case of *Rhizobium* the bacteria are located within root nodules of certain plants, such as pea and clover. The relationship is symbiotic (mutually beneficial) because the enzymatic machinery for nitrogen fixation is produced by the bacterium and the host plant provides a suitable environment for nitrogen fixation. A micrograph of a root nodule on pea plants is shown in Figure 3.2.

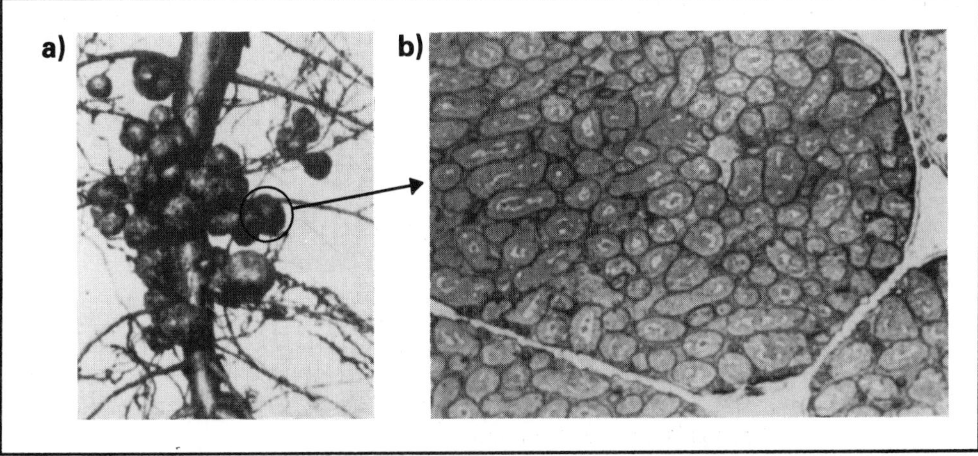

Figure 3.2 Root nodules on plant root system. a) root nodules on a pea plant; b) micrograph of a cell within the nodule packed with nitrogen fixing bacteria. The bacteria within the nodules take on a much enlarged morphology and become known as bacteriods. They are members of the genus *Rhizobium*.

All nitrogen-fixing organisms assimilate the ammonia formed from nitrogen by the low ammonia concentration route (Section 3.3).

requirements for nitrogen fixation

There are five essential requirements for biological nitrogen fixation which are:

- a source of reducing potential;
- suitable electron carriers;
- active nitrogenase (the nitrogen-fixing enzyme);
- a source of energy;
- protection against oxidation by molecular oxygen.

The various physiological groups of nitrogen fixers fulfil these requirements in different ways. However, the relationship between these requirements is essentially the same in all organisms. This relationship is illustrated in Figure 3.3 and represents part of a unifying concept of biological nitrogen fixation.

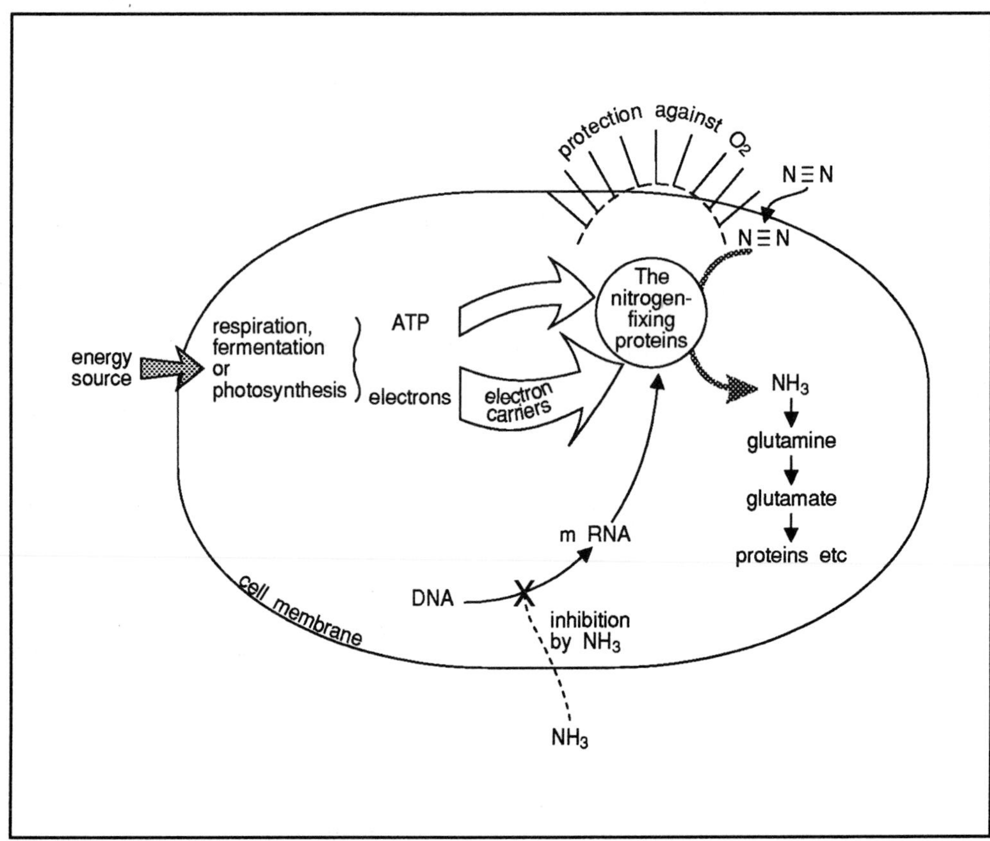

Figure 3.3 A unifying concept of biological nitrogen fixation.

We shall now examine each of the five essential requirements of nitrogen fixation in more detail. As you proceed you might find it useful to refer back to the unifying concept (Figure 3.3) to remind yourself of the overall scheme.

3.5.2 The biological reductant

electron donors in N_2 fixation

Nitrogen fixation is basically a reductive process requiring a continuous source of strong reducing agent. The origin and nature of the electron donors used in nitrogen fixation vary among the different physiological groups of nitrogen fixers. In anaerobic systems, pyruvate, hydrogen or formate are used and in aerobic or symbiotic nitrogen fixers NADH or NADPH are the electron donors.

1) Pyruvate, hydrogen or formate.

These three electron donors are strong reducers and will support nitrogen fixation in anaerobes but not aerobes.

Pyruvate undergoes a phosphoroclastic reaction involving inorganic phosphate and the released electrons are passed on to a low redox potential carrier called ferredoxin (Fd_{ox}).

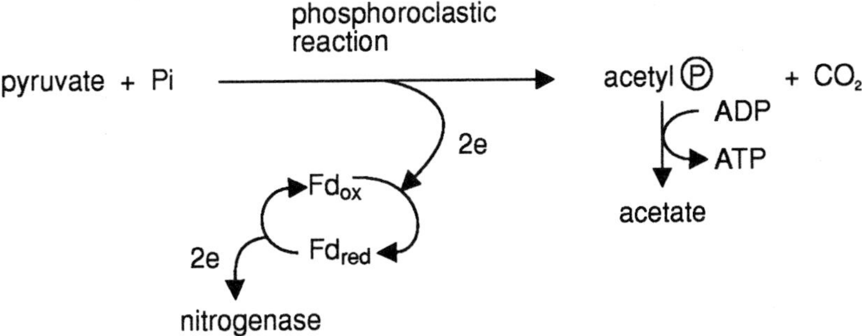

The acetyl-P can be used for the formation of ATP via acetyl kinase. In this way nitrogen is supplied with both reducing potential (Fd_{red}) and ATP. Since more ATP than Fd_{red} is required for nitrogen fixation the cells must provide extra ATP from other reactions.

SAQ 3.4

Does this extra ATP come from substrate level phosphorylations or from the respiratory chain?

Formate and hydrogen can also support nitrogen fixation in anaerobes by donating electrons directly to nitrogen fixation.

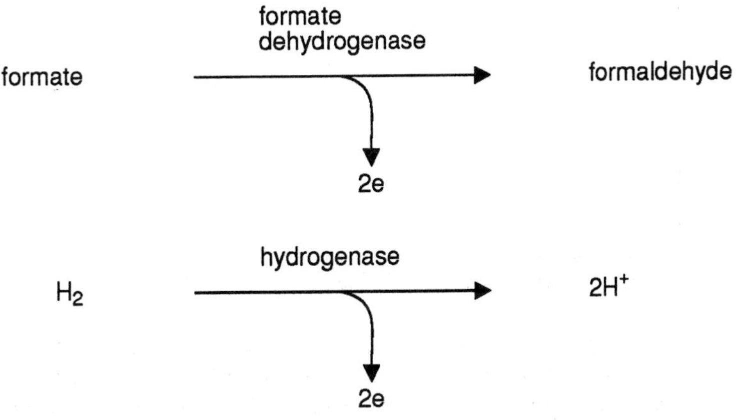

2) NADH or NADPH

There is no phosphoroclastic type of reaction in free living aerobic or symbiotic nitrogen fixers. Both NADH and NADPH have been also shown to function as electron donors.

Π The standard redox potentials (E_o) for $NADP^+/NADPH$ and ferredoxin are as follows: $NADP^+/NADPH$ -0.32 V, Fd_{ox}/Fd_{red} -0.42 V. What does this suggest to you about the transfer of electrons?

The redox potentials suggest that electrons will be transferred from ferredoxin to $NADP^+$ since ferredoxin is the stronger reductant. We should remember that the redox potential is, however, dependent upon relative concentrations of the oxidised and reduced forms.

Consider the reaction:

$$A_{red} \rightarrow A_{ox} + ne^{-1}$$

If the standard oxidation reduction potential for the reaction is E_o then we can write:

$$E = E_o + \frac{RT}{nF} \ln \frac{[A_{ox}]}{[A_{red}]}$$

where R is the gas constant, T is the temperature, n is the number of electrons transferred and F is the Faraday constant.

Thus the actual redox potential depends upon the ratio of $[A_{ox}]$ to $[A_{red}]$.

To go back to our problem concerning $NADP^+$ and ferredoxin, providing the ratio of $NADPH/NADP^+$ is high and Fd_{red}/Fd_{ox} is low, then:

$$E = E_o + RT \ln \frac{[NADP^+]}{[NADPH]}$$

will give a greater negative value than:

$$E = E_o + RT \ln \frac{[Fd_{ox}]}{[Fd_{red}]}$$

thus NADPH will be capable of reducing Fd_{ox}.

In cells the ratio of $NADP^+$:NADPH is often 1:100 or greater.

3.5.3 The electron carriers

flavodoxin

Electron donors do not pass their electrons directly to nitrogenase but use two types of electron carrier to achieve this. These are ferredoxin, which we have already met, and flavodoxin (Fl), which we have not. They are both relatively small proteins, (relative molecular mass from 6,000 to 24,000), which have the common property of existing in oxidised and reduced forms which can easily be interconverted. The reduced forms are strong reducing agents capable of reducing many other biological molecules and usually react with air, becoming oxidised.

structure of ferredoxins

You will recall that ferredoxins are involved in a variety of biological processes, including photosynthesis in plants and pyruvate metabolism in anaerobic bacteria. They are electron carriers for nitrogen fixation in most organisms. Although there are many different types, those concerned with nitrogen fixation belong to the 'four iron' class, which means that they have at least one cluster of four iron and four sulphur atoms in the molecule.

Changes in the oxidation state of the iron atoms in the clusters are responsible for the special oxidation-reduction properties of the ferredoxins.

In some organisms flavodoxin replaces ferredoxin as the primary reductant to nitrogenase. In others flavodoxin accepts electrons from ferredoxin and then serves as the primary reductant for nitrogenase. Flavodoxins do not contain iron atoms; their oxido-reducible centre is a yellow, fluorescent molecule called a flavin.

3.5.4 The nitrogenase enzyme complex

structure of nitrogenase

Nitrogenase complexes from different sources are very similar. They consist of two completely distinct metal-containing proteins.

Nitrogenase (also called component I or MoFe-protein) is a large protein composed of four subunits. It has a molecular weight of 220,000 and contains two molybdenum atoms, 22 to 34 iron atoms and 30 sulphur atoms.

Nitrogenase reductase (component II or Fe-protein) is a small protein with a molecular weight of 64,000 and has four iron and four sulphur atoms.

Nitrogenase and nitrogenase reductase combine to form a functional nitrogenase complex. Although biochemists do not know the exact nature of the nitrogenase complex they do now have some idea how the enzyme binds and then fixes nitrogen by reducing it to ammonia; the mechanism is shown in Figure 3.4.

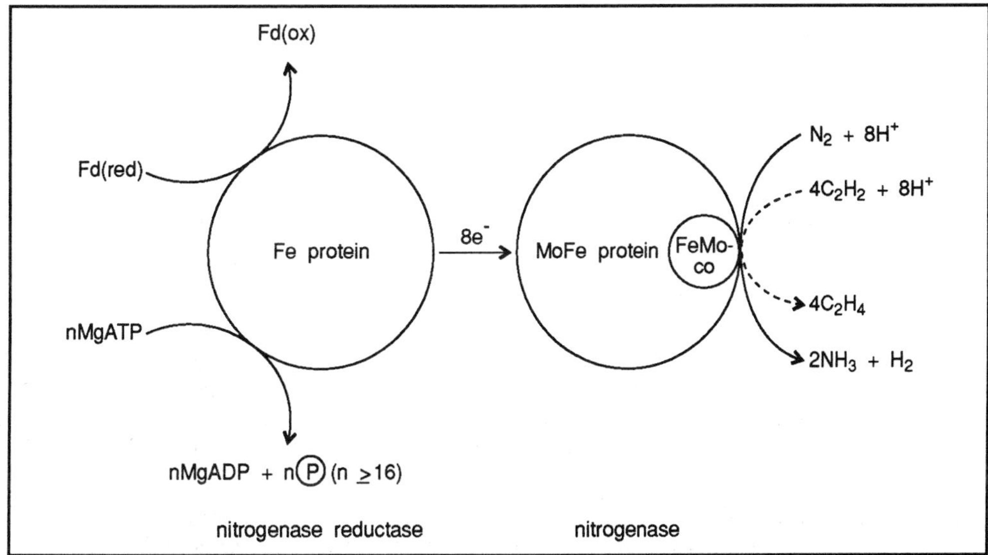

Figure 3.4 Structure and function of the nitrogenase complex.Note the dotted line at the right edge of this figure. Nitrogenase is not specific, it will reduce some compounds in addition to dinitrogen. In the example shown, acetylene is being reduced to ethylene. This is discussed in more detail in the text.

Electron transfer occurs in three stages within the complex:

- electrons are transferred through either ferredoxin or flavodoxin to a Mg-ATP complex in nitrogenase reductase;

- electrons are transferred to nitrogenase at the same time as hydrolysis of molecules of ATP occurs at the Mg-ATP complex;

- reduction of N_2 to NH_3 and H^+ to H_2 occurs. The active site at which reduction occurs is occupied by a special cofactor, called MoFe-co, containing molybdenum atoms as well as iron and sulphur atoms. Recent studies have shown that some species of *Azotobacter* use vanadium in place of molybdenum in the active site.

nitrogenase has low specificity

Nitrogenase has low substrate specificity, which means that the enzyme will reduce not only N_2 but also several other compounds, such as cyanide (CN^-) and acetylene ($CH{\equiv}CH$). Some of these reductions involve the transfer of only two electrons rather than the six required to reduce N_2. The proposed mechanism of the reduction suggests that two such electron reductions should proceed at three times the rate of reduction of N_2, and in most cases this is true.

Π Describe how you would measure the activity of nitrogenase?

assay of nitrogenase

You might have considered measuring the rate of ammonia formation with time. However, this method would lack sensitivity and is not specific for the nitrogenase complex since other enzymes can also form ammonia. Measuring the rate of removal of nitrogen gas from an assay vessel also lacks sensitivity. So how is the enzyme assayed?

A suitable alternative is to assay the enzyme using acetylene as substrate, which is reduced to ethylene. The reduction of acetylene by nitrogenase is shown in Figure 3.4. Ethylene concentrations can be easily quantified, and the reaction is a highly specific one since no enzyme system other than nitrogenase can catalyse this reduction. A sample, which may be soil, water, a culture, or a cell extract is incubated with acetylene under appropriate conditions. The reaction mixture is later analysed by gas chromatography for production of the gaseous substance ethylene.

Since the introduction of acetylene as assay substrate for the enzyme many claims for N_2 fixation in micro-organisms have been proved to be false.

Π Can you think of a reason why the growth of an organism in a medium to which no nitrogen compounds have been added does not mean that the organism is fixing N_2 from the air.

The reason for this is that traces of nitrogen compounds often occur as contaminants in the ingredients of culture medium or drift into the medium in gaseous form or as dust particles. Even distilled water may be contaminated with ammonia.

3.5.5 The need for energy

competition for ATP

The nitrogenase enzyme complex utilises a lot of ATP; it must be supplied continuously by an ATP-generating system because the enzyme is inhibited by excess free ATP. The source of ATP is the normal metabolism of the micro-organism. If the ATP was not used for nitrogen fixation, it would be used for growth and multiplication. It follows that cultures of cells which are fixing nitrogen are less efficient at converting food into cell material than are cultures which are using fixed nitrogen. We can see from Figure 3.4 that 16 ATP molecules are required to reduce N_2 to $2NH_3$.

Kinetic experiments have shown that two ATP molecules are hydrolysed to ADP each time an electron is transferred to the nitrogenase protein (stage 2 of the electron transfer). We already know that six electrons are required to reduce N_2 to NH_3, equivalent to 12 ATP molecules.

This requirement for ATP is purely kinetic - to activate the movement of electrons in the nitrogenase complex. So the involvement of ATP in nitrogen fixation, unlike its involvement in many other enzymes, is not to overcome a thermodynamic barrier. Although the minimum free-energy input (ΔG) for the reaction $N_2 + 6H^+ \rightarrow 2NH_3$ is 658kJ mol^{-1} N_2 the reaction from ferredoxin is thermodynamically downhill.

We have seen that six electrons are theoretically required to reduce N_2 to NH_3 and 2 ATP are used to transfer each pair of electrons. However, the nitrogenase complex consumes 16 or more ATP for every N_2 reduced biologically. What are the extra 4ATP used for? They are required for the transfer of two electrons that are transferred to $2H^+$ forming one molecule of H_2. The formation of H_2 always accompanies the reduction of N_2.

So why is hydrogen formed at all? Evidence suggest that the MoFe-co site, before it can bind N_2, must bind two H^+ ions as hydride groups; these can then be displaced by N_2. It is thought that reduction of N_2 then proceeds as follows:

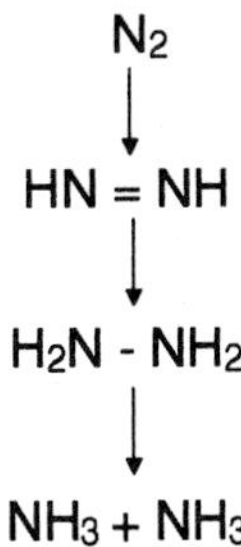

Chemists have actually detected the partially reduced dinitrogen hydride (HN= NH) as an intermediate in the reduction of N_2 to NH_3.

Π Write a balanced equation to describe biological nitrogen fixation, involving N_2, H^+, ATP and electrons as reactants.

The equation:

$$N_2 + 8H^+ + 8e^- + 16ATP \rightarrow 2NH_3 + H_2 + 16ADP + 16Pi$$

requires eight electrons and eight H^+. Although only six are required for the reduction of N_2 to $2NH_3$, H_2 is also formed and this accounts for the extra electrons.

This equation shows that nitrogenase is an expensive enzyme in terms of biological energy. For this reason nitrogen-fixing organisms regulate its activity and its synthesis very precisely.

ATP/ADP ratio controls activity of nitrogenase

- Regulation of nitrogenase activity. This is not understood in detail, but ADP, the product of utilisation of ATP by nitrogenase, inhibits the enzyme. ATP is essential for the enzyme to work at all, so the ratio of ATP to ADP in the neighbourhood of the enzyme can determine the rate at which the enzyme works; this conserves ATP when cellular ATP levels are low. In several nitrogen fixing organisms, addition of ammonia to the population also causes immediate 'switching off' of nitrogenase activity; the mechanism is still obscure but is again thought to conserve ATP.

regulation of nitrogenase synthesis

- Regulation of nitrogenase synthesis. The genes for nitrogen fixation, called nif genes, have been studied intensively in the soil micro-organism *Klebsiella pneumoniae.* There are 17 nif genes and they are regulated in a complex manner which is not entirely understood; a detailed examination of this is beyond the scope of this chapter. Three of the genes, called nif H, nif D and nif K provide the codes for the polypeptides making up nitrogenase.

At a descriptive level, the regulation of nitrogenase synthesis is straightforward:

ammonia suppresses nitrogenase synthesis

- Ammonia suppresses the synthesis of nitrogenase: if the environment contains sufficient ammonia for the organism's needs, it does not make any nitrogenase. If there is less than sufficient, but still some, the organism makes only sufficient nitrogenase to satisfy its requirements for fixed nitrogen.

O_2 inactivates nitrogenase

- Nitrogenase is inactivated by O_2 and some organisms repress nitrogenase synthesis when O_2 is present in the environment. In the following section, we shall examine the effect of O_2 on nitrogen fixation in more detail.

3.5.5 Methods of oxygen protection

This is necessary because the nitrogenase complex is extremely sensitive to O_2 and can be irreversibly inactivated by even low concentrations. The problem can be overcome in many different ways which we will now consider.

- Obligate anaerobes (eg *Clostridium pasteurianum*). O_2 is not present in the environments in which these organisms grow and so they do not need to protect their nitrogenase. There is no special O_2 protection process in these organisms.

- Facultative anaerobes (eg *Klebsiella pneumoniae*). These bacteria can grow in the presence or absence of O_2 but can only fix nitrogen when O_2 is absent. This is because nitrogenase synthesis is switched off in the presence of O_2.

- Aerobes. These have special physiological and/or morphological features to protect their nitrogenase from O_2.

Respiratory protection. Respiratory protection is found in *Azotobacter vinelandii.* This organism possesses two types of electron transport chain. Type 1 operates at both high and low O_2 tensions and is coupled to ATP production. This is 'normal' aerobic respiration. Type 2 operates at high oxygen tensions only and is not coupled to oxidative phosphorylation. This dramatically reduces respiratory control and 'burns off' oxygen thus keeping the oxygen tension low.

leghaemoglobin

Root nodules. The root nodules of leguminous plants contain a protein called leghaemoglobin (also referred to as legumohaemoglobin) which is produced by the plant. It is similar to the haemoglobin of mammalian blood but it has an even higher affinity for O_2. The leghaemoglobin prevents the accumulation of high concentrations of free oxygen while at the same time providing the oxygen necessary for the metabolism of *Rhizobium*, which inhabits the nodules.

heterocysts

Heterocysts. Certain blue-green algae carry out O_2 evolving photosynthesis and fix dinitrogen. O_2 protection is achieved by physically separating photosynthesis and nitrogen fixation by confining nitrogen fixation to specialised cells called heterocysts. Within these cells the photosystem involved with O_2 evolution (photosystem 2) is absent. Although air can diffuse into heterocysts to some extent, it is required for aerobic respiration which can be regarded as a low capacity form of respiratory protection for the nitrogenase. Figure 3.5 shows heterocysts and the metabolic exchange between them and vegetative cells.

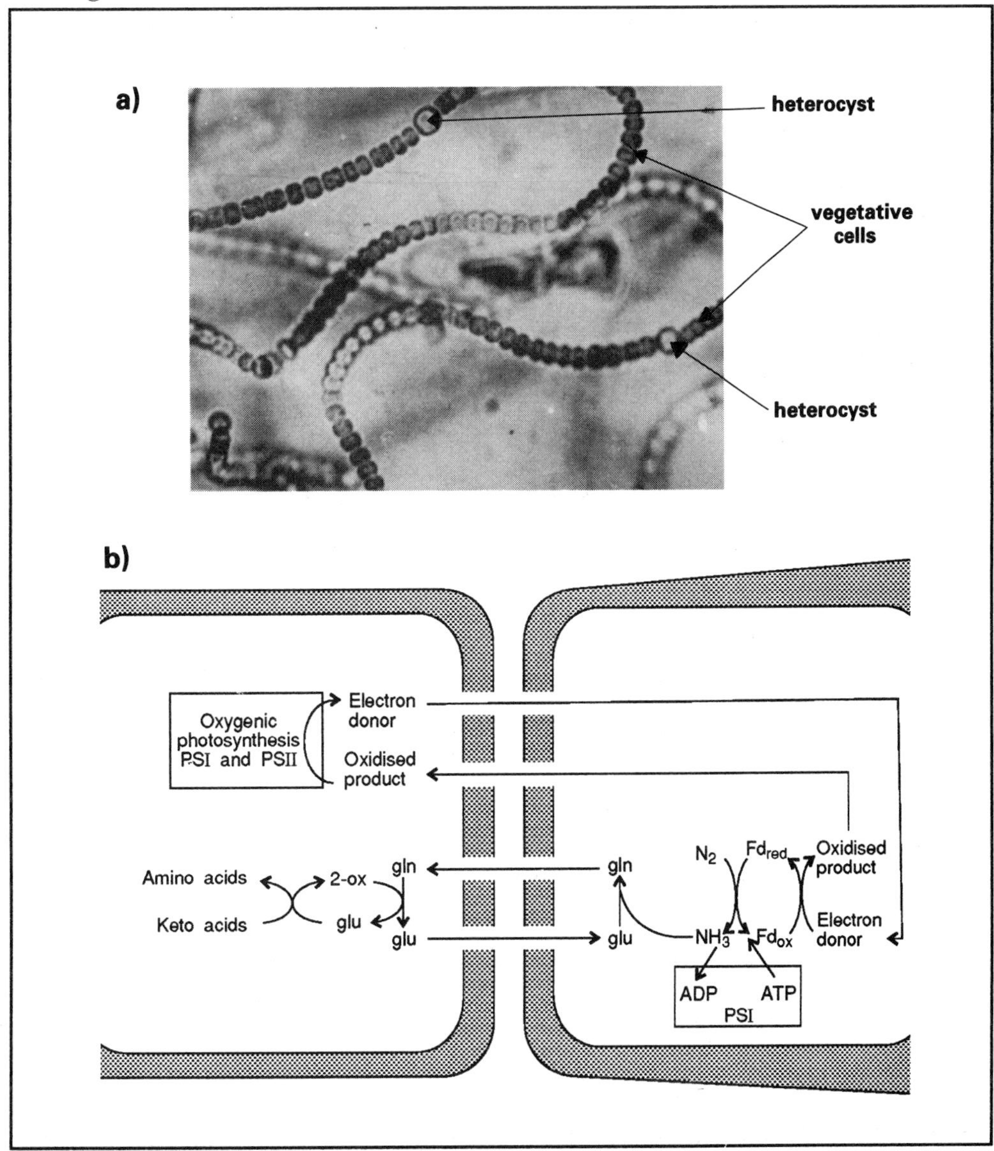

Figure 3.5 a) Photomicrograph of *Anabena cylindrica* showing enlarged cells (heterocysts) on the filaments. b) Metabolic exchange between vegetative cells and heterocysts. (Abbreviations: PSI, photosystem 1; PSII, photosystem 2; glu, glutamate; gln, glutamine; 2-ox, 2-oxoglutarate).

SAQ 3.5

Identify whether each of the following statements is true or false and justify your decisions.

1) Nitrogen fixation by facultative anaerobes only occurs during anaerobic growth.

2) Aerobic nitrogen-fixing bacteria can use NAD(P)H as a source of reducing power.

3) The ammonia formed during nitrogen fixation is assimilated via glutamate dehydrogenase.

4) Symbiotic nitrogen fixers have a respiratory mechanism for protecting their nitrogenase.

5) Binding of nitrogen to the nitrogenase complex occurs at the MoFe-co site of nitrogenase reductase.

6) Depletion of carbon source in *Azotobacter* species will inhibit nitrogen fixation.

7) Reduction of acetylene to ethylene by nitrogenase is likely to proceed at a rate three times faster than the reduction of N_2.

8) Molybdenum accepts and donates electrons within nitrogenase.

SAQ 3.6

For each of the organisms shown below select the appropriate responses from each of the other three lists. (Figure 3.1 might help you - but do not cheat!)

Organism; *Rhizobium*; *Azotobacter*; *Frankia*; *Klebsiella*; Purple bacteria; Cyanobacteria.

Physiological group

1) Free-living/aerobic/phototrophic

2) Free-living/aerobic/chemotrophic.

3) Free-living/anaerobic/phototrophic.

4) Free-living/anaerobic/chemotrophic.

5) Symbiotic/aerobic/phototrophic.

6) Symbiotic/aerobic/chemotrophic.

7) Symbiotic/anaerobic/phototrophic.

8) Symbiotic/anaerobic/chemotrophic.

Source of reducing power

a) Pyruvate, hydrogen or formate.

b) NADH or NADPH.

Special mechanism of O_2 protection

i) None.

ii) Heterocysts.

iii) Leghaemoprotein.

iv) Respiratory protection.

SAQ 3.7

Bacteria from the genus *Klebsiella* (a facultative anaerobe) were incubated under the different conditions shown below, labelled 1) to 5). For each of these conditions identify which of the enzymes, labelled a) to e), will function in the assimilation of nitrogen.

1) medium containing nitrate incubated under aerobic conditions.

2) medium containing ammonia incubated under anaerobic conditions.

3) medium containing dinitrogen incubated under aerobic conditions.

4) medium containing dinitrogen incubated under anaerobic conditions.

5) medium containing ammonia and dinitrogen incubated under anaerobic conditions.

a) glutamate dehydrogenase; b) nitrogenase; c) nitrate reductase; d) GOGAT and glutamine synthetase; e) nitrite reductase.

3.6 The importance of nitrogen fixation in nature

nitrogen cycle

All living things on this planet continually recycle the chemical elements of which they are composed. The most important, from both ecological and economic viewpoints, is the nitrogen cycle. Figure 3.6 is a simple representation of the nitrogen cycle which shows the transformations undergone by the element nitrogen (N) as a consequence of biological activity.

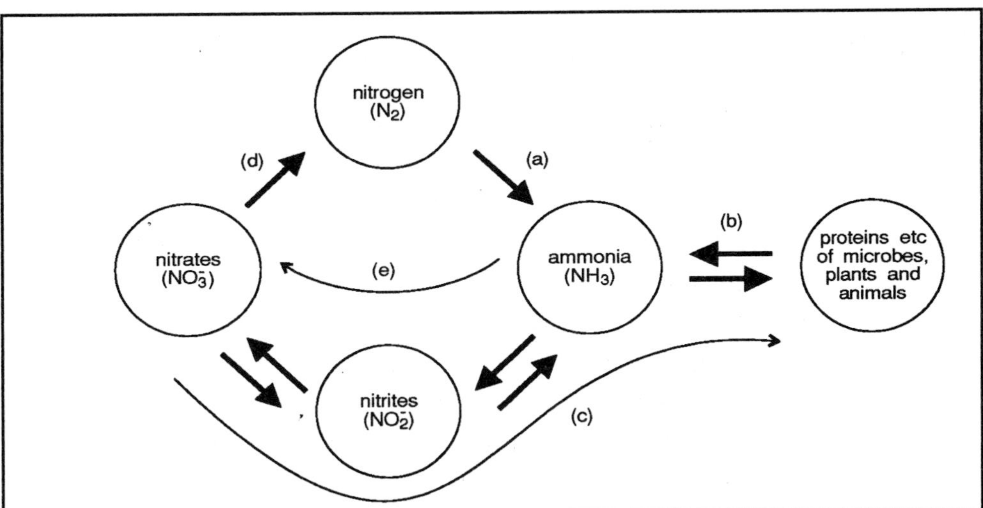

Figure 3.6 The nitrogen cycle.

There are five processes operating in the cycle:

1) nitrogen assimilation; 2) nitrogen fixation; 3) nitrification; 4) denitrification; 5) ammonification.

Π See if you can match each of the processes named above with the transformations labelled a) to e) in the nitrogen cycle of Figure 3.6.

Transformation [ ] is nitrogen assimilation

Transformation [ ] is nitrogen fixation

Transformation [ ] is nitrification

Transformation [ ] is denitrification

Transformation [ ] is ammonification

The letters should be inserted into the statements above in the following order - c), a), e), d) and b).

We have examined the assimilation and nitrogen fixation reactions in detail in this chapter. You may recall from your previous studies that the nitrification reactions are used by certain chemoautotrophic bacteria to obtain energy and reducing power, whereas the denitrifying reactions are used by certain other bacteria as ways to accept electrons in anaerobic respiration. The final process, ammonification, is one in which inorganic nitrogen is returned to the cycle from organic matter as a result of autolysis, decay and putrefaction of biological material, and the principal form in which it appears is ammonia.

crop productivity

An important general point about the cycle is that, in agricultural areas of this planet, biological productivity is largely determined by the availability of inorganic nitrogen in the soil. This means that the rate at which the cycle turns determines biological productivity, with the exception of highly sophisticated agricultural communities. Nitrogen fixation is the rate-limiting step in the cycle and increasing the rate of nitrogen fixation would increase productivity of the biosphere.

Haber process for making NH_3

Many different processes contribute to the rate-limiting step in the cycle. Nitrogen-based fertilisers, for example, are produced by chemical nitrogen fixation processes. (Here we are using nitrogen fixation in the sense of nitrogen being combined with other elements. The chemical products may be assimilated, directly or indirectly into biological processes.) The most important of these is the Haber process, in which hydrogen and nitrogen are reacted at high temperature and pressure in the presence of an iron catalyst.

$$N_2 + 3H_2 \xrightarrow[\text{Fe catalyst}]{\substack{450^\circ C \\ 200 \text{ atmospheres}}} 2NH_3$$

This process uses a lot of energy so the product is expensive. Because of this the use of fertilisers is generally limited to rich countries.

Nitrogen fixation can also occur by purely physical processes, such as electrical discharge in the atmosphere (lightning). During such discharges, the high energy released is sufficient to separate the two nitrogen atoms and oxides of nitrogen are produced. These are then hydrated by water vapour and carried to the earth as nitrates or nitrites. It has been estimated that 2kg NO_3^- hectare^{-1} year^{-1} is produced by lightning.

By far the most important process, however, is biological nitrogen fixation. This fixes an estimated 2 to 10 x 10^8 tonnes of nitrogen every year, which is about 100 times that fixed by physical processes and by the Haber process combined.

Symbiotic leguminous nitrogen-fixers, of which *Rhizobium* species are the major group, play the most important ecological role. They fix an estimated 30 times more nitrogen than all the other groups. The problem of the availability of fixed nitrogen to crops has been tackled by the use of chemical fertilisers and crop rotation. In crop rotation legume plants eg peas and clover are grown once every four or five years. Much of the fixed nitrogen is retained in the roots and so boosts the level of nitrogen in the soil.

crop rotation

Considerable research effort is now being devoted to extending the range of plant species that can harbour nitrogen-fixing bacteria.

Although the range of nitrogen-fixing plants is wide, it does not include the world's major food crops, wheat and rice, nor the major forage crop, grass. Thus there is a very big challenge here for research scientists.

SAQ 3.8

For statements 1) to 5) shown below select appropriate nitrogen fixing organisms from the list provided:

1) This organism contributes most to the rate-limiting step in the nitrogen cycle.

2) This organism contributes to the nitrogen cycle.

3) The organism is the one most likely to fix dinitrogen in freshly dug top-soil.

4) This organism is the one most likely to fix dinitrogen in deep swamp mud.

5) This organism is the one most likely to fix dinitrogen in shallow fresh water.

Frankia; Klebsiella; Anabaena; Azotobacter; Rhizobium

Summary and objectives

We have seen that sulphur is assimilated as hydrogen sulphide and nitrogen as ammonia and more oxidised sources of these elements must be reduced before they can be incorporated into organic material. In the case of sulphate, reduction proceeds via sulphite to hydrogen sulphide which is then used to form the sulphur containing amino acid cysteine. In the case of ammonia there are three assimilation reactions resulting in the formation of the amino acids glutamic acid, asparagine and glutamine.

The main route of ammonia assimilation depends upon the availability of ammonia. At relatively high concentrations of ammonia, glutamate dehydrogenase is used but at low concentrations glutamine synthetase and GOGAT are used. The assimilation of nitrate proceeds via nitrite and involves two reductase enzymes and a truncated electron transport chain. Assimilation of dinitrogen, called biological nitrogen fixation, is carried out only by certain bacteria. Nitrogen fixation is a reductive and energy-demanding process catalysed by the two proteins of the nitrogenase complex. ATP is required to move electrons from one protein to the other in the complex and the binding of dinitrogen to the enzyme requires the simultaneous binding of hydrogen ions. The activity and synthesis of the nitrogenase complex is under strict control and this acts to conserve ATP. The ammonia formed is assimilated via the 'low ammonia route'. Nitrogenase is inactivated by molecular oxygen and aerobic bacteria have special mechanisms to avoid this. Nitrogen fixation is the rate-limiting step in the geological nitrogen cycle and *Rhizobium*, a symbiotic leguminous nitrogen-fixer, contributes most to this step.

Now that you have completed this chapter you should be able to:

- describe assimilation reactions of common sources of sulphur and nitrogen;
- calculate the number of electrons required for the assimilation of different sources of sulphur or nitrogen;
- correlate assimilation reactions with named enzymes;
- explain how availability of ammonia affects the route of ammonia assimilation;
- describe how the five essential requirements of nitrogen fixation interrelate and explain how different physiological groups of bacteria satisfy these requirements;
- describe the structure and mode of action of the nitrogenase complex. Explain the need for its regulation and how this may be achieved;
- illustrate the geological nitrogen cycle by means of a labelled flow diagram and explain the importance of assimilation reactions to soil fertility.

Amino acid and nucleotide biosynthesis

Introduction 66

4.1 A brief review of amino acids; their general formula and importance in the diet 66

4.2 Derivation of biosynthetic precursors of amino acids 70

4.3 The assimilation of nitrogen 71

4.4 The biosynthesis of amino acids 72

4.5 The recycling and biosynthesis of nucleic acid precursors 94

4.6 The metabolic links between amino acid biosynthesis and the biosynthesis of purines and pyrimidines 106

Summary and objectives 108

Amino acid and nucleotide biosynthesis

Introduction

In this chapter we shall study the biosynthesis of amino acids and then study the biosynthesis of purines and pyrimidines and their subsequent incorporation into nucleotides. This chapter briefly reviews the chemical nature of the twenty amino acids and identifies those essential to the diet of higher organisms. The derivation of carbon skeletons (precursor metabolites) and the assimilation of nitrogen follows. The biosynthesis of the amino acids is then presented in some detail; this may appear a laborious task at the outset - the aim however is to indicate the major trends on the one hand and to stress the important aspects of the metabolic demands and controls of these major biosynthetic processes on the other. We hope you find this approach less daunting.

Then a section follows where the biosynthesis of purines and pyrimidines is studied. Three sources of material are identified; nitrogenous base salvage pathways followed by prokaryotic use of exogenous nitrogenous bases and nucleosides and finally *de novo* biosynthesis.

Throughout the chapter, links with other metabolic pathways have been highlighted and the final section examines the link between histidine and purine biosynthesis.

4.1 A brief review of amino acids; their general formula and importance in the diet

Amino acids have the general formula:

$$
\begin{array}{c}
\quad\quad\quad COO^- \\
\quad\quad\quad | \\
H_3N^+ - CH \\
\quad\quad\quad | \\
\quad\quad\quad R
\end{array}
$$

amino acid structure

where R may be one of a variety of groups from hydrogen (as in glycine) through to complicated side chains including aromatic rings. All except proline have the amino group on carbon two, the α –carbon, thus the compounds are called α-amino acids.

Remember that convention dictates that we write the amino group to the left of the molecule signifying an L-amino acid. These amino acids are far more common in nature than the D-amino acids - being the only amino acid found in higher animals and constituting the vast majority of the amino acids in other organisms. D-amino acids have specific roles to play in prokaryotic systems, for example D-alanine and D-glutamate are present in all bacterial cell walls.

Π There are twenty common amino acids. Just pause for a moment and try to write down as many as you can and indicate any special features about the individual ones.

Figure 4.1 shows the amino acids with their structures.

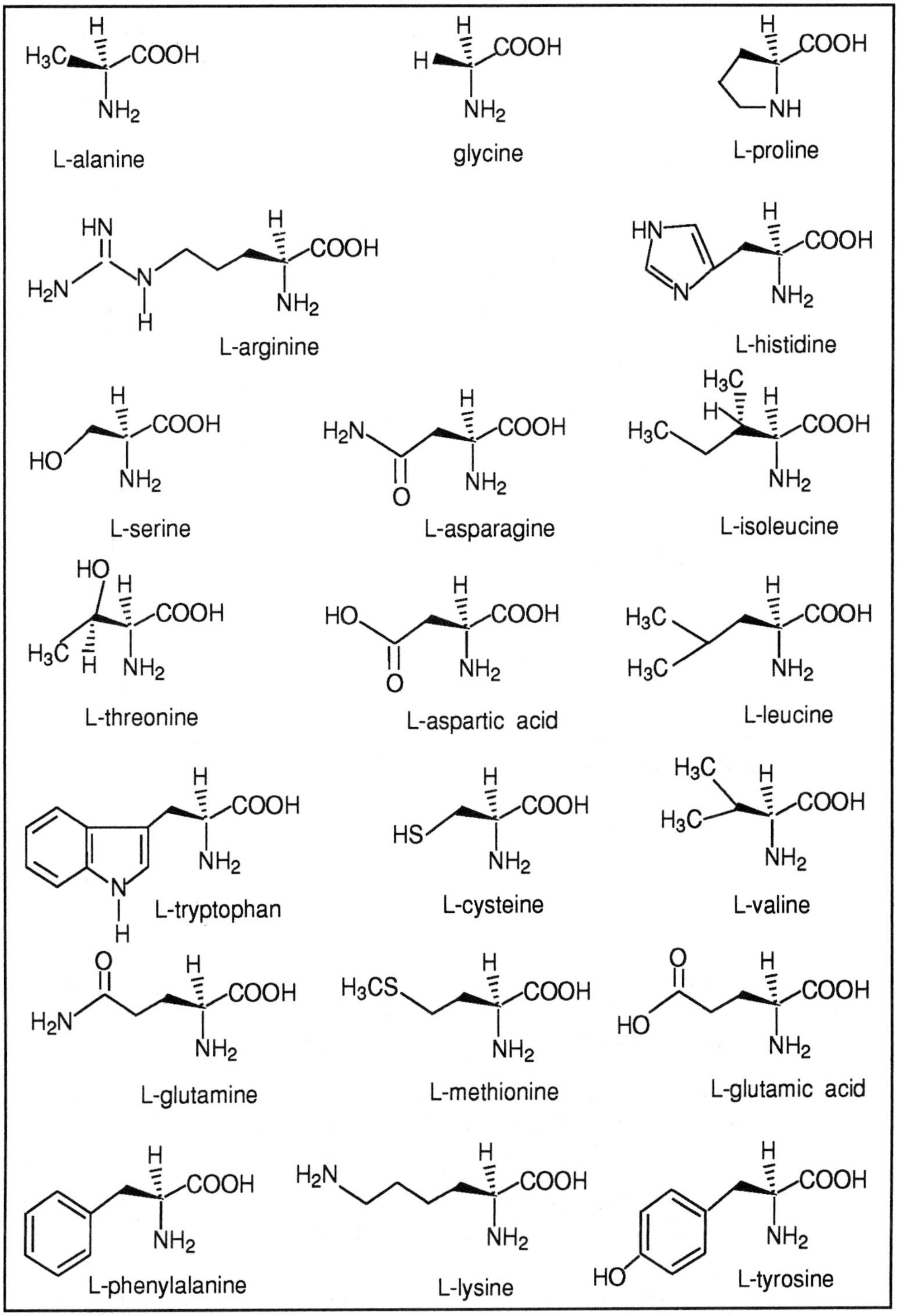

Figure 4.1 The names and structures of the twenty common amino acids. The figure shows simplified structures. It might be sensible to draw these out for yourself putting in all of the carbons and hydrogens and to pin it up somewhere where you can see it.

Π Which of the amino acids are: 1) dicarboxylic acids; 2) contain sulphur; 3) contains an imidazole ring; 4) diamino acids; 5) contain a guanidino group; 6) aromatic amino acids?

Your answer should be: 1) glutamic and aspartic acid; 2) cysteine and methionine; 3) histidine; 4) lysine; 5) arginine; 6) phenylalanine, tyrosine and tryptophan; (note that in tryptophan the aromatic ring is part of an imidazole moiety). The structures shown in Figure 4.1 are in their acid forms. You should remember, however, that at physiological pHs, the acids are usually in their zwitterions. For example, glutamic acid is usually present as glutamate. Thus:

$$R-\underset{{}^{+}NH_3}{\overset{H}{\underset{|}{\overset{|}{C}}}}-COO^-$$

Amino acids are the building blocks from which proteins are made. The acids are added on to each other by peptide bonds in the order predetermined by the nuclear DNA. It is not the purpose of this chapter to investigate protein synthesis but as the order and identity of amino acids within proteins must be exact, it follows that all of the amino acids must be produced in sufficient quantity to satisfy protein synthesis. This is an awesome task as up to 50% of the dry weight of cells may be protein and in rapidly growing cells or organisms the demand for amino acids is high and constant.

high demand for amino acids

There are basically three possible ways in which cells can obtain supplies of amino acids:

- from degradation of endogenous protein;
- from their diet;
- by synthesising them from metabolic precursors.

sources of amino acids

For a variety of reasons protein is constantly being degraded to free amino acids by all organisms and it is not surprising that the majority of this, 80% for example in mammalian systems, is recycled by the amino acids being incorporated into new protein. This occurs in all living systems irrespective of whether any amino acids are dietarily essential compounds or not.

The second source of amino acids is from the diet. All organisms will absorb amino acids from the diet even if they are capable of synthesising them. It is obviously beneficial to an organism to do this because it then does not have to expend energy to synthesise them. The question which needs answering here is what constitutes an ideal diet? This will obviously vary from organism to organism and the likelihood of getting a perfect diet, that is, one in which all amino acids are supplied in exactly the right quantities, is remote indeed. For example, Man needs 3.5 times the weight of leucine compared to tryptophan in an ideal diet (14 g/kg/day compared to 4.8 g/kg/day). If the protein source contains very little tryptophan then either a protein supplement is required or sufficient protein must be consumed to take in the minimum requirement of tryptophan. This would inevitably mean the absorption of excess leucine which would have to be metabolised away.

absorption of amino acids

Amino acids are taken into cells by specific carrier systems; due to their polar nature they cannot enter by simple diffusion. This entry can in fact be a rate-limiting step and certain organisms prefer to take in small peptides. Each peptide or amino acid enters at the same rate, thus with peptides more amino acids enter the cell per unit time.

dietarily essential amino acids

The third source of amino acids is the individual biosynthesis by the cell. Great variation occurs in the capacity of living systems to synthesise each of the twenty amino acids. The phrase 'dietarily essential' amino acids is often used in the literature and indicates amino acids which cannot be synthesised by a particular organism. However the meaning is not always precise as we can see if we study the amino acid requirements in humans. Table 4.1 shows a total of twelve amino acids. The first eight, in order of amounts required, are strictly dietarily essential in that they cannot be synthesised at all by humans. The next two, tyrosine and cysteine, can only be synthesised from phenylalanine and methionine respectively, themselves dietarily essential. Finally histidine and arginine are included although they can be synthesised by humans. Synthesis of both in babies and in the young generally is grossly inadequate to meet the requirements for growth. There is now evidence that the adult ability to synthesise histidine is also inadequate. Thus only eight of the twenty amino acids are synthesised totally and adequately by humans - the remainder must be supplied in the diet and we should remember that little, if any, is supplied by the gut flora.

leucine	phenylalanine	cysteine
valine	methionine	tyrosine
isoleucine	lysine	arginine
threonine	tryptophan	histidine

Table 4.1 Amino acids which are 'dietarily essential' for humans.

plants make their own amino acids

Higher plants are more versatile and can synthesise all of their amino acids.

Micro-organisms vary widely in their capacity to synthesise amino acids. For example the bacterium *Escherichia coli* can synthesise all of its amino acids from a simple salts medium (basically from glucose and ammonium sulphate) although it will preferentially use amino acids if provided. Other bacteria are termed nutritionally demanding, which means that they must be supplied with all or most of their amino acids and vitamins. One extreme example is *Leuconostoc mesenteroides*, strains of which have to be provided with sixteen or more of the amino acids in a growth medium. As a general comment, micro-organisms including bacteria, fungi and protozoa can usually synthesize their full complement of amino acids.

large amounts of amino acids required by bacteria

It is not surprising that bacteria in particular and micro-organisms generally can synthesise their own amino acids. Their requirements on a weight-for-weight basis are phenomenal compared to higher organisms due to their elevated growth rate. For example the biosynthetic rate for protein in elderly adults, young adults, infants and newly-born infants respectively is 1.9g; 3g; 6.9g and 17.4g /kg/day. Bacteria can double their numbers every thirty minutes or less under ideal conditions. If the protein content of bacteria is 50% of dry weight and the dry weight is 12% of their total weight, then 60g of every kg will be protein: thus 1 kg of bacteria will produce a further kilogram after thirty minutes and this 2 kg will produce a further 2 kg in the next thirty minutes. Overall three kilogram of bacteria have been produced in the one-hour period indicating a net synthesis of 180g protein /kg/hour. Obviously great industry is required!

4.2 Derivation of biosynthetic precursors of amino acids

In the next few sections of this chapter we shall follow the biosynthesis of each of the twenty amino acids noting whether or not they are essential for given organisms and occasionally noting alternative pathways for a particular amino acid.

precursor metabolites

Before looking at individual pathways we need to determine the nature of the precursors required for synthesis. These are twelve compounds termed precursor metabolites from which all biosyntheses begin. These are listed below:

glucose-6-phosphate	ribose-5-phosphate
fructose-6-phosphate	erythrose-4-phosphate
glyceraldehyde-3-phosphate	acetyl CoA
3-phosphoglycerate	α-oxoglutarate
phosphoenol pyruvate	succinyl CoA
pyruvate	oxaloacetate

These compounds should be familiar to you by name and you should be able to identify the pathways in which they are involved.

The amino acids conveniently divide into families based on the nature of the precursor molecules required. Effectively there are six families as shown in Table 4.2 and a total of seven precursor metabolites are required. We now need to identify how each of these precursor metabolites is provided in living cells.

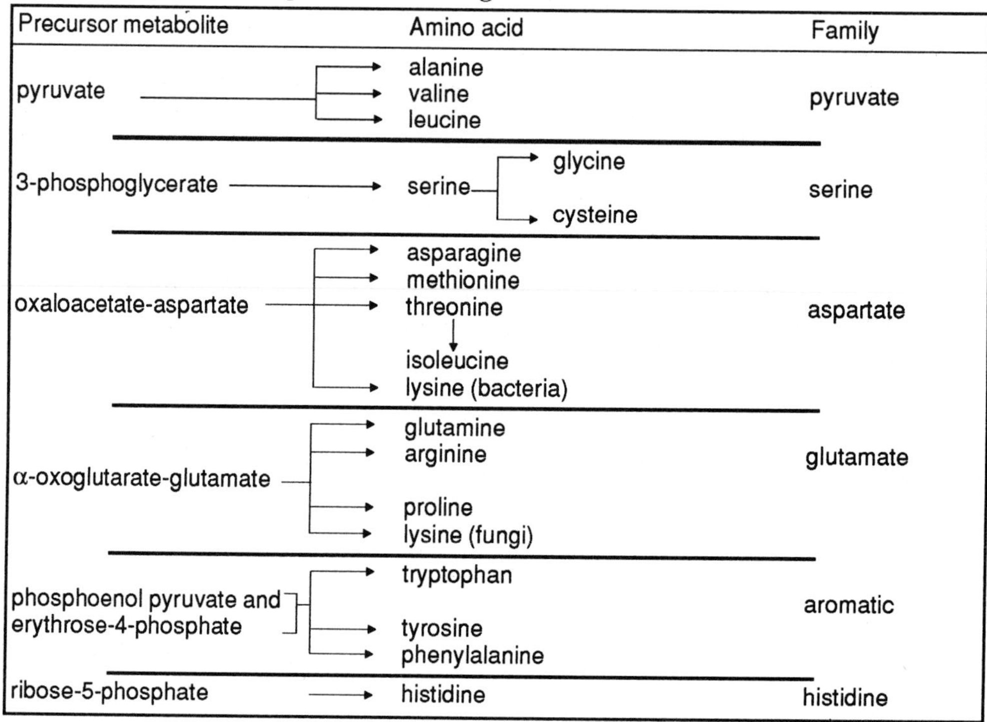

Precursor metabolite	Amino acid	Family
pyruvate	alanine valine leucine	pyruvate
3-phosphoglycerate	serine → glycine serine → cysteine	serine
oxaloacetate-aspartate	asparagine methionine threonine → isoleucine lysine (bacteria)	aspartate
α-oxoglutarate-glutamate	glutamine arginine proline lysine (fungi)	glutamate
phosphoenol pyruvate and erythrose-4-phosphate	tryptophan tyrosine phenylalanine	aromatic
ribose-5-phosphate	histidine	histidine

Table 4.2 The biosynthetic families of amino acids.

Π Can you recall in which pathway(s) each of the seven precursor metabolites is found? Look back at Table 4.2 and see if you can do this and then check your answers withour list below.

Precursor metabolite	Pathway
3-phosphoglycerate phosphoenol pyruvate pyruvate	the Embden Meyerhof pathway
α-oxoglutarate oxaloacetate	the TCA cycle
erythrose-4-phosphate ribose-5-phosphate	pentose phosphate pathway

The pathways, either in part or in their entirety are found in all living systems. Remember that the Embden Meyerhof pathway or the pentose phosphate pathway produce pyruvate which is fed via acetyl CoA into the TCA cycle. Thus a starting material of glucose for example would be sufficient to produce all of the precursor metabolites.

4.3 The assimilation of nitrogen

sources of nitrogen

One extra requirement is needed before we can proceed to study the pathways and that is a suitable source of nitrogen for the amino groups. Higher animals utilise ammonium ions for synthesis of their dietarily non-essential amino acids, they cannot use alternatives such as nitrate, nitrite or atmospheric nitrogen. Plants can utilise ammonium or nitrite ions but prefer nitrate and cannot themselves fix nitrogen. Bacteria and fungi are much more varied in their requirements. The majority prefer ammonium ions but some will accept nitrate or nitrite. A restricted number will 'fix' or incorporate atmospheric nitrogen converting it into ammonia before incorporating it into organic molecules. This topic was covered in great detail in the previous chapter.

The previous paragraph indicates that whatever nitrogen source is used, incorporation occurs at the level of ammonium ions. There are two key ammonium assimilation routes used by animals, plants and micro-organisms. Firstly reductive amination of an α-oxo acid and secondly the formation of an amide of a carboxylic acid.

glutamate dehydrogenase

The key enzyme in reductive amination is glutamate dehydrogenase which has the following mode of action:

$$\underset{\alpha\text{-oxoglutarate}}{COO^- - C(=O) - CH_2 - CH_2 - COO^-} + NH_4^+ + NAD(P)H + H^+ \longleftrightarrow \underset{\text{glutamate}}{COO^- - CH(NH_3^+) - CH_2 - CH_2 - COO^-} + NAD(P)^+ + H_2O$$

Plants or animals have at least one enzyme which uses either NADH + H^+ or NADPH + H^+ whereas bacterial and fungal systems have an enzyme which uses one or the other cofactor, not both. The enzymes in animals and plants are freely reversible whereas those in bacteria and fungi generally only operate from right to left as above for ammonium assimilation.

Π Bacteria also have a relatively active alanine dehydrogenase; can you draw the reaction for this enzyme?

$$\underset{\text{pyruvate}}{\begin{array}{l} COO^- \\ | \\ C=O \\ | \\ CH_3 \end{array}} + NH_4^+ + NAD(P)H + H^+ \longleftrightarrow \underset{\text{alanine}}{\begin{array}{r} COO^- \\ | \\ H_3N^+-CH \\ | \\ CH_3 \end{array}} + NAD(P)^+ + H_2O$$

The formation of amide bonds revolves almost entirely around the formation of glutamine from glutamate though a small amount of asparagine may be formed from aspartate. The enzymes which catalyse these reactions are similar in mode of action and are ATP-requiring, for example:

$$\text{glutamate} + NH_3 + ATP \xrightarrow[\text{glutamine synthetase}]{} \text{glutamine} + AMP + PPi$$

NB do not forget that glutamine is technically misnamed for it is an amide not an amine.

4.4 The biosynthesis of amino acids

The previous sections have shown us that some amino acids can be synthesised by all organisms and some by relatively few, generally micro-organisms and plants. Each of the twenty amino acids can be placed into one of six groups depending on their precursor metabolite. We shall now look at each of these families in turn to study amino acid biosynthesis. Some of the pathways are very long and have metabolites in them which have not been encountered before. You will find the pathways complicated. Try not to get disheartened by this, you are not expected to learn them by heart. You should aim to understand what is going on, to identify certain types of reaction and get an overall grasp of the tasks facing an organism.

pathways are branched and multi-functional

You will find that some of the pathways are complicated because they are branched and because they are used for both biosynthesis and, occasionally, degradation. In addition they are used to make other constituents of the cell, for example:

- the glutamate family pathway is also used to make polyamines;
- diaminopimelate and dipicolinate are produced from the aspartate family pathway;
- the serine family pathway is used to make purines and porphyrins;
- pantothenic acid is produced from the pyruvate family pathway;

- para-hydroxybenzoate and para-aminobenzoate are produced from the aromatic family pathway.

These aspects will be pointed out at the appropriate points in the following sections.

4.4.1 Biosynthesis of the pyruvate family - alanine, valine and leucine

The production of alanine is a simple, one-step procedure catalysed by alanine dehydrogenase which was described in Section 4.3. A reductive amination occurs during which pyruvate is converted to alanine directly.

valine biosynthesis

Biosynthesis of valine from pyruvate is a four-reaction sequence, the first three reactions of which are common to the leucine biosynthetic pathway. Leucine is a six-carbon compound whereas valine is only a five-carbon compound and the insertion of an extra carbon and subsequent rearrangement in leucine biosynthesis means a seven stage pathway is required.

The first reaction, common to both pathways, is the condensation of two pyruvate molecules during which the carboxyl group of one pyruvate is lost as carbon dioxide. The coenzyme thiamine pyrophosphate (TPP) is required during the reaction.

$$CH_3-\overset{O}{\overset{\|}{C}}-COO^- + \begin{array}{c}CH_3\\|\\C=O\\|\\COO^-\end{array} \longleftrightarrow CH_3-\overset{O}{\overset{\|}{C}}-\begin{array}{c}CH_3\\|\\C\\|\\COO^-\end{array}-OH \quad + CO_2$$

pyruvate pyruvate acetolactate carbon dioxide

The enzyme is called an acetohydroxy acid synthetase. It is relatively non-specific, also functioning in the biosynthesis of isoleucine. In the next reaction a 'reductive isomerisation' occurs as follows:

$$H_3C-\overset{O}{\overset{\|}{C}}-\begin{array}{c}CH_3\\|\\C\\|\\COO^-\end{array}-OH + NADPH + H^+ \longrightarrow \begin{array}{c}CH_3\\|\\H_3C-C-OH\\|\\HC-OH\\|\\COO^-\end{array} \quad + NADP^+$$

acetolactate α, β-dihydroxyisovalerate

The enzyme is called acetohydroxy acid reducto-isomerase.

The reaction is reductive because a keto (oxo) group has been reduced to a hydroxyl group and as the methyl group is now a β-methyl not an α-methyl, isomerisation has taken place.

There are now only two stages left to convert the α, β-dihydroxyisovalerate to valine.

SAQ 4.1

The following incomplete diagram shows the structures of α, β-dihydroxyisovalerate and valine. From your knowledge to date fill in the intermediate and as much extra information as you can.

If you get stuck look at the following clues one by one:

1) the last step is usually a transamination;

2) glutamate is the usual amino donor;

3) amino groups are transferred to keto (oxo) groups.

$$H_3C-\overset{\displaystyle CH_3}{\underset{|}{\overset{|}{C}}}-OH \quad HC-OH \quad COO^- \longrightarrow \quad ? \quad \longrightarrow \quad H_3C-\overset{\displaystyle CH_3}{\underset{|}{\overset{|}{CH}}} \quad H_3N^+-CH \quad COO^-$$

α, β-dihydroxyisovalerate ? valine

Study the answer at the back of the book and if you identified all parts - well done! The first reaction could have been a dehydrogenation but was likely to be a dehydration when comparing dihydroxyisovalerate with valine. The intermediate must be an isovalerate backbone and you should be able to recognise α-oxo groups.

leucine biosynthesis

In the valine pathway we saw two three-carbon compounds condensed to a five-carbon compound with the loss of carbon dioxide. The five-carbon α-oxoisovalerate is a precursor for the six-carbon amino acid leucine. In some ways the four reaction sequence leading to leucine is similar to the four reaction sequence leading to valine. Here a five-carbon molecule condenses with acetyl from acetyl CoA to give a seven-carbon compound. Isomerisation is followed by decarboxylation and then aminotransfer to give leucine. The reactions are shown in Figure 4.2. The reactions are repetitive but the formulae are new and somewhat complicated.

As indicated earlier some control of the enzymes of these pathways must be exercised in order to avoid overproduction of the amino acids. The topic of the regulation of metabolism will be dealt with in depth in a later chapter but certain points will be raised here. In addition to the control of production of alanine, valine and leucine there is an extra complication. The first four reactions in valine and leucine production (pyruvate to α-oxoisovalerate) are chemically identical to four reactions in isoleucine production (α-oxobutyrate to α-oxo-β-methyl valerate). You might like to try to draw these out for yourself. Begin with α–oxobutyrate:

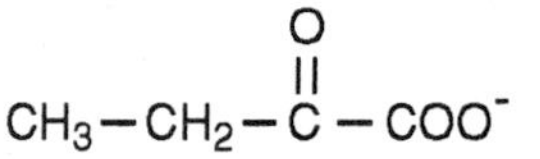

$$CH_3-CH_2-\overset{\displaystyle O}{\overset{\|}{C}}-COO^-$$

and then follow the reaction sequence described above. Remember that isoleucine has the structure:

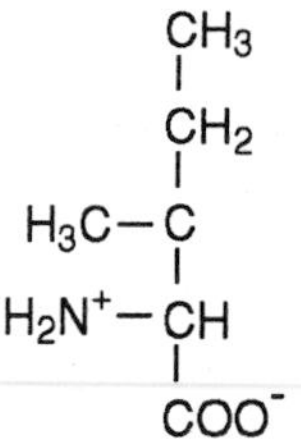

$$\begin{array}{c} CH_3 \\ | \\ CH_2 \\ | \\ H_3C-C \\ | \\ H_2N^+-CH \\ | \\ COO^- \end{array}$$

The names of the compounds are given in Figure 4.7. The only difference is that the intermediates in the latter pathway contain an extra methyl group. So similar are they that it is likely that all organisms utilise the same enzymes for the two pathways. Competition between the pathways is avoided by regulating the availability of the entering precursor.

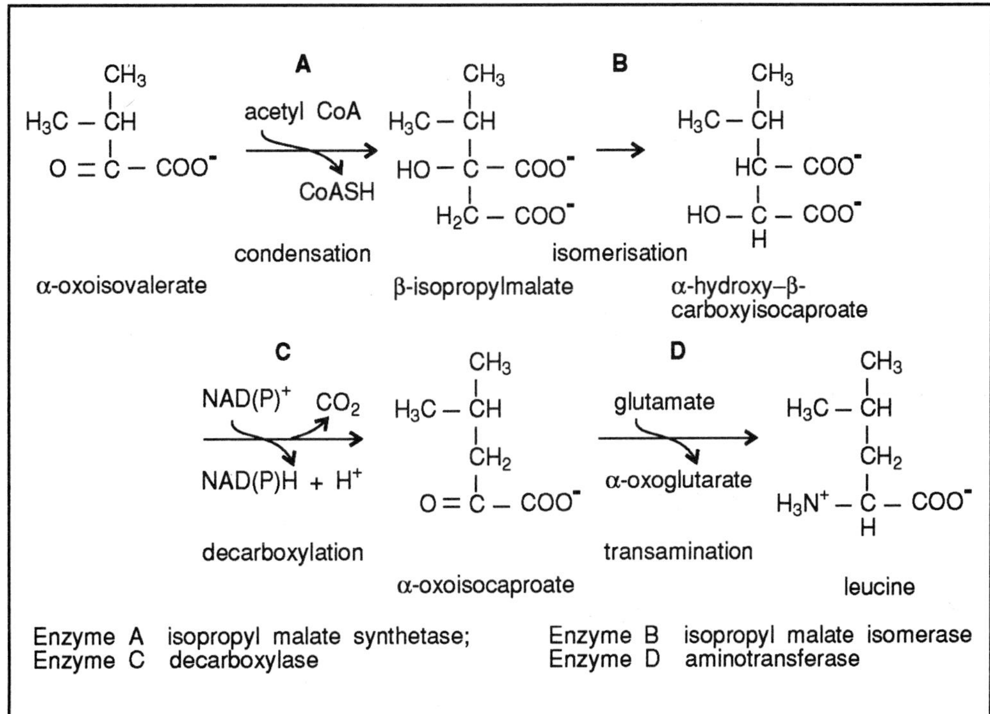

Figure 4.2 Biosynthesis of leucine from isovalerate.

regulation of production of the pyruvate family

To control the availability of pyruvate would be impractical because of its widespread importance thus the availability of α-oxobutyrate must be controlled. α-Oxobutyrate is formed from threonine by the reaction of threonine deaminase. In cells which can synthesise isoleucine two separate threonine deaminases are present; the synthesis of one of these - a biosynthetic enzyme - is severely inhibited in the presence of isoleucine. Thus whilst pyruvate is always present at a relatively constant concentration, α-oxobutyrate availability varies considerably and will influence the rate of isoleucine production. All enzymes mentioned in this section are membrane bound and in eukaryotes are mitochondrial. Acetohydroxy acid synthetase activity is inhibited and its synthesis repressed by valine in bacteria and by valine and leucine in fungi. α-oxoisovalerate is obviously the branch point of valine and leucine biosynthesis. Isopropyl malate synthetase is sensitive to inhibition by leucine and no further regulation occurs in this pathway. The meaning and use of the terms inhibition and repression of enzymes will be explained more fully in a later chapter.

inhibition and repression

The last point before closing this section on the biosynthesis of the pyruvate family is that two of the three, namely valine and leucine, are dietarily essential amino acids in higher organisms whereas all living systems can synthesise alanine.

SAQ 4.2

Produce a single simplified flow diagram of the reactions from pyruvate to alanine, valine and leucine and write an overall equation for each of the three parts of the pathway.

After completing this task the pathway should now look far less daunting even though it provided many new names to cope with. The purpose of the overall reaction is to indicate how little energy is required for amino acid biosynthesis once the precursor metabolites have been provided. No ATP / ADP changes occur throughout and only one redox imbalance occurs, in valine biosynthesis, and this reaction is cancelled out during leucine biosynthesis. Two molecules of pyruvate are expended however and each of these could be combusted to yield 15ATP. In one sense this is a significant loss but 'loss' is not really the correct word as the amino acid formed could be later degraded to yield pyruvate (from alanine), succinyl CoA (from valine) or acetyl CoA (from leucine) and each of them is capable of yielding energy. We will now move on to the serine family.

4.4.2 The biosynthesis of the serine family - serine, glycine and cysteine

Serine, glycine and cysteine may be produced from 3-phosphoglycerate by a relatively simple sequence of reactions.

early part of pathway is common

The first three reactions are common to each of the three amino acids and, as will be seen in Figure 4.3, begins with an oxidation of the α-hydroxyl group to yield an α-oxo group. This is followed by a transamination and finally removal of the phosphate which remains from the 3-phosphoglycerate.

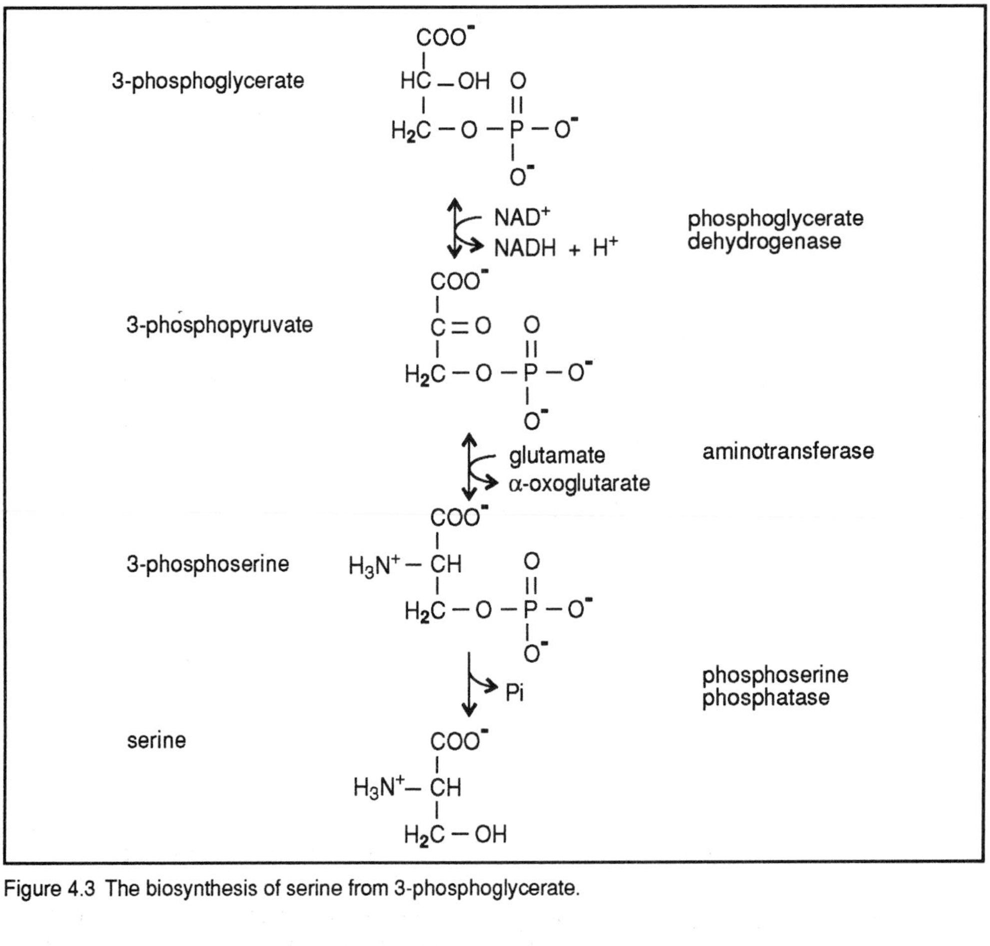

3-phosphoglycerate: COO^- / $HC-OH$ / $H_2C-O-P(=O)(O^-)-O^-$

NAD^+ → $NADH + H^+$ — phosphoglycerate dehydrogenase

3-phosphopyruvate: COO^- / $C=O$ / $H_2C-O-P(=O)(O^-)-O^-$

glutamate → α-oxoglutarate — aminotransferase

3-phosphoserine: COO^- / H_3N^+-CH / $H_2C-O-P(=O)(O^-)-O^-$

→ Pi — phosphoserine phosphatase

serine: COO^- / H_3N^+-CH / H_2C-OH

Figure 4.3 The biosynthesis of serine from 3-phosphoglycerate.

3-Phosphoglycerate is an intermediate of the Embden Meyerhof pathway which, on its way to pyruvate, is usually converted to 2-phosphoglycerate by phosphoglycerate mutase. Phosphoglycerate dehydrogenase is inhibited by serine thus regulating the overall biosynthetic pathway. This pathway is called the phosphorylated pathway as all of the intermediates contain phosphate. An alternative non-phosphorylated pathway via glycerate to serine is found in some fungi and bacteria but is not quantitatively important.

phosphorylated pathway

Glycine is produced from serine in a single stage reaction involving tetrahydrofolic acid (THF).

$$H_3N^+ - \underset{CH_2OH}{\overset{COO^-}{C}} - H \xrightleftharpoons[\text{serine hydroxymethyl transferase}]{THF \rightarrow N^5N^{10}\text{ methylene THF}} H_3N^+ - \underset{H}{\overset{COO^-}{C}} - H$$

serine — serine hydroxymethyl transferase — glycine

Glycine and serine are intimately linked and are major contributors to the metabolic one-carbon pool via THF acid. They are in fact, the major contributors of methylene THF for nucleotide biosynthesis.

glycine also produced by threonine breakdown

An alternative source of glycine which should be mentioned is by the cleavage of threonine to glycine and acetaldehyde. Certain Clostridia have difficulty in producing sufficient serine for their metabolic needs and in such cases there would obviously be no serine available for conversion to glycine so an alternative pathway is required.

Cysteine is usually produced from serine in a two-stage reaction in which the beta-hydroxyl group is replaced by a beta-sulphydryl group.

$$H_3N^+ - \underset{CH_2OH}{\overset{COO^-}{C}} - H \xrightarrow[\text{serine acetyl transferase}]{\text{acetyl CoA} \rightarrow CoASH} H_3N^+ - \underset{CH_2OCOCH_3}{\overset{COO^-}{C}} - H$$

serine — serine acetyl transferase — O-acetyl serine

$$\xrightarrow[\text{serine sulphydrase}]{H_2S \rightarrow \text{acetate}} H_3N^+ - \underset{CH_2SH}{\overset{COO^-}{C}} - H$$

cysteine

serine sulphydrase

Remember that whilst glycine and serine can be produced by most living systems cysteine can only be produced by bacteria and plants. The synthesis of serine acetyl transferase is severely repressed when bacteria are grown in cysteine-rich media. As we saw in Section 4.1 cysteine can be synthesised from methionine in higher animals but this process is not favoured in micro-organisms.

Π Can you derive the overall reactions for synthesis of each of the amino acids from 3-phosphoglycerate? Try this on a piece of paper and check your answers with the overall reactions writtenout below.

1) 3-phosphoglycerate + glutamate + NAD^+ →

serine + α-oxoglutarate + NADH + H^+ + Pi

2) 3-phosphoglycerate + glutamate + NAD^+ + THF →

glycine + α-oxoglutarate + NADH + H^+ + N^5N^{10} methylene THF

3) 3-phosphoglycerate + glutamate + NAD^+ + acetyl CoA + H_2S →

cysteine + α-oxoglutarate + NADH + H^+ + acetate + CoASH

SAQ 4.3

Now you have mastered the overall reactions see if you can put together a simple flow diagram showing the formation of serine, glycine and cysteine from 3-phosphoglycerate.

Drawing the reactions out as a single diagram shows that there is only a single redox reaction and no direct energy loss in terms of ATP utilisation showing that, as with the pyruvate family, once the precursor metabolites are available there is little energy required for the process of biosynthesis to occur.

4.4.3 Biosynthesis of the aspartate family - aspartate, asparagine, lysine, methionine, threonine and isoleucine

The biosynthesis of three members of this family are relatively straightforward but the other three involve quite complicated pathways, particularly that of lysine. Only aspartate and asparagine can be synthesised by all living systems, the other four are synthesised by higher plants and micro-organisms except fungi. Owing to the complexity of the biosynthesis within the family it is advantageous to produce a simplified flow diagram just to demonstrate the overall pathways. This diagram is shown in Figure 4.4. Each arrow represents a reaction step - thus threonine to isoleucine includes five steps catalysed by different enzymes.

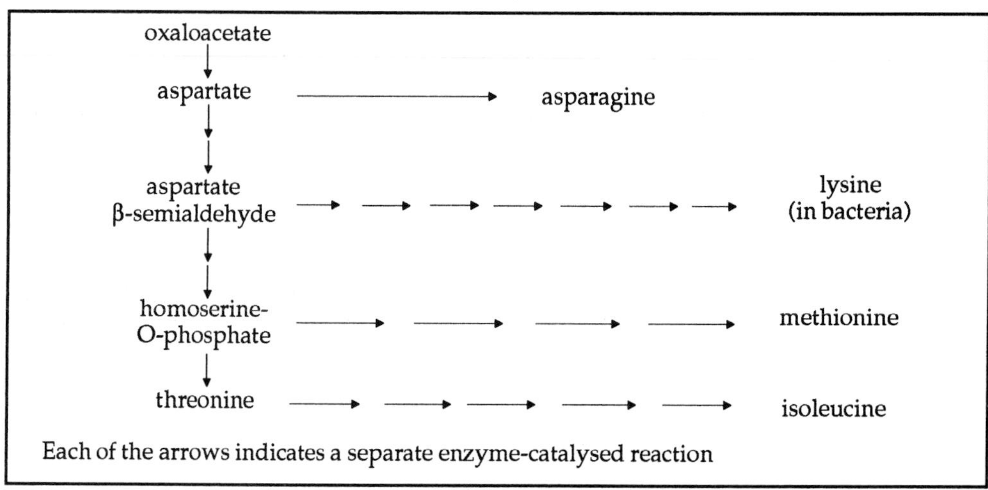

Figure 4.4 A simplified flow diagram of the biosynthesis of the aspartate family of amino acids.

Our task when looking at Figure 4.4 looks very daunting due to the sheer complexity and new compounds which will be encountered. Although all reactions are equally important to the well-being of the cell, we shall concentrate on highlighting only a few to explain aspects of metabolic control.

The first part of the diagram in Figure 4.4 is very familiar to us and the following questions should elicit ready answers.

SAQ 4.4

1) What is the source of oxaloacetate in living cells?

2) What is the formula of oxaloacetate?

3) From 2) can you therefore write down the structure of aspartate?

4) What enzyme is used to convert oxaloacetate to aspartate?

5) What is the formula of asparagine (it is an amide of aspartate)?

6) Can you write the equation of the reaction to produce asparagine from aspartate?

regulation of the pathway

Before we look at the individual reactions of each of the pathways let us consider what sort of control processes are employed. The cell must be able to regulate production of each of the amino acids as required but this will be against a background of the continuous presence of oxaloacetate. In addition aspartate will be present in relatively high concentrations due to the oxaloacetate $\rightleftarrows$ aspartate and aspartate $\rightleftarrows$ asparagine reactions employed in nitrogen assimilation and conservation.

SAQ 4.5

From the discussion above and looking at Figure 4.4 can you decide where control will be exercised in the pathways and why? Do not go into detail about the individual reactions. If after considerable thought you are stuck, look at the clues below one by one.

1) rarely are all enzymes controlled;

2) in the figure there are five key control points;

3) there is generally a 'master switch' which can turn on or off all or most of the pathway if required;

4) individual arms of the pathway could usefully be controlled.

The answer to SAQ 4.5 is fairly lengthy but it is worth spending time to grasp the principles of control. Metabolism is very logical in its processes and the elucidation of control is a fascinating task.

Continuing with our story, aspartate is converted to aspartate-β-semialdehyde via β-aspartyl phosphate as follows:

$$H_3N^+{-}CH(COO^-){-}CH_2{-}COO^- \xrightarrow{ATP \rightarrow ADP} H_3N^+{-}CH(COO^-){-}CH_2{-}C(=O){-}O{-}PO_3^{2-} \xrightarrow{NADPH + H^+ \rightarrow NADP^+ + Pi} H_3N^+{-}CH(COO^-){-}CH_2{-}HC{=}O$$

aspartate β-aspartyl phosphate aspartate-β-semialdehyde

The first enzyme is aspartokinase (remember a kinase involves ATP formation or utilisation).

Π Can you name the second enzyme?

Note the use of NADPH + H^+ indicating the use of a hydrogenation/dehydrogenation reaction: the compound being dehydrogenated is aspartate semialdehyde, thus the enzyme is aspartate semialdehyde dehydrogenase.

lysine biosynthesis

From this point the pathways diverge, and we will next examine the pathway to lysine found in higher plants, most algae and all prokaryotes. This pathway, shown in Figure 4.5, is complex but is of interest not only for the production of lysine but also of two intermediates which are important to many bacterial cells. Examine the diagram carefully and, knowing the names of the compounds then the enzyme names are not too difficult. The two intermediates which are important for biosynthesis of other compounds are dihydropicolinic acid (dihydropicolinate) and mesodiaminopimelate. The latter is an intermediate of the murein component of the cell wall of many bacteria; more of this in a later chapter of this book. Dihydropicolinate is converted to dipicolinate which is a major component of bacterial spores (largely in members of the genera *Bacillus* and *Clostridium*).

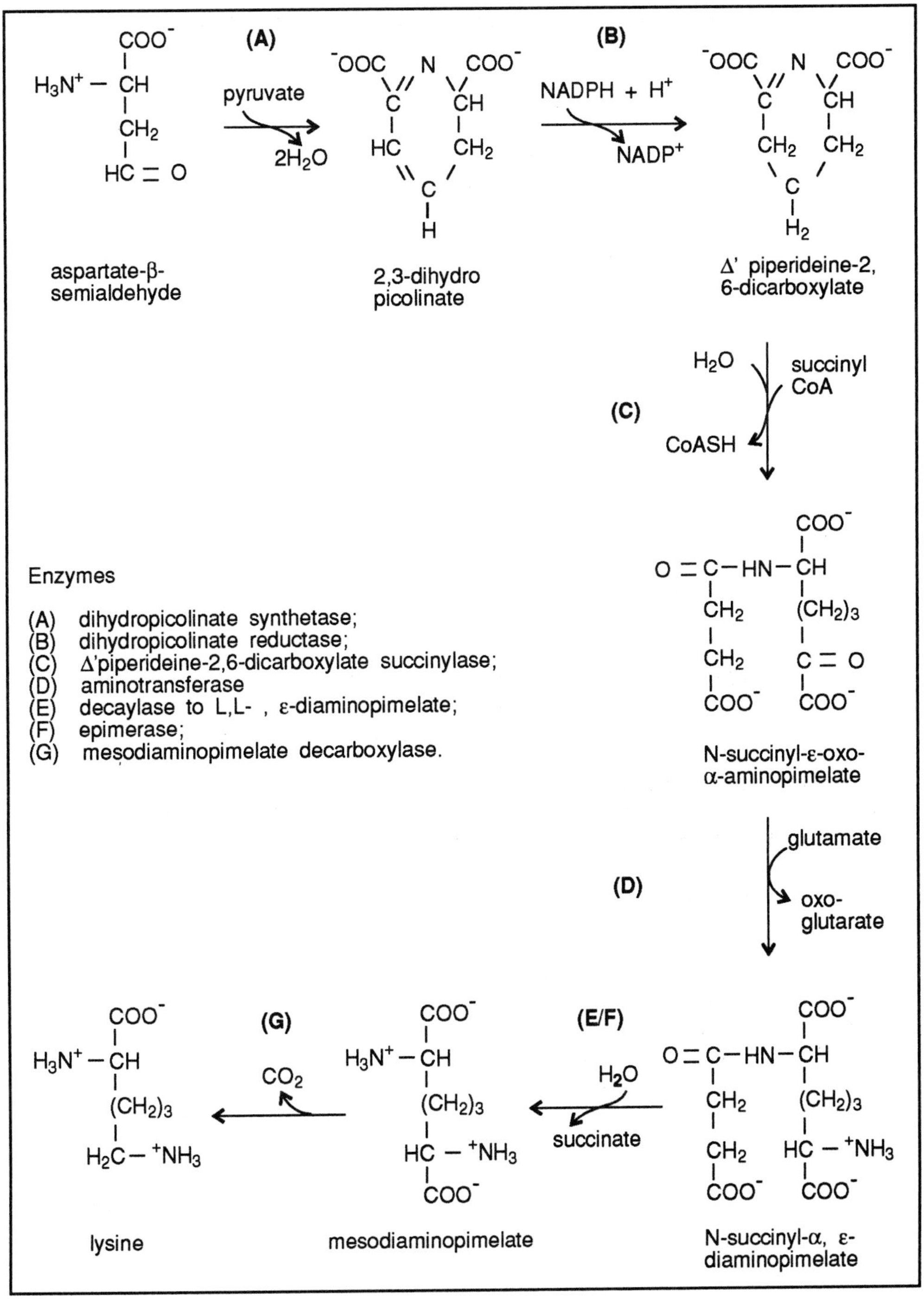

aspartate-β-semialdehyde

(A) pyruvate → $2H_2O$

2,3-dihydro picolinate

(B) NADPH + H^+ → $NADP^+$

Δ' piperideine-2, 6-dicarboxylate

(C) H_2O, succinyl CoA → CoASH

N-succinyl-ε-oxo-α-aminopimelate

(D) glutamate → oxo-glutarate

N-succinyl-α, ε-diaminopimelate

(E/F) H_2O → succinate

mesodiaminopimelate

(G) → CO_2

lysine

Enzymes

(A) dihydropicolinate synthetase;
(B) dihydropicolinate reductase;
(C) Δ'piperideine-2,6-dicarboxylate succinylase;
(D) aminotransferase
(E) decaylase to L,L- , ε-diaminopimelate;
(F) epimerase;
(G) mesodiaminopimelate decarboxylase.

Figure 4.5 The diaminopimelate pathway for the production of lysine.

There are two reactions in common from aspartate-β-semialdehyde when considering the biosynthesis of methionine, threonine and isoleucine. Firstly a reduction (hydrogenation) occurs to form homoserine followed by a phosphorylation to yield homoserine-O-phosphate as follows:

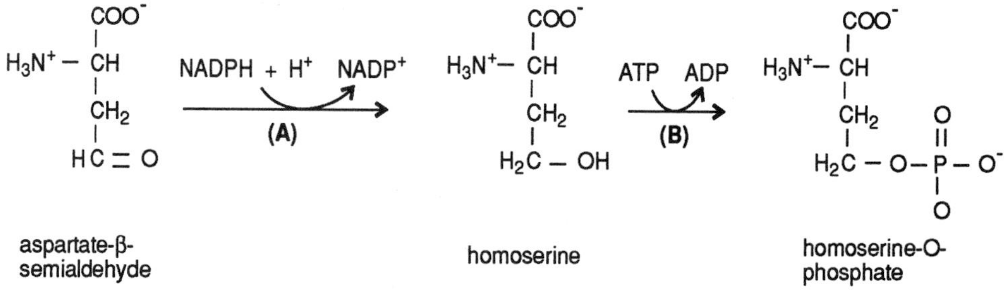

aspartate-β-semialdehyde —(A): NADPH + H⁺ → NADP⁺→ homoserine —(B): ATP → ADP→ homoserine-O-phosphate

Π Can you name the enzymes in this sequence?

The first enzyme (A) is clearly a dehydrogenase but be careful, it is homoserine dehydrogenase not aspartate-β-semialdehyde dehydrogenase. In naming the enzyme you need to note which of the two compounds is actually dehydrogenated. Enzyme (B) is a kinase, hence homoserine kinase.

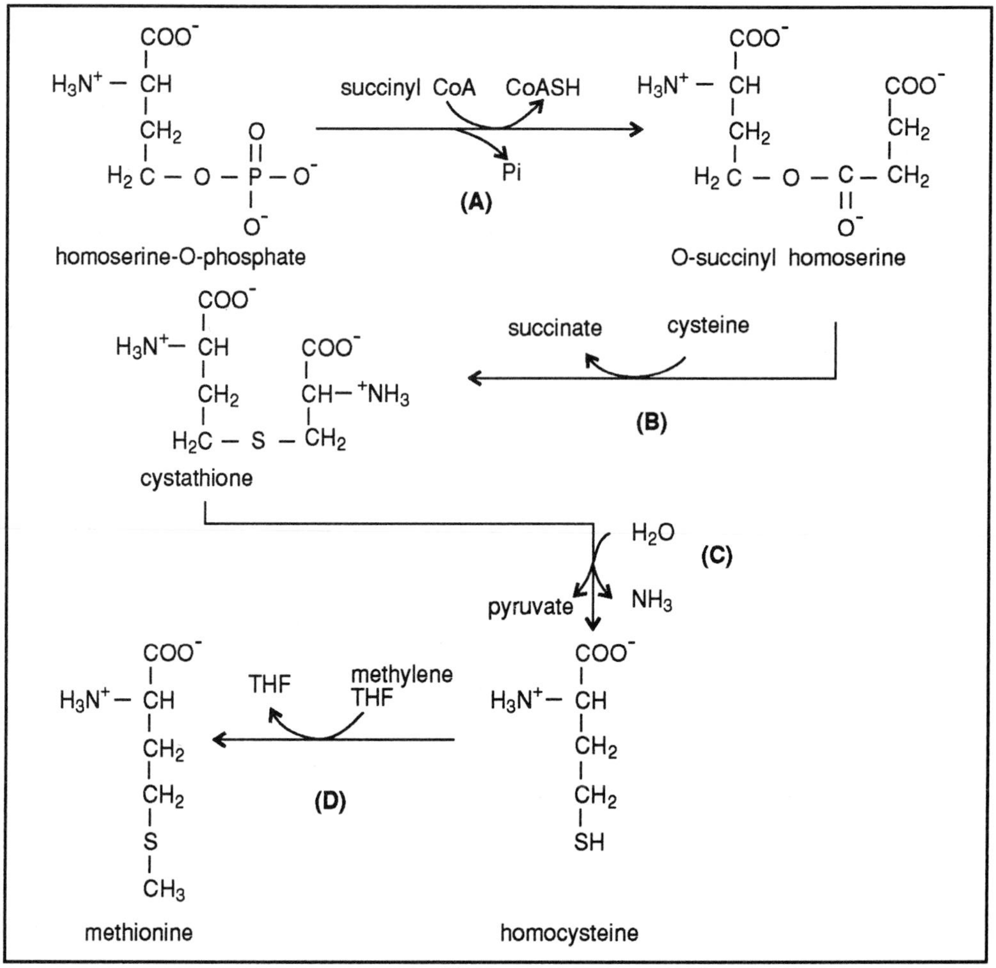

homoserine-O-phosphate —(A): succinyl CoA → CoASH, Pi→ O-succinyl homoserine —(B): cysteine → succinate→ cystathione —(C): H_2O → pyruvate, NH_3→ homocysteine —(D): methylene THF → THF→ methionine

Figure 4.6 The synthesis of methionine from homoserine-O-phosphate. Key to enzymes: (A) homoserine succinyl transferase; (B) cystathione synthetase; (C) cystathionase; (D) homocysteine methyl transferase.

Four steps are required to produce methionine from homoserine-O-phosphate and this part of the pathway is quite demanding in that succinyl CoA and cysteine are required (though succinate and pyruvate are released) together with methylene THF. The sequence is given in Figure 4.6.

isoleucine biosynthesis

The sequence of reactions from homoserine-O-phosphate to isoleucine via threonine will not be given in detail but only as a flow diagram (Figure 4.7). This is because, as indicated in Section 4.4.1, the biosynthesis of isoleucine occurs in an entirely analogous manner to that of valine. Following deamination of threonine the product is α-oxobutyrate and this compound differs from pyruvate only in that it contains an extra methylene group. The end-products of the four reaction sequence, isoleucine and valine respectively, differ from each other only in that isoleucine contains an extra methyl group. As explained earlier the same set of enzymes is considered to catalyse both sets of reactions.

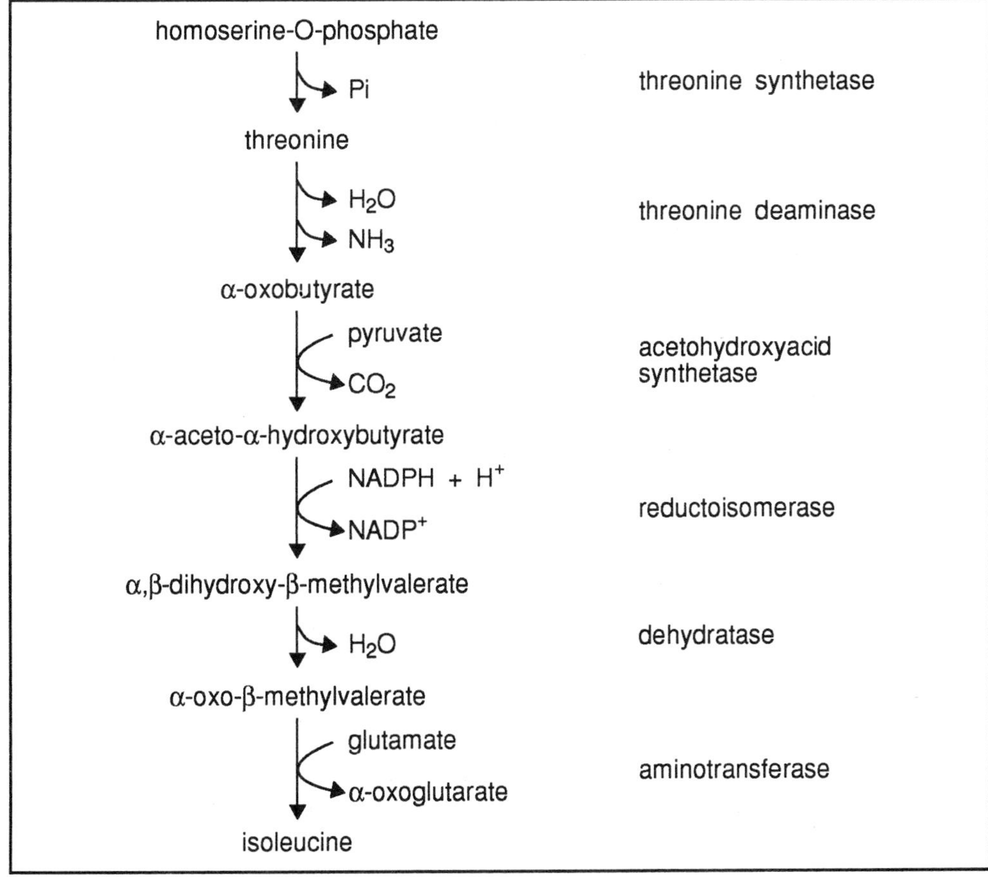

Figure 4.7 Biosynthesis of isoleucine from homoserine-O-phosphate.

To conclude our studies of the aspartate family of amino acids study the flow diagram in Figure 4.8.

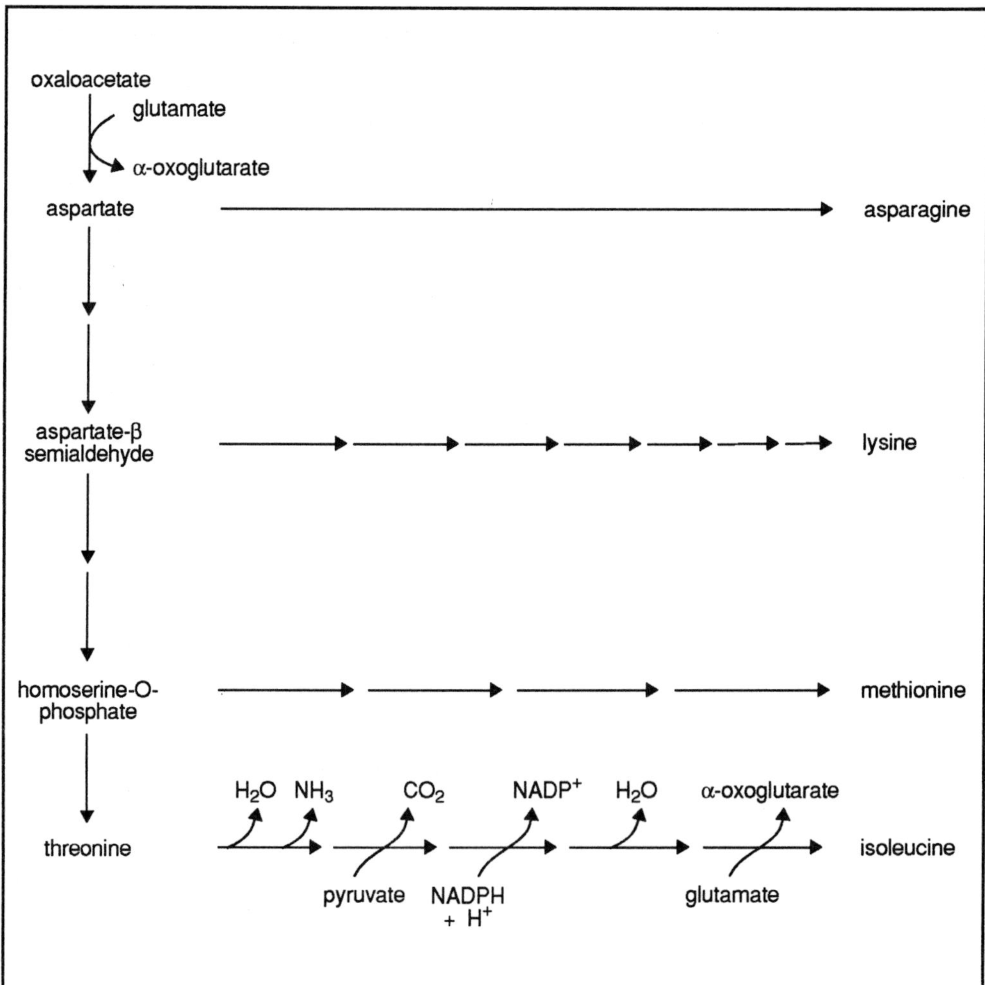

Figure 4.8 The biosynthesis of the aspartate family of amino acids - a flow diagram for SAQ 4.6.

SAQ 4.6

Complete Figure 4.8 by indicating only the overall changes which occur during each section of the pathways. To help you along, the overall changes are shown for the oxaloacetate to aspartate stage and for the threonine to isoleucine reactions.

Π From the information given try to work out the overall equations for the conversion of oxaloacetate to lysine, to methionine, to threonine and finally to isoleucine.

1) oxaloacetate + 2 glutamate + succinyl CoA + pyruvate + ATP + 2NADPH + $2H^+$

→ lysine + 2α-oxoglutarate + succinate + CoASH + CO_2 + ADP + Pi + $2NADP^+$

2) oxaloacetate + glutamate + succinyl CoA + cysteine + methylene THF + 2ATP + 2NADPH + $2H^+$ + H_2O

$\rightarrow$ methionine + α-oxoglutarate + succinate + CoASH + pyruvate + THF + 2ADP + 2Pi + NH_3 + $2NADP^+$

3) oxaloacetate + glutamate + 2ATP + 2NADPH + $2H^+$

$\rightarrow$ threonine + α-oxoglutarate + 2ADP + 2Pi + $2NADP^+$

4) oxaloacetate + 2 glutamate + pyruvate + 2ATP + 3NADPH +$3H^+$

$\rightarrow$ isoleucine + 2α-oxoglutarate + CO_2 + 2ADP 2Pi + $3NADP^+$ + $2H_2O$ + NH_3

The experience gained in carrying out exercise SAQ 4.6 and the ITE above is useful in many ways. Firstly continued working with the pathways makes them seem less awe-inspiring or complicated and you should gradually get more confident in tackling them.

general points about the aspartate family

Secondly they gradually begin to appear as 'variations on a similar theme' where cells are carrying out repetitive processes to accomplish a particular set of tasks. Finally we should study the overall equations for a few minutes to identify any generalisations possible. Note the following:

- the repeated use of glutamate as the amino group donor;
- the relatively small requirement for ATP throughout the pathways;
- the alternative use of succinyl CoA as an energy source. This allows the addition of CoASH to a compound without the involvement of ATP; the compound-CoA bond when broken yields energy for a metabolic reaction;
- the use of NADPH + H^+ in biosynthetic reactions rather than NAD^+ which is used for catabolism, a theme which we will develope throughout many chapters.

4.4.4 The biosynthesis of the glutamate family - glutamate, glutamine, arginine, proline and lysine.

The conversion of α-oxoglutarate to glutamate and glutamine are simple and have already been discussed. Proline and arginine are produced from glutamate by four and eight-stage reactions respectively. These four amino acids are not dietarily essential in higher animals but lysine, the fifth amino acid in this group, is. In the previous section we studied the biosynthesis of lysine via the diaminopimelate pathway in higher plants and prokaryotes. In fungi, however, an alternative, the α-aminoadipic acid pathway, is used to synthesise lysine.

α-Oxoglutarate is, as we know, a TCA cycle intermediate and therefore will be readily available for amino acid biosynthesis.

We studied the assimilation of nitrogen by living organisms in Section 4.3 and noted that the principal way is reductive amination by glutamate dehydrogenase which produces glutamate from α-oxoglutarate.

importance of glutamine

The formation of glutamine from glutamate was also discussed in Section 4.3. Glutamine is important in metabolism not only to take up excess amino groups and then to distribute them as circumstances dictate but also as an amino group donor during purine and pyrimidine biosynthesis as we shall see later.

proline biosynthesis

In proline biosynthesis glutamate is phosphorylated to glutamyl phosphate which is then reduced and dephosphorylated to glutamic semialdehyde. A non-enzymic spontaneous ring closure occurs giving Δ-pyrroline-5-carboxylate and finally this compound is reduced to proline. The process is shown in Figure 4.9.

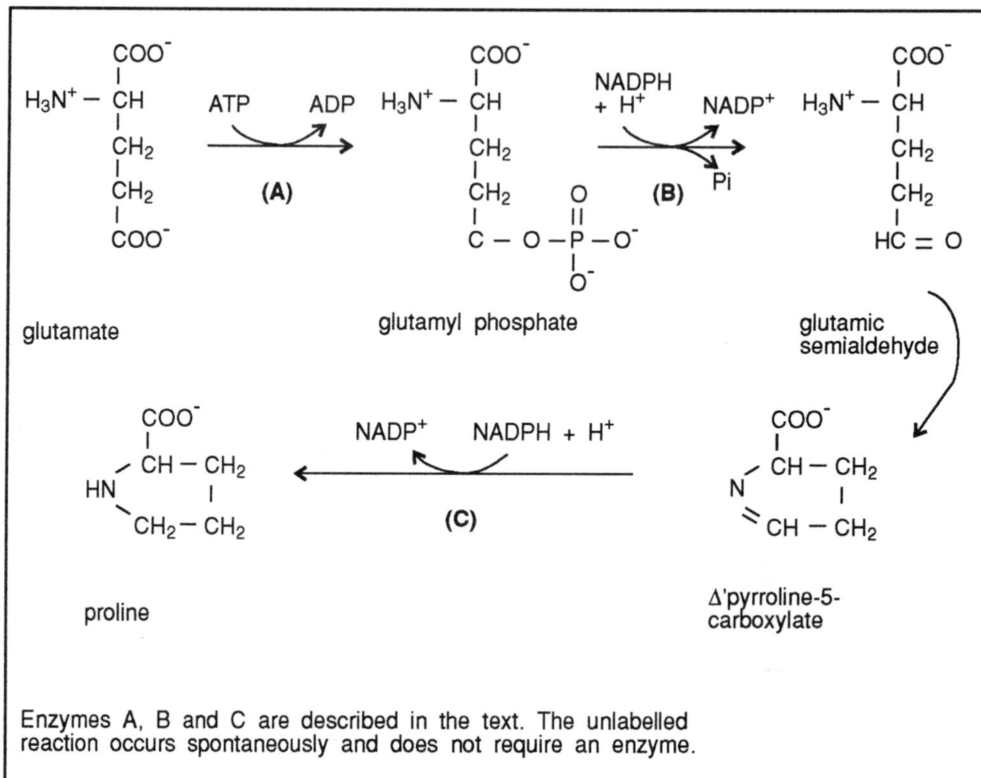

Figure 4.9 Flow diagram to show the biosynthesis of proline from glutamate.

Π Can you name the enzymes A, B and C shown in Figure 4.9?

Enzyme A is glutamate kinase; enzyme B is glutamyl phosphate reductase and enzyme C is pyrroline-5-carboxylate reductase. Note that the cyclisation of glutamic semialdehyde to form Δ'pyrroline-5-carboxylate occurs spontaneously and no enzyme is required.

You should recognise the kinases because of the utilisation of ATP. Enzymes B and C are reductases rather than dehydrogenases as use of the term dehydrogenase would imply that the back reaction is possible but these two reactions are essentially irreversible.

The overall reaction for the process is:

$$\text{glutamate} + \text{ATP} + 2\text{NADPH} + 2\text{H}^+ \rightarrow \text{proline} + \text{ADP} + \text{Pi} + 2\text{NADP}^+ + \text{H}_2\text{O}$$

regulation of the pathway

In most micro-organisms the first enzyme glutamate kinase is subject to end-product inhibition by proline. In addition the final enzyme, pyrroline-5-carboxylate reductase, is repressed by the presence of proline in the medium.

arginine biosynthesis

urea cycle

Arginine biosynthesis is a more complex process than proline biosynthesis, largely because arginine is a C_6N_4 containing compound whereas glutamate is only a C_5N_1 compound. The pathway is extremely important in ureotelic organisms, that is those which produce urea as nitrogenous waste. Higher animals belong to this group. They cannot remove the toxic ammonia quickly enough so they produce urea and excrete it via the kidneys. This process, called the urea cycle, is essentially a cyclical process involving the last four intermediates of arginine biosynthesis and is of little if any importance in micro-organisms. Figure 4.10 is a simplified diagram describing arginine biosynthesis from glutamate. The enzyme which is present in higher organisms to complete the urea cycle is shown in Figure 4.10 by a dotted line.

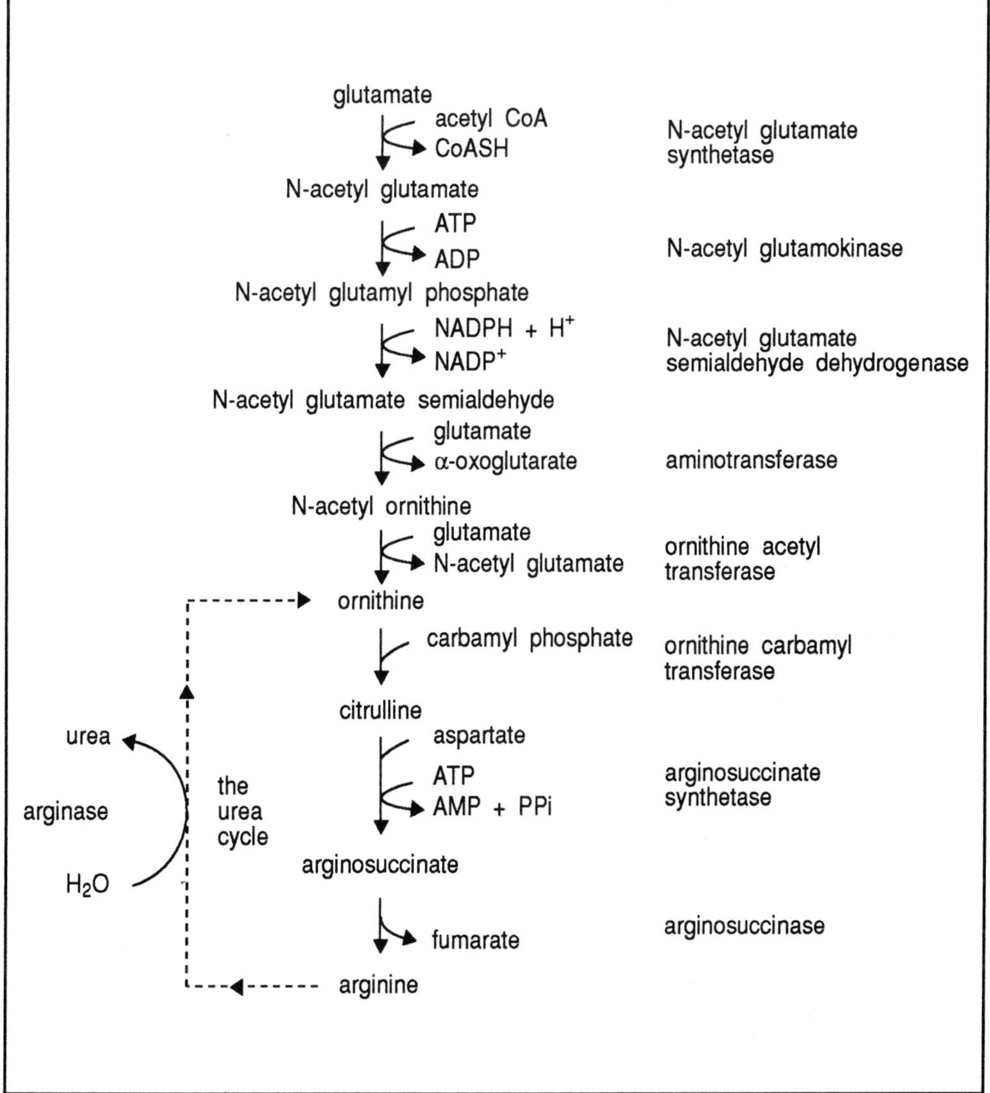

Figure 4.10 Diagram to show the biosynthesis of arginine and to demonstrate the urea cycle.

The overall reaction for arginine production is:

3 glutamate + acetyl CoA + aspartate + carbamyl phosphate + 2ATP + NADPH + H^+

→ arginine + N-acetyl glutamate + fumarate + α-oxoglutarate + CoASH + ADP + AMP + Pi + PPi + $NADP^+$

A complicated sequence indeed though the N-acetyl glutamate will be recycled via step two of the pathway.

polyamines

As expected the N-acetyl glutamine synthetase, the first enzyme of the pathway, is inhibited by the end-product, arginine. In addition to being used in proteins, arginine is also used in the biosynthesis of the polyamines, putrescine, spermidine and spermine. Polyamines are often present in substantial amounts: for example 4% of the total nitrogen of *Azotobacter vinelandii* may be spermidine and putrescine. Not only are these compounds important in prokaryotes, they are present in all living systems, being closely associated with DNA and RNA (probably neutralising their formidable negative charges) and helping to control the intracellular ionic strength. Either ornithine or arginine can act as the starting point as shown below.

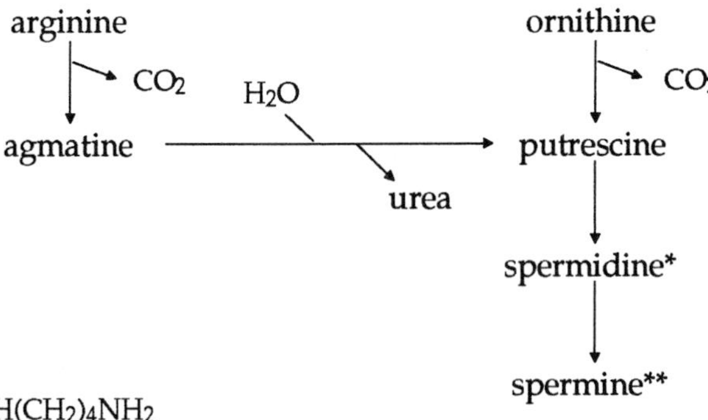

* $NH_2(CH_2)_3NH(CH_2)_4NH_2$
** $NH_2(CH_2)_3NH(CH_2)_4NH(CH_2)_3NH_2$

lysine biosynthesis

The last amino acid in this group is lysine. Its biosynthesis occurs from α-oxoglutarate by a seven-step sequence, the first three being unfamiliar to us in terms of nomenclature but recognisable in terms of the chemistry. In the TCA cycle an acetyl moiety from acetyl CoA condenses onto oxaloacetate to give citrate which is converted to α-oxoglutarate via isocitrate. In lysine biosynthesis there is an extra methyl group present on an identical series of reactions, from α-oxoglutarate to α-oxoadipic acid. Again familiar reactions follow: this six-carbon dicarboxylic acid is aminated and the terminal carboxyl group reduced via the semialdehyde before again being aminated from glutamate. The whole α-aminoadipic acid pathway is shown in Figure 4.11.

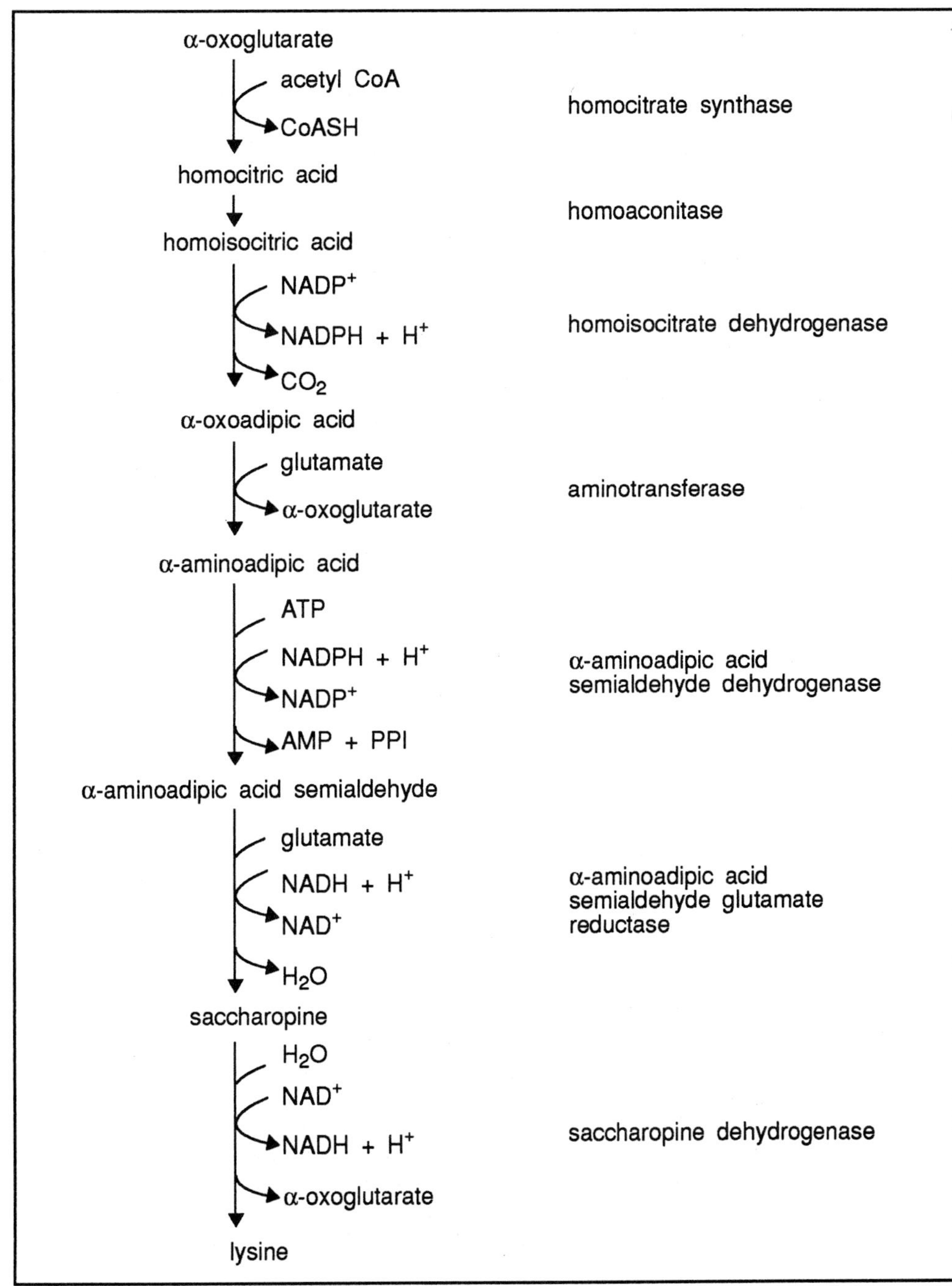

Figure 4.11 The biosynthesis of lysine in fungi via the α-aminoadipic acid pathway.

Control is exercised on the first two enzymes of the pathway which are repressed (but not inhibited) by high lysine concentrations.

4.4.5 The biosynthesis of the aromatic family phenylalanine - trosine and tryptophan

early pathway is common

importance of aromatic compounds

We have already established that essentially all three of these amino acids are dietarily essential for higher animals; tyrosine can be synthesised but only by hydroxylation of phenylalanine. The pathways involved are common to all three amino acids for the seven reactions leading to chorismic acid. From here simple three-step reactions yield either phenylalanine or tyrosine and a separate five-step sequence to tryptophan is needed. Inevitably the control will be somewhat complex due to the late branching but a further complication arises due to the tremendous importance of the intermediates of aromatic amino acid biosynthesis as precursors for other compounds and the use of the amino acids in non-protein components. The flow diagram (Figure 4.12) is a very useful starting point to show the pathway and to indicate the branch points to other metabolites.

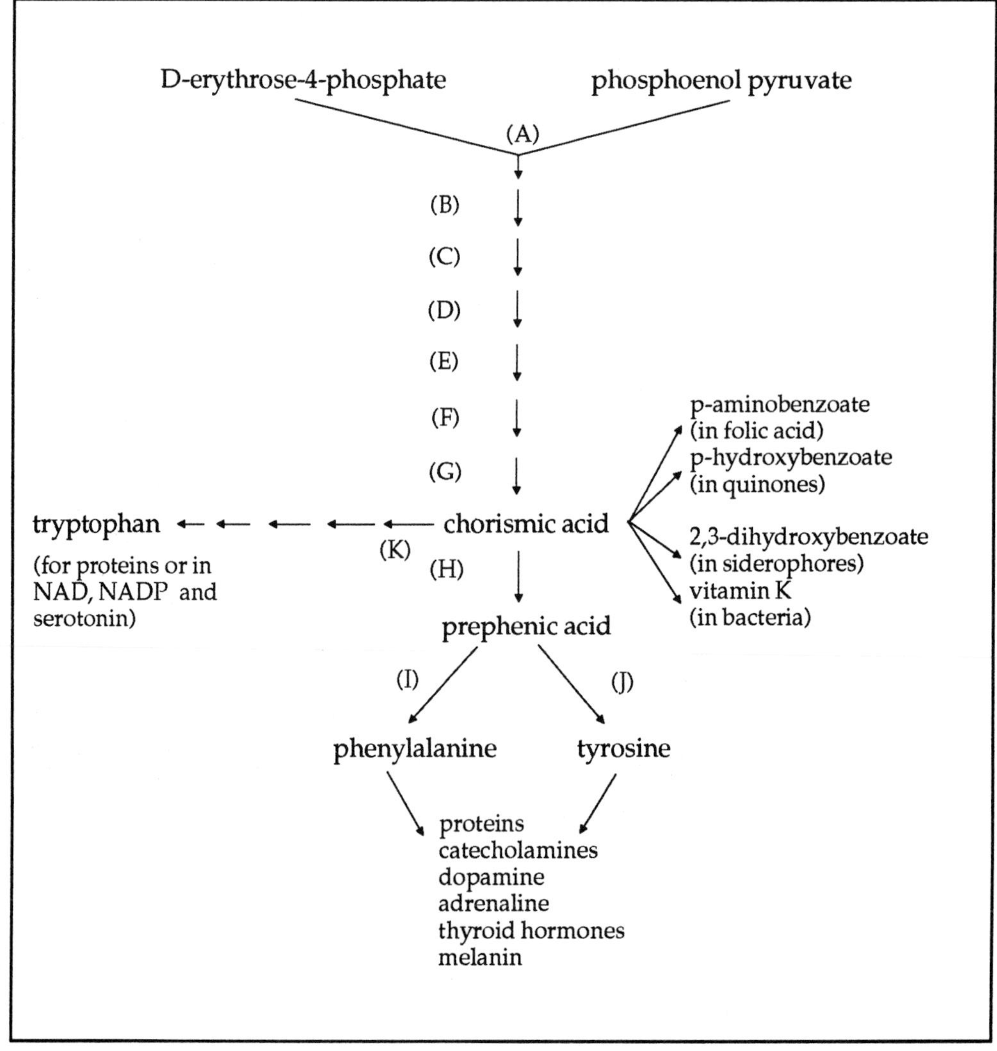

Figure 4.12 Flow diagram of the biosynthetic pathway for aromatic amino acids and the branch points for other metabolites. (Letters relating to enzymes are referred to in the text).

Π In terms of control, which enzyme (s) do you think are inhibited and/or repressed?

The answer is enzymes A, H (I), H (J) and K.

Obviously enzyme A, the first enzyme of the whole pathway. In most organisms there are three separate enzymes, one inhibited by each of the three amino acids. Enzymes B to F in some fungi, for example *Neurospora crassa* are in a multi-enzyme complex but this is not the case in most bacterial systems.

Enzyme H is generally present as two isoenzymes, one associated with enzyme I and inhibited by phenylalanine and a second associated with enzyme J and inhibited by tyrosine. Interestingly excess of tryptophan stimulates enzyme H thus drawing chorismic acid to the other two amino acids.

Enzyme K is strongly inhibited by an excess of tryptophan.

The intermediates in these three pathways are complicated because they lead up to the three amino acids with the highest molecular weights of all. Much of the effort in the first seven reactions is the production of a six-carbon ring structure which, by the time chorismic acid is reached, has two, not three, double bonds. Enzymes I, J and K each in their own way produce compounds which are genuinely aromatic in preparation for the amino acids themselves. Undoubtedly as we have mentioned these biosynthetic schemes are important and perusal of the structures in text books could be rewarding. We will not describe them further here.

4.4.6 The biosynthesis of histidine

Histidine stands alone in being the sole product of a complex unbranched pathway from the five-carbon compound phosphoribosyl pyrophosphate (PRPP). All five of the carbon atoms in PRPP are found in histidine, three constitute the side chain and the other two are two of the carbons within the imidazole ring (a ring containing three carbons and two nitrogens). Thus the source of one carbon and two nitrogens still needs to be clarified. One nitrogen is derived from glutamine and, curiously, the carbon and nitrogen remaining are together derived from the purine nucleus of ATP. This utilisation of a C-N pair of atoms from a purine nucleus is unique and looks particularly surprising in terms of energy conservation. However the remainder of the ATP molecule after it loses the C-N fragment, called aminoimidazole carboxamide ribotide, (AICAR) is not wasted but is in fact a precursor of purine biosynthesis, as we shall see later in this chapter. Thus rapid recycling occurs without great loss of energy or carbons. The early reactions are concerned with forming the imidazole ring, the remainder with altering the side chain (Figure 4.13).

C-N taken from a purine

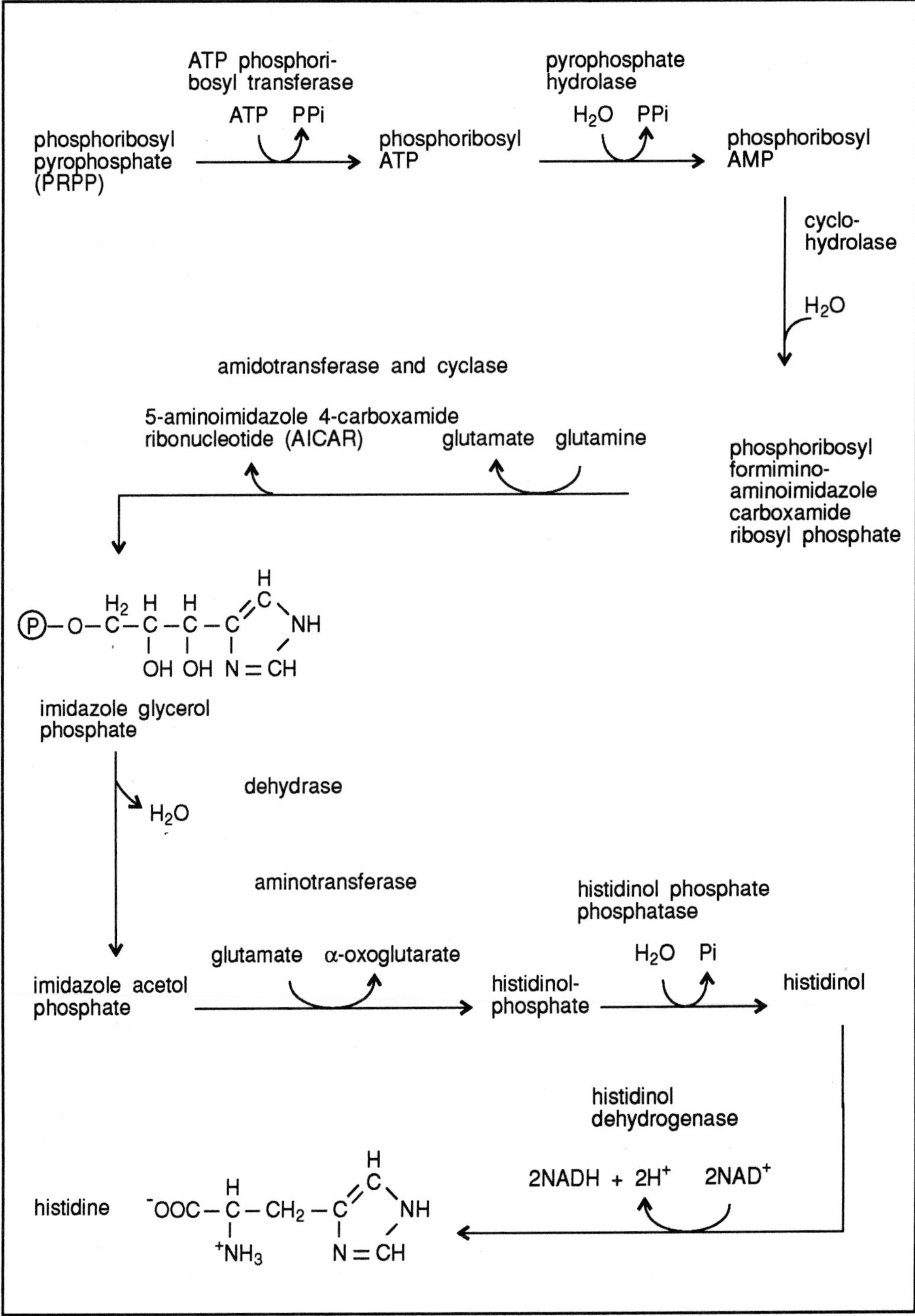

Figure 4.13 The biosynthesis of histidine.

As there are no branch points control is relatively simple, the ATP phosphoribosyl transferase being strongly inhibited by histidine.

SAQ 4.7

Study the partially completed diagram in Figure 4.14. It is a simplified overall diagram showing the biosynthesis of amino acids by micro-organisms. Can you:

1) identify by underlining, the seven precursor metabolites and name the pathways from which they are derived;

2) fill in the compounds missing from the spaces which are denoted by boxes?

This exercise is worthwhile in that it helps to give that essential overview of the processes involved in amino acid biosynthesis. In many ways the reactions are chemically simple but the sheer numbers of the intermediates involved makes learning of sequences a virtually impossible task and also an unnecessary one.

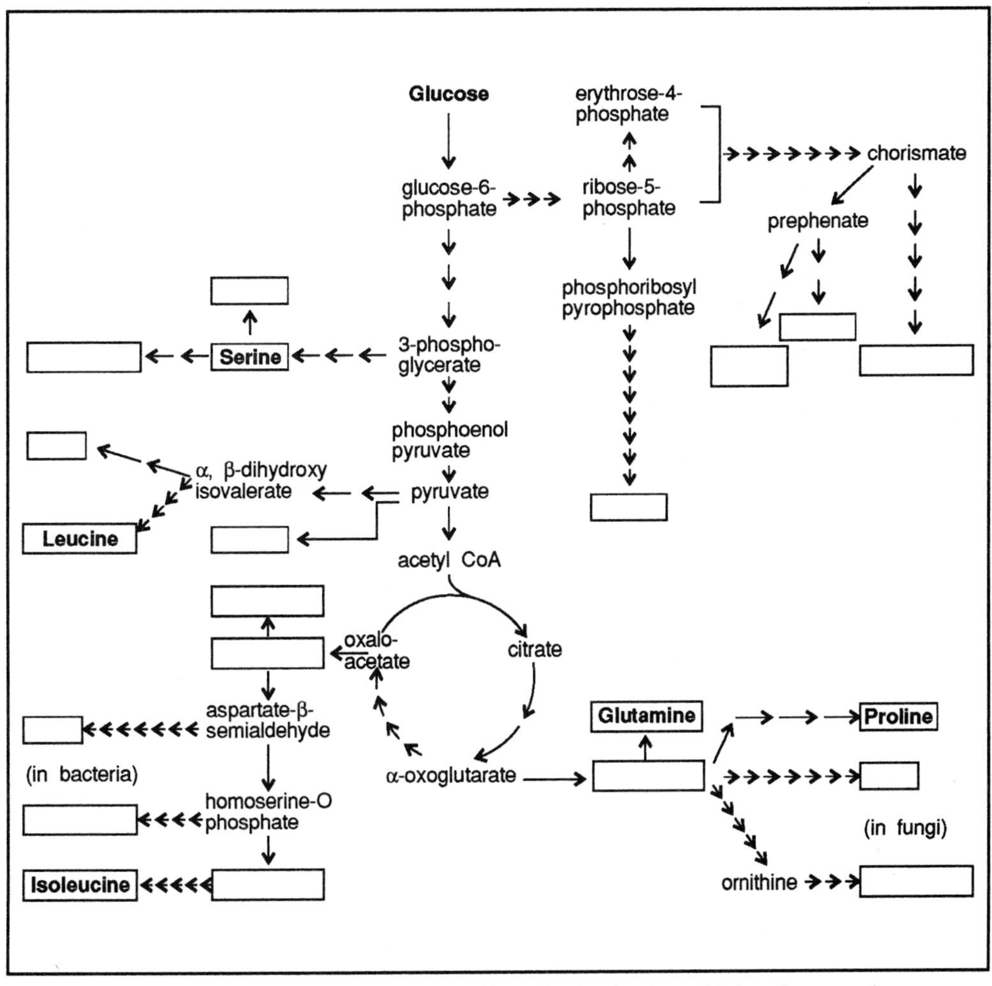

Figure 4.14 Simplified overall diagram showing the biosynthesis of amino acids by micro-organisms. (Note SAQ 4.7 refers to this figure).

It is worth pausing for a time merely to study your completed Figure 4.14. In addition to giving a feeling of rapport with the metabolic processes, you should get a feeling of satisfaction, possibly 'relieved satisfaction' at having completed a long and difficult task.

4.5 The recycling and biosynthesis of nucleic acid precursors

Remember that the nucleic acids form 5-15% of the dry weight of the living cells and some nucleic acid species, for example messenger RNA, have a very rapid turnover. The purine and pyrimidine components of nucleic acids are structurally quite complex. Most eukaryotes and some prokaryotes have problems with purine degradation - in higher animals, for example they are degraded only to uric acid which is excreted in urine. This constitutes a considerable loss of N atoms. Whatever the degradative pathway there is little or no benefit in terms of energy or intermediates and excretion is an inevitable burden. Pyrimidines are simpler and their degradation is more rewarding energetically in that the vast majority of living systems produce the metabolically useful succinyl CoA from them.

succinyl CoA is a useful degradation product of pyrimidine

It is therefore sensible and desirable for all cells to recycle the degradative products of nucleic acids. This is in fact what happens, the vast majority of such degradation products being recycled to produce new nucleic acid material. Remember that nucleic acids are broken down to nucleotides (nitrogenous bases plus sugar plus phosphate) which are themselves broken down to nucleosides (nitrogenous base plus sugar) and then to individual base and sugar. The recycling or biosynthesis of sugars will not be considered further in the chapter. The metabolism of ribose and deoxyribose is considered in some depth in the Biotol text, 'The Principles of Cell Energetics', Chapter 6. It is sufficient to remind ourselves that any free ribose or deoxyribose would simply return to the pentose metabolic pool.

There is a third possibility which we will consider briefly - the utilisation of exogenous nitrogenous bases or nucleosides by some bacterial species.

We shall look first at the recycling of breakdown products of endogenous nucleic acids followed by an examination of the use of exogenous bases and nucleosides by various bacteria. Finally *de novo* synthesis of purines and pyrimidines will be considered.

4.5.1 Recycling of endogenous nucleic acid metabolites

The purines adenine and guanine are degraded in nucleotide form from nucleic acids, that is adenosine monophosphate (AMP) and guanosine monophosphate (GMP). These compounds can obviously be utilised as such though use would need to be fairly rapid as much of the cellular control depends on the maintenance of AMP: ADP: ATP ratios and could not tolerate a massive influx of AMP (see Chapters 6 & 7).

maintaining nucleotide ratios

One problem which arises is that nucleic acid degradation may yield AMP and GMP in ratios which are different to cellular requirements. The solution to this problem and the description of the way in which cells partially regenerate AMP and GMP are shown as follows.

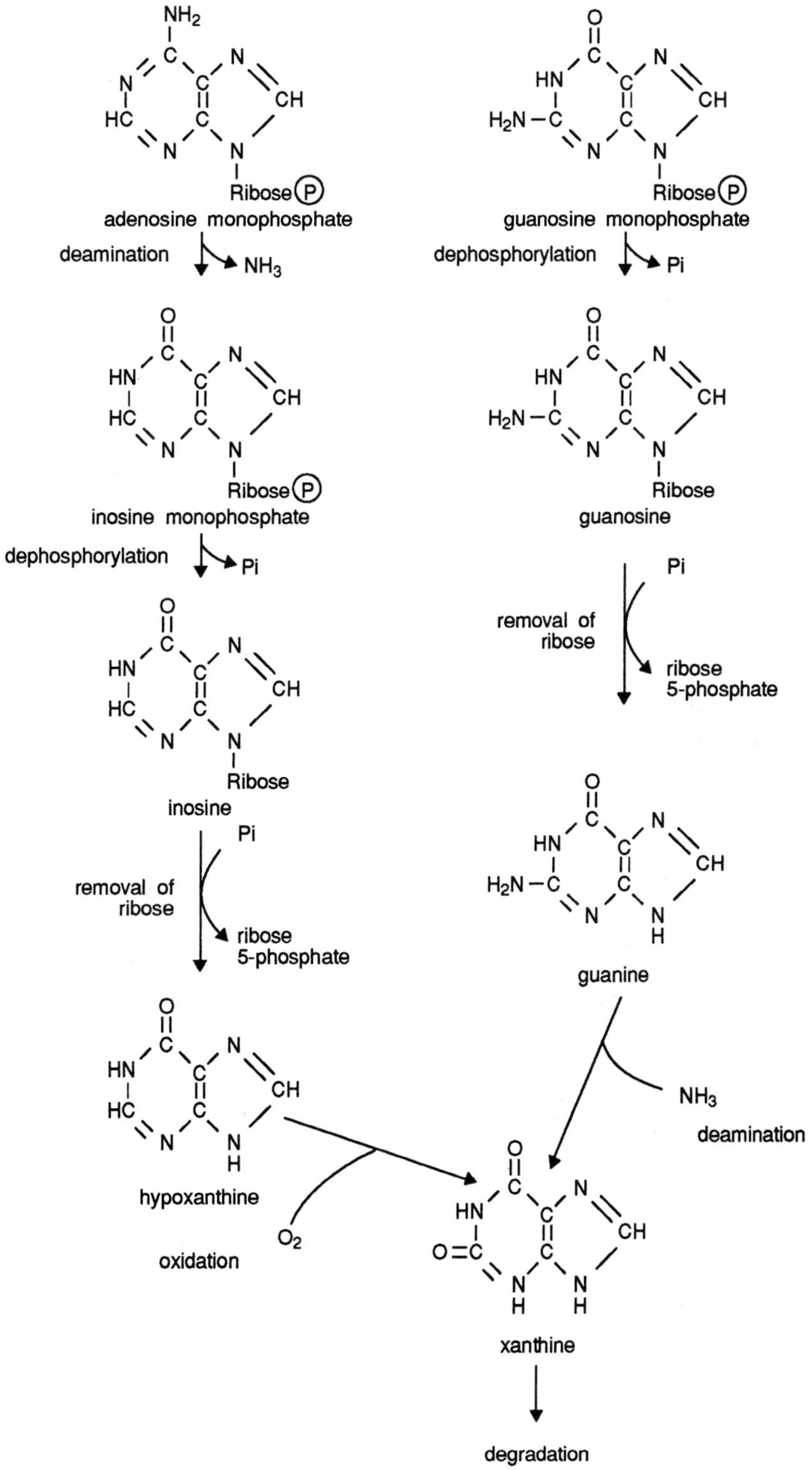

The diagram shows the principal breakdown route of purine catabolism. The balance of AMP:GMP is maintained largely by the inhibition of AMP deaminase by GMP.

If we draw this in the form of a flow diagram we can simplify the reactions to:

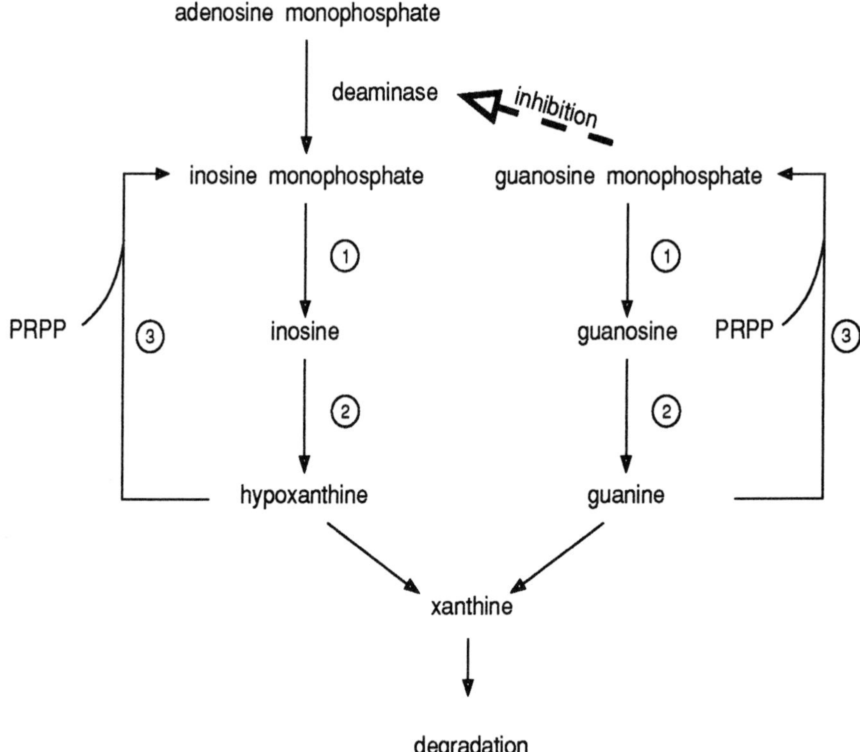

Note that PRPP is phosphoribosyl pyrophosphate. The enzymes listed are: 1) phosphomonoesterase, 2) purine nucleoside phosphorylase and 3) hypoxanthine xanthine phosphoribosyl transferase.

Pyrimidines are simpler molecules than purines thus their biosynthesis demands less energy. In addition pyrimidine degradation, unlike purine degradation, yields a useful central metabolite, succinyl CoA, thus salvage pathways are less important. There is much interconversion possible within the biosynthetic processes and discussion of these aspects will follow in Section 4.5.3.

4.5.2 Utilisation of exogenous nucleosides, purine or pyrimidine bases by prokaryotes.

Much variation occurs with respect to the use of exogenous bases within the prokaryotes. The following examples shown in Figure 4.15 are typical of the reactions occurring in the majority of the *Enterobacteriaceae*.

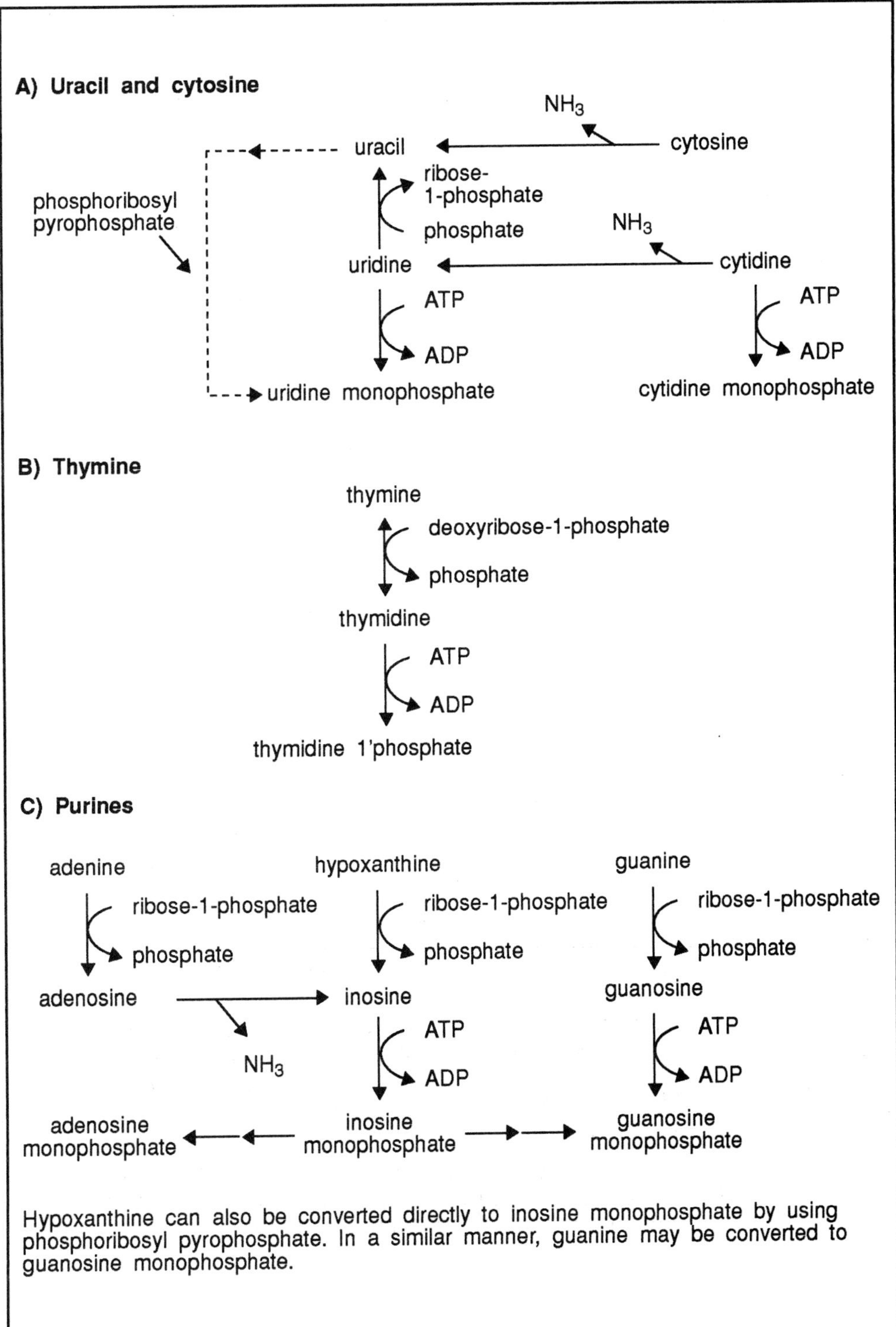

Figure 4.15 The mechanisms in *Enterobacteriaceae* by which exogenous purines and pyrimidines are utilised.

4.5.3 The biosynthesis of purines

biosynthesis proper starts at PRPP

Purine biosynthesis is considered to start properly from the ribosyl moiety of phosphoribosyl pyrophosphate (PRPP). By successive additions of amino groups and small one- or two-carbon fragments the nine membered purine molecule is built up. We can introduce the pathway simply as follows:

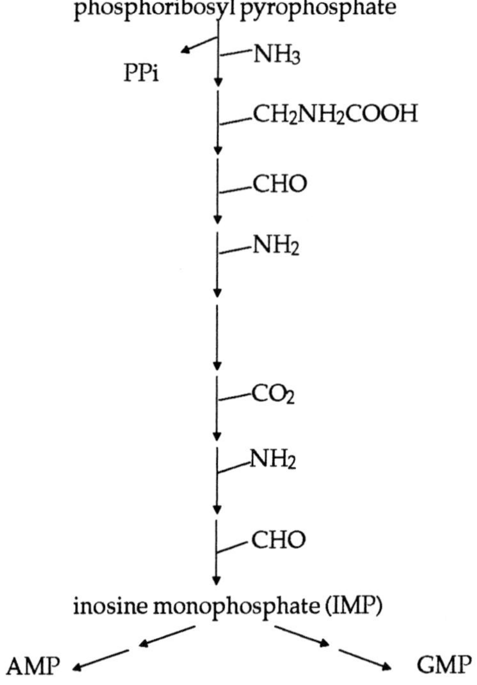

SAQ 4.8

Draw the structure of PRPP and write out the equation for its formation from ribose-5-phophate. Name the enzyme involved and state the source of the starting compound.

purines exert coarse control

PRPP is involved in the biosynthesis of pyrimidines and the committal step for purine synthesis *de novo* is the next reaction, which is the amination of PRPP using glutamine as donor. Although PRPP synthetase is inhibited by purine products, this occurs only when these compounds are present at high concentrations so it is a coarse and not a fine control.

Thus the first reaction unique to purine biosynthesis is the transfer of an amino group from the amide moiety of glutamine. Let us now proceed to building a purine molecule.

Π Try to complete the reaction shown below naming the products and the enzyme. Take care in naming the product, the clue is in the full name of PRPP. One necessary piece of information: PRPP has an α-linkage, the product is β-1-linked.

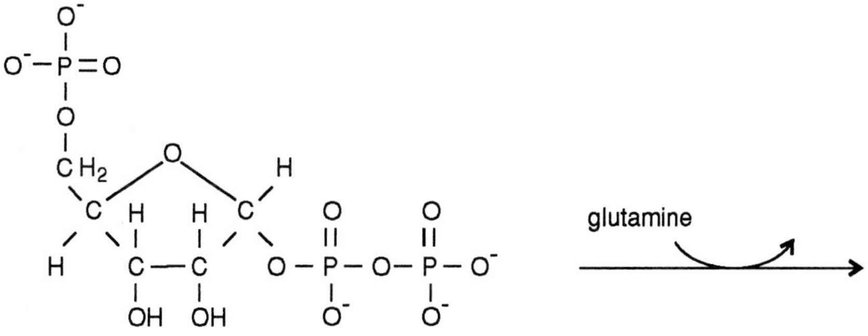

glutamine

PRPP (5-phosphoribosyl-1-pyrophosphate)

The right hand side of the equation is:

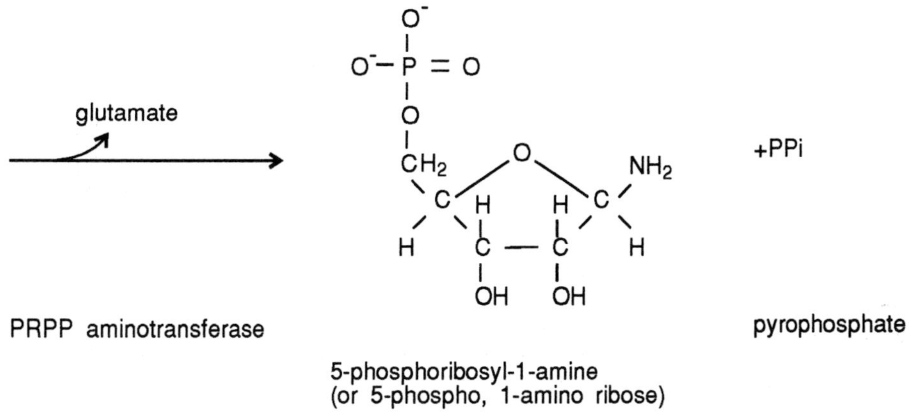

regulation of the purine pathway

PRPP aminotransferase has two inhibition sites, one sensitive to AMP and one to GMP. There is no further metabolic control of the following reactions until IMP is reached and then AMP and GMP regulate their own biosynthesis.

The detailed pathway of inosine monophosphate biosynthesis is shown in Figure 4.16. Note that the purine molecule is progressively built up on the amide group of the 5-phosphoribosyl-1-amine compound. For simplicity this compound is indicated throughout Figure 4.16 as:

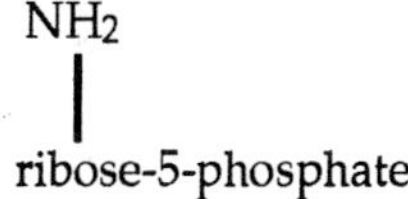

The chemistry involved in Figure 4.16 is relatively simple, the chemical names are complex but informative. Study the diagram and attempt to identify chemical groups from the individual names. You may find it easier to do than you think. Note that the purine molecule is built up slowly, the 2-carbon unit of glycine being the largest single addition.

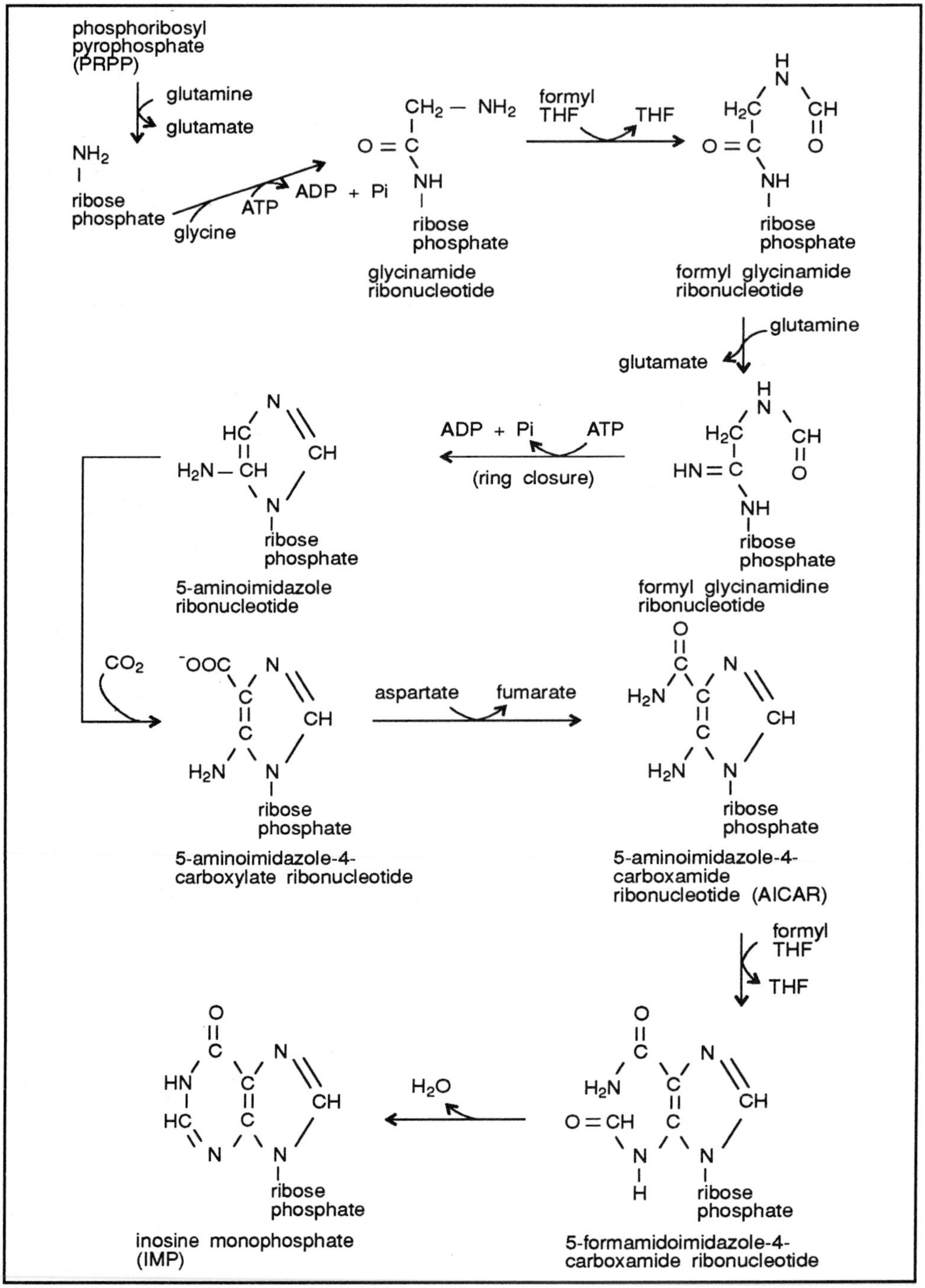

Figure 4.16 The biosynthesis of inosine monophosphate (IMP).

Π Can you produce the rather complicated overall reaction from Figure 4.16? The answer is given below but if you cannot get to the same answer write down all of the reactions individually and cancel as indicated below.

PRPP + 2 glutamine + 2 formyl THF + glycine + aspartate + CO_2 + 2ATP
→ IMP + 2 glutamate + 2THF + fumarate + 2ADP + 2Pi + PPi + H_2O

PRPP + 2 glutamine → 5-phosphoribosyl-1-amine + glutamate + PPi

5-phosphoribosyl-1-amine + glycine + ATP → glycinamide ribonucleotide + ADP + Pi

glycinamide ribonucleotide + formyl THF → formyl glycinamide ribonucleotide + THF

formyl glycinamide ribonucleotide + glutamine
→ formyl glycinamidine ribonucleotide + glutamate

formyl glycinamidine ribonucleotide +ATP
→ 5-aminoimadazole ribonucleotide + ADP + Pi

5-aminoimidazole ribonucleotide + CO_2
→ 5-aminoimadazole-4-carboxylate ribonucleotide

5-aminoimadazole-4-carboxylate ribonucleotide + aspartate
→ 5-aminoimadazole-4-carboxamide ribonucleotide + fumarate

5-aminoimidazole-4-carboxamide ribonucleotide + formyl THF
→ 5-formamidoimidazole-4-carboxamide ribonucleotide + THF

5-formamidoimidazole-4-carboxamide ribonucleotide → inosine monophosphate + H_2O

PRPP + 2 glutamine + 2 formyl THF + glycine + aspartate + CO_2 + 2ATP
→ IMP + 2 glutamate + 2THF + fumarate + 2ADP + 2Pi + PPi + H_2O

The details of AMP and GMP biosynthesis from IMP are shown in Figure 4.17. Adenylosuccinate synthetase (enzyme A) and IMP dehydrogenase (enzyme B) are inhibited by AMP and GMP respectively, examples of product inhibition. Note that GTP is used in the conversion of IMP to AMP, but ATP for the conversion of IMP to GMP. This is considered to aid the ATP: GTP balance.

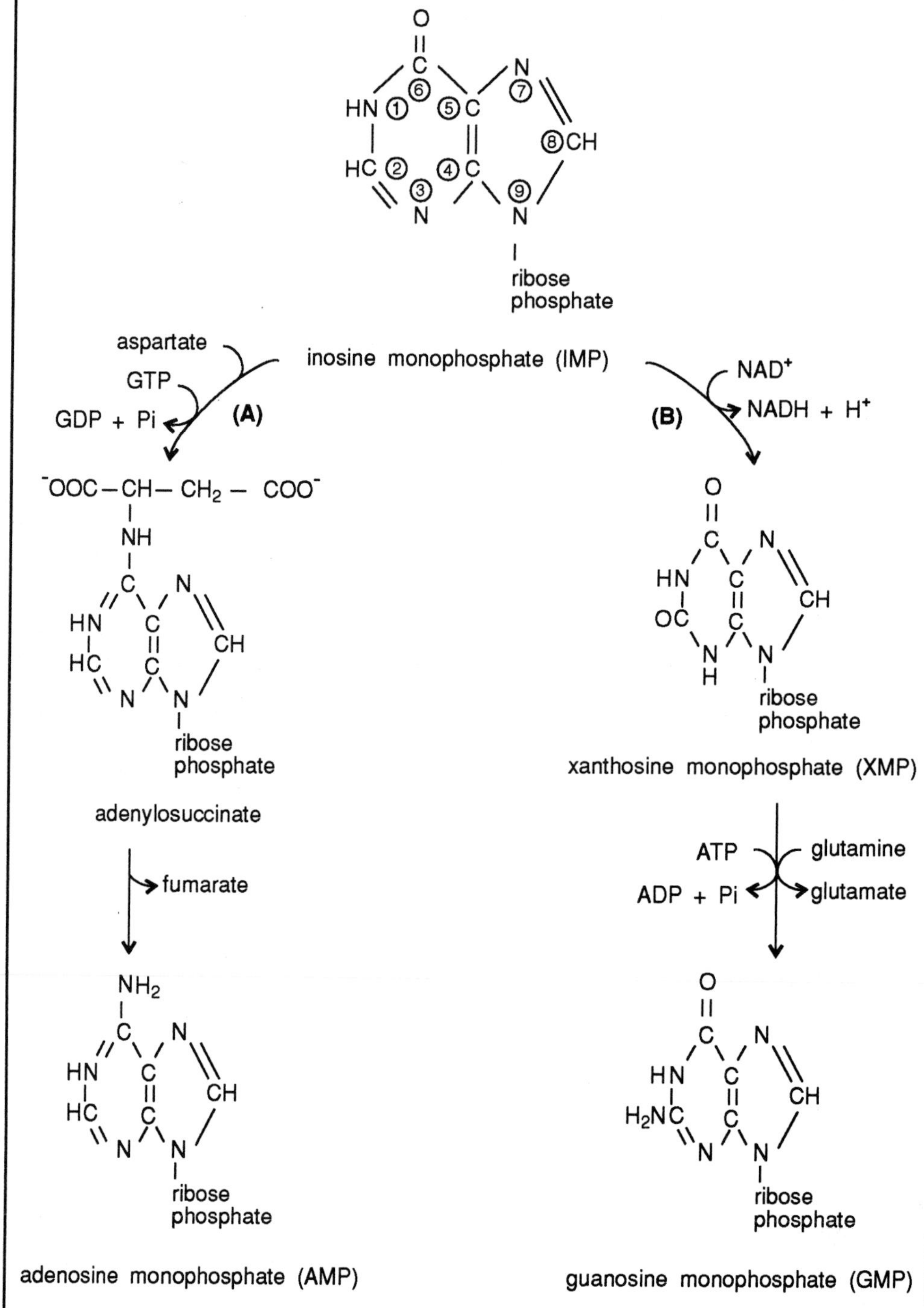

The numbering system for the atoms of purines is shown for IMP

Figure 4.17 The biosynthesis of AMP and GMP from IMP. (A) = adenylosucciate synthetase; (B) - inosine monophosphate dehydrogenase.

SAQ 4.9

Match up on the purine backbone below the parts indicated 1-7 with the compounds labelled A-E. Note that some compounds will need to be used twice.

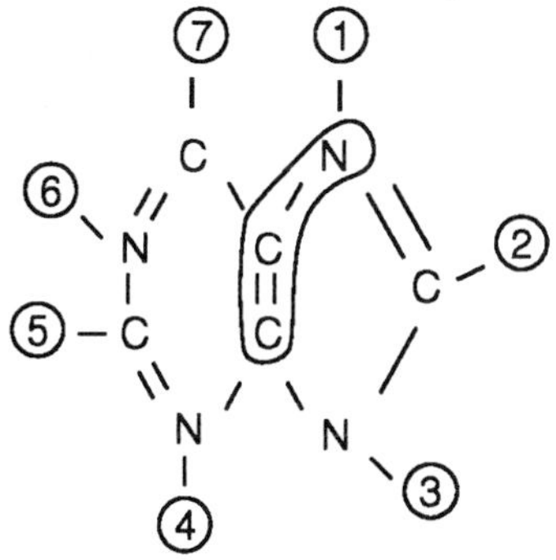

A) formyl THF; B) glycine; C) glutamine; D) aspartate; E) CO_2.

3.5.4 The biosynthesis of pyrimidines

two precursors used to make pyrimidines

The biosynthesis of pyrimidines is distinct from the biosynthesis of purines and the process is much less complicated. The basic six membered pyrimidine ring is synthesised from two components in only three reactions. Simple redox reactions or substitution reactions then occur to produce the required pyrimidine bases.

One of the precursors is carbamyl phosphate, a compound prominent in the urea cycle of higher animals. These organisms have two carbamyl phosphate synthetases; enzyme I which is involved in the urea cycle is mitochondrial and utilises ammonia whereas enzyme II is cytosolic and uses glutamine as amine donor. The two enzymes are regulated separately. Micro-organisms do not have a urea cycle (they can excrete ammonia) and hence have no requirement for such regulation.

The pyrimidine biosynthetic pathway is shown in Figure 4.18.

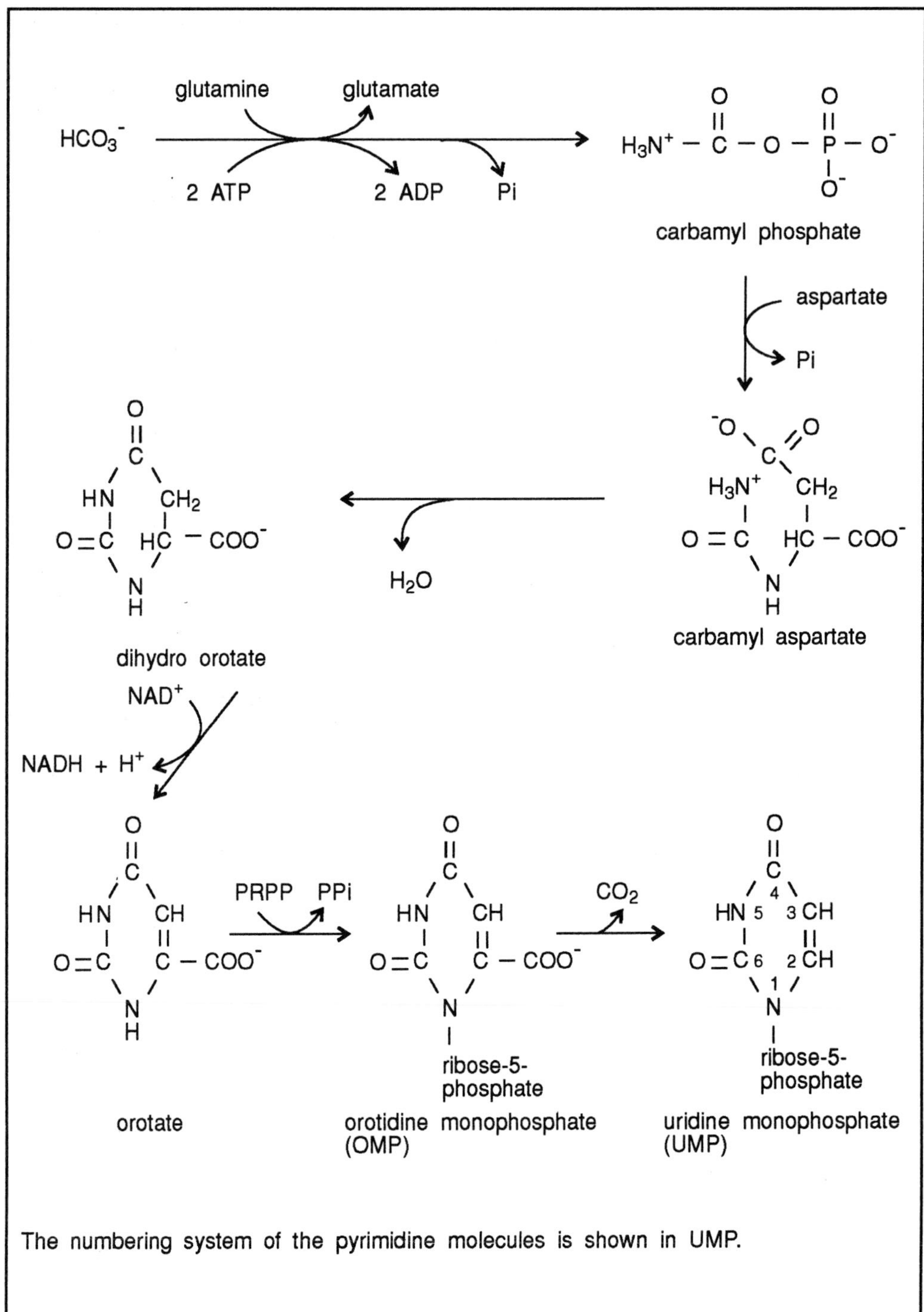

Figure 4.18 The biosynthesis of pyrimidines.

regulation of the pathway

The precursor molecules are aspartate and carbamyl phosphate but note that again PRPP is required. PRPP also has a marked effect on the first reaction, catalysed by carbamyl phosphate synthase, as PRPP acts as an activator of this enzyme.

4.5.5 Production of ribonucleoside di- and tri-phosphates and of deoxyribonucleoside derivatives

In Section 4.5.3 we studied the biosynthesis of AMP and GMP and in Section 4.5.4 the biosynthesis of UMP. We will now examine the biosynthesis of the di- and tri-phosphate nucleotides.

Production of ADP and GDP are brought about by specific kinases, for example:

$$\text{AMP} + \text{ATP} \underset{\text{adenylate kinase}}{\rightleftharpoons} \text{ADP} + \text{ADP}$$

Nucleoside di- and tri-phosphates are interconverted by nucleoside diphosphate kinase, an enzyme with broad specificity, for example:

$$\text{ZDP} + \text{YTP} \rightleftharpoons \text{ZTP} + \text{YDP}$$

One remaining base, cytosine, has not yet been mentioned and is required in both RNA and DNA. Cytosine has an amino group instead of an oxo group at carbon 4 and this is donated by glutamine. It is synthesised from UTP by the reaction:

$$\text{UTP} + \text{glutamine} + \text{ATP} \rightarrow \text{CTP} + \text{glutamate} + \text{ADP} + \text{Pi}$$

The nitrogen donor in many bacteria is NH_4^+ rather than glutamine.

formation of deoxy sugars

Finally the production of deoxyribonucleosides should be mentioned. Adenosine, guanosine and cytidine nucleosides remember, are found in both RNA and DNA. In DNA deoxythymidine is found rather than uridine ribonucleoside, thymine having a methyl substitution at carbon 3. The interconversions are usually at the levels shown below:

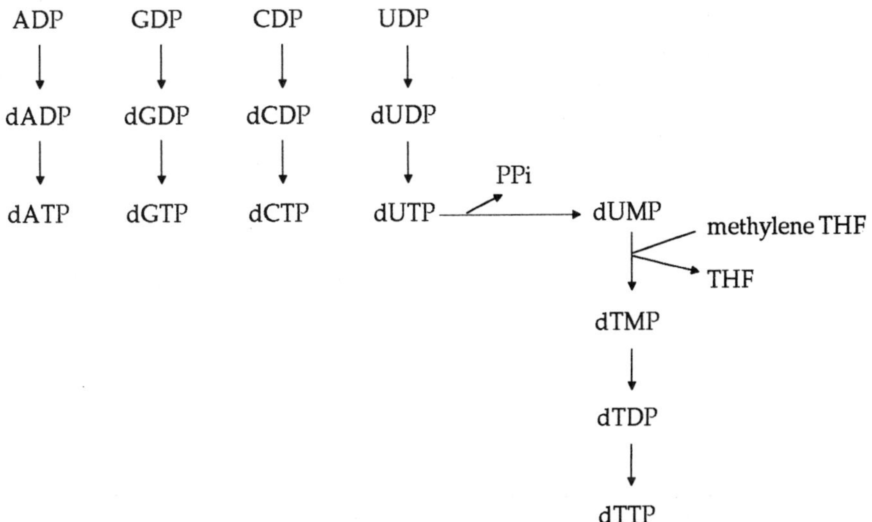

Ribonucleosides prefixed by d indicate deoxyribonucleotides. All conversions of the type dZDP → dZTP are catalysed by nucleoside diphosphokinase whereas dTMP → dTDP requires a specific kinase.

SAQ 4.10

Decide whether the statements below are true or false and give your reasons.

1) Two nitrogen atoms are donated by aspartate molecules during the synthesis of AMP.

2) Glutamine donates two nitrogen atoms to the structure of GMP.

3) Two nitrogen atoms are donated from glutamine during CTP biosynthesis in animals but only one in bacteria.

4.6 The metabolic links between amino acid biosynthesis and the biosynthesis of purines and pyrimidines

At a general level we have seen that three of the four nitrogen atoms of inosine monophosphate are derived from glutamine (2) and aspartate (1). In addition the fourth nitrogen and two of the carbons are provided by glycine. A further nitrogen from aspartate is required during the IMP to AMP conversion and from glutamine in the synthesis of GMP from IMP.

Pyrimidine biosynthesis requires an amino group from glutamine and the amino group together with all four carbons of an aspartate molecule. Finally UTP to CTP conversion requires an amino group and this comes from glutamine or NH_4^+.

The final relationship, which was mentioned in Section 4.4.6, is the linkage between purine and histidine biosynthesis. Simplified extracts of Figures 4.12 and 4.15 are shown in Figure 4.19. This figure highlights the relationship between the two pathways and shows that the apparently serious loss of an AMP molecule to provide only a C-N group for histidine biosynthesis is misleading. The waste product from histidine biosynthesis, AICAR, is an intermediate only two stages before IMP itself during IMP biosynthesis.

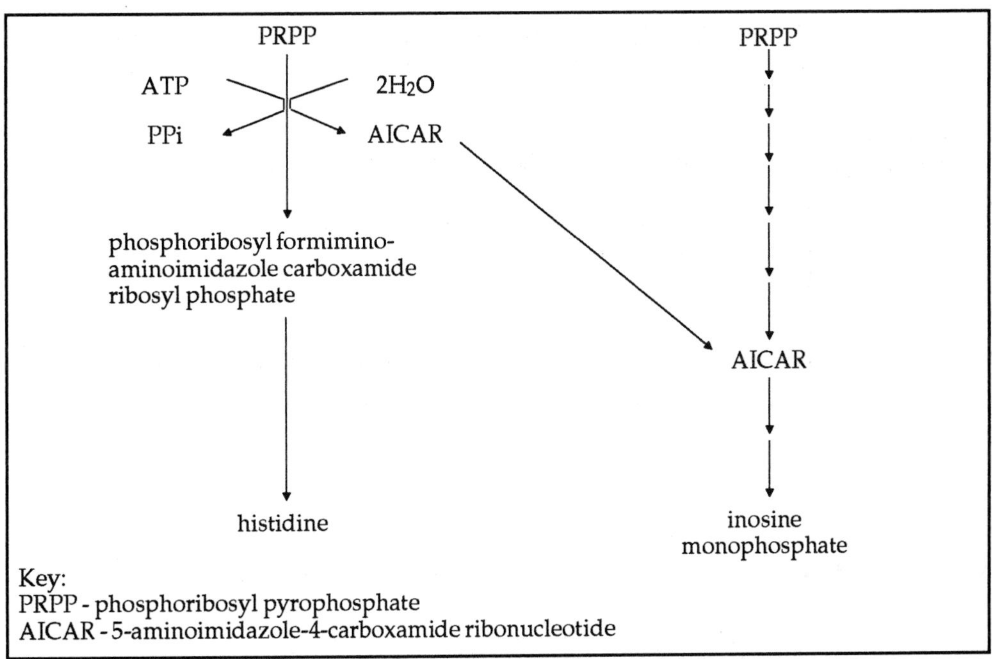

Figure 4.19 Diagram to show the relationship between histidine and purine biosynthesis.

SAQ 4.11

This chapter has introduced a lot of new compounds many of which are referred to by initials. It is obviously important that you remember these, hence this SAQ.

Expand the following abbreviations:

ATP; PRPP; IMP; THF; TTP; TPP; dUTP; CDP; GDP and there is a special prize if you get the last one right; AICAR.

Summary and objectives

The majority of the amino acids released following endogenous protein degradation are recycled to make new protein.

De novo biosynthesis can be divided into six groups or families of amino acids depending on the nature of the precursor metabolites required.

Ten of the twenty amino acids are dietarily essential to higher animals, such living systems relying on the plants and micro-organisms for adequate dietary supply.

The seven precursor metabolites required are all central metabolites and the other requirement, nitrogen, is usually supplied as an ammonium salt though nitrate and occasionally atmospheric nitrogen will suffice.

Each amino acid biosynthetic pathway is under some form of control, generally either inhibition of enzyme activity or repression of enzyme synthesis by the end-product.

The majority of the products of nucleic acid metabolism, the nucleotides, are recycled into new nucleic acid. Some bacteria can utilise exogenous nucleosides or nitrogenous bases. *De novo* biosynthesis of purines occurs via inosine monophosphate and of pyrimidines via orotate.

Links between amino acid biosynthetic pathways, nitrogenous base biosynthesis and other biosynthetic pathways are common. There is a particularly strong link between histidine and purine biosynthesis.

Now you have completed this chapter you should be able to:

- list the common amino acids, describe their general formula and indicate which are dietarily essential and which are non-essential;
- discuss the way in which amino acids are grouped into families depending on their anabolic starting point;
- explain the ways in which sufficient quantities of precursors can be provided to satisfy amino acid requirements;
- list the reactions involved in the biosynthesis of selected examples of the principal amino acid families;
- discuss the control mechanisms used in the regulation of amino acid biosynthesis;
- describe the biosynthesis of nitrogenous compounds derived from branches of amino acid biosynthetic pathways;
- demonstrate a knowledge of the biosynthesis of purine and pyrimidine nucleotides;
- assess the value of recycling of nucleotides and discuss the relationship between the recycling process and the metabolic control of nucleotide biosynthesis;
- discuss the value of having the reactions in common for histidine biosynthesis and adenine and guanine biosynthesis.

The biosynthesis of lipids

Introduction 110

5.1 The classes of lipids 110

5.2 The function and occurrence of lipids 114

5.3 Biosynthesis versus degradation of fatty acids 115

5.4 Provision of precursors for fatty acid biosynthesis 116

5.5 The biosynthesis of fatty acids 119

5.6 Biosynthesis of triglycerides 128

5.7 Biosynthesis of phospholipids 129

5.8 Biosynthesis of lipids from isoprene derivatives 133

5.9 Microbial biotransformations (bioconversions) of steroids 140

Summary and objectives 142

The biosynthesis of lipids

Introduction

As a generalisation the lipids of prokaryotes are more varied than those of eukaryotes and this necessitates a concomitant versatility in both degradative and biosynthetic pathways. In this chapter we shall review briefly the lipid types before revising their distribution and function in all living systems. The majority of the chapter will then be devoted to a study of the biosynthesis of lipids including reference to the subcellular distribution and control of relevant pathways. Several examples of biosynthesis of lipids unique to prokaryotes will be mentioned. Finally a section on the way in which micro-organisms are industrially important in lipid biotransformations will be presented.

5.1 The classes of lipids

We have defined lipids as a heterogeneous group of substances which may be classified as compounds insoluble in water but soluble in a variety of non-polar solvents such as ether, benzene and chloroform. They rarely form polymers and they generally have molecular weights below 1000; they are, however, constructed from a wide variety of precursors.

neutral lipids

The first class of lipids, the neutral lipids, are composed of the three-carbon compound glycerol esterified to one, two or three fatty acids.

Π Using the general formula for a fatty acid (R.COOH) for the moment can you draw a diagram of a triglyceride?

The answer is:

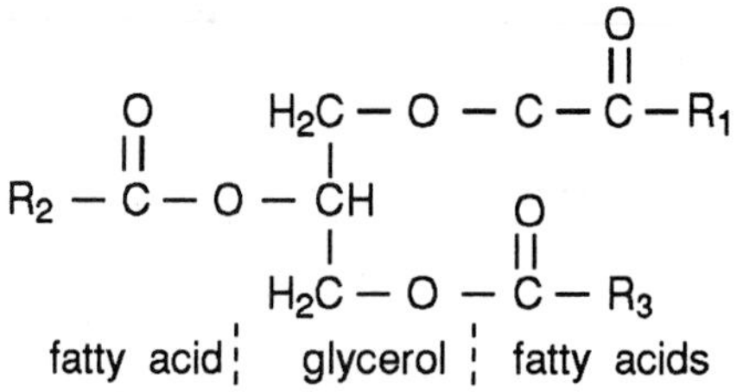

```
                                        O
                                        ||
            O       H2C - O - C - C - R1
            ||       |
   R2 - C - O - CH          O
                      |           ||
                    H2C - O - C - R3
     fatty acid |  glycerol  |  fatty acids
```

Remember that the -COOH of the fatty acid links to the -OH of glycerol forming an ester bond as shown above. Each bond results in the loss of one molecule of water; a hydrogen from the fatty acid and a hydroxyl from glycerol.

Although they are rare in prokaryotes, triglycerides are the principal storage form of energy in many eukaryotic cells. The presence of mono- and diglycerides in prokaryotes are thought to be transient by-products in the biosynthesis or degradation of phospholipids.

Phospholipids are lipids containing phosphates: they are composed of phosphatidic acid which is linked to an alcohol group. We remind you of some of the components of the phospholipids.

The core structure is phosphatidic acid:

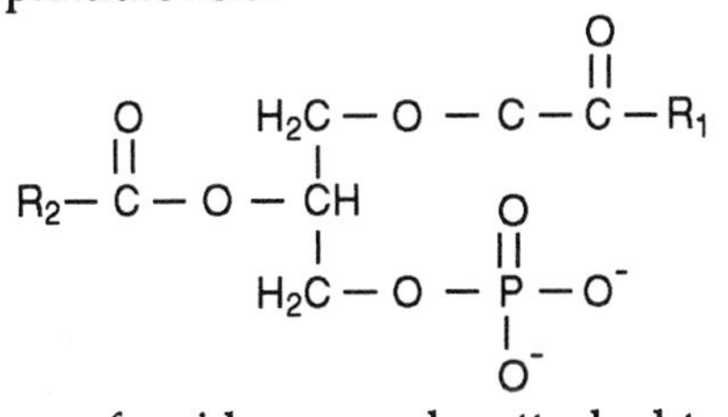

$$
\begin{array}{l}
\mathrm{H_2C-O-C(=O)-R_1} \\
\mathrm{R_2-C(=O)-O-CH} \\
\mathrm{H_2C-O-P(=O)(O^-)-O^-}
\end{array}
$$

A number of different types of residues may be attached to the phosphate moiety of phosphatidic acid. The most common are:

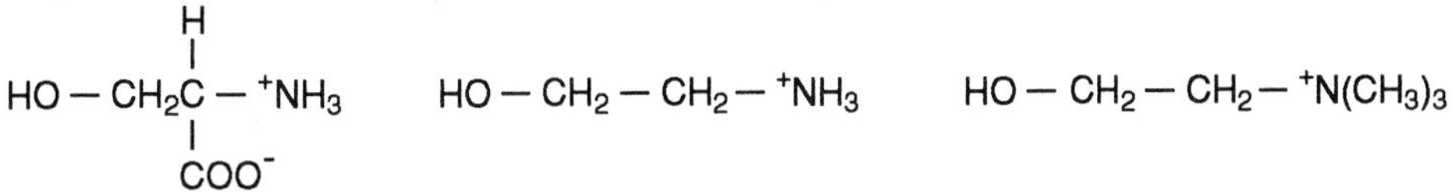

$$\mathrm{HO-CH_2CH(COO^-)-{}^+NH_3} \quad \text{serine}$$

$$\mathrm{HO-CH_2-CH_2-{}^+NH_3} \quad \text{ethanolamine}$$

$$\mathrm{HO-CH_2-CH_2-{}^+N(CH_3)_3} \quad \text{choline}$$

Polyols (eg inositol) may also be attached to phosphatidic acids. Thus phospholipids have the general structure:

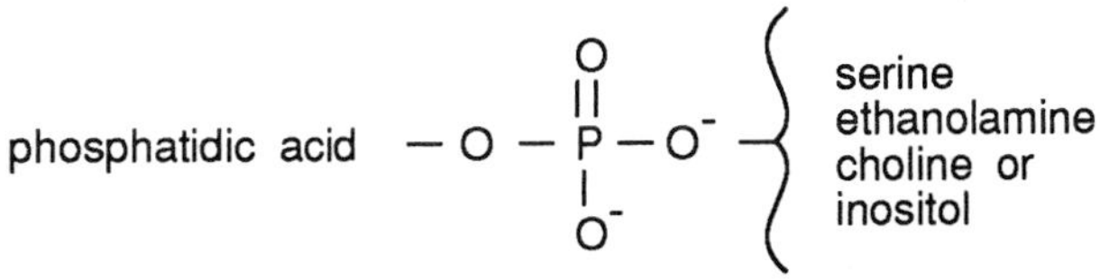

$$\text{phosphatidic acid} \; \mathrm{-O-P(=O)(O^-)-O^-} - \left\{ \begin{array}{l} \text{serine} \\ \text{ethanolamine} \\ \text{choline or} \\ \text{inositol} \end{array} \right.$$

These compounds are found in all living systems as they are essential membrane components. In addition to what may be termed the conventional phospholipids there are two other groups. Firstly the plasmalogens which generally contain sixteen-carbon to nineteen-carbon fatty acids, which, at glycerol carbon one, are bound by an ether rather than an ester group. One carbon of the ether group is always involved in a double bond with an adjacent carbon. These compounds appear to be widespread in anaerobic organisms replacing conventional phospholipids.

plasmalogens

archaebacterial phospholipids

Secondly a group of phospholipids unique to archaebacteria have been characterised. These compounds contain two long-chain alcohols, instead of fatty acids, both linked to glycerol by ether bonds. This ether bond is a feature shared only with the plasmalogens. Archaebacterial phospholipids are however unique in that the hydrophobic long-chain hydrocarbon is branched being derived from the five-carbon compound isoprene. Generally the twenty-carbon derivatives, phytane, or the forty-carbon derivatives, biphytane, are used, the latter exhibiting cyclopentane rings in thermoacidophiles (heat-loving and acid-loving bacteria). Also important are the lipids based on sphingosine (eg the sphingomyelins) and the diphosphatidyl glycerol lipids. We will not enlarge on their structure here. The reader is referred to the Biotol text, 'The Molecular Fabric of Cells'.

SAQ 5.1

Study Figure 5.1. Firstly label a), b) and c) with one of the following; phosphatidyl serine; a choline plasmalogen; an archaebacterial phospholipid.

Now insert the correct labels onto the diagram from the list below:

an ester bond; glycerol; phosphate; serine; choline; fatty acid (twice); ether bonds (twice); long chain alcohol; phosphatidic acid; glycerol phosphate (twice).

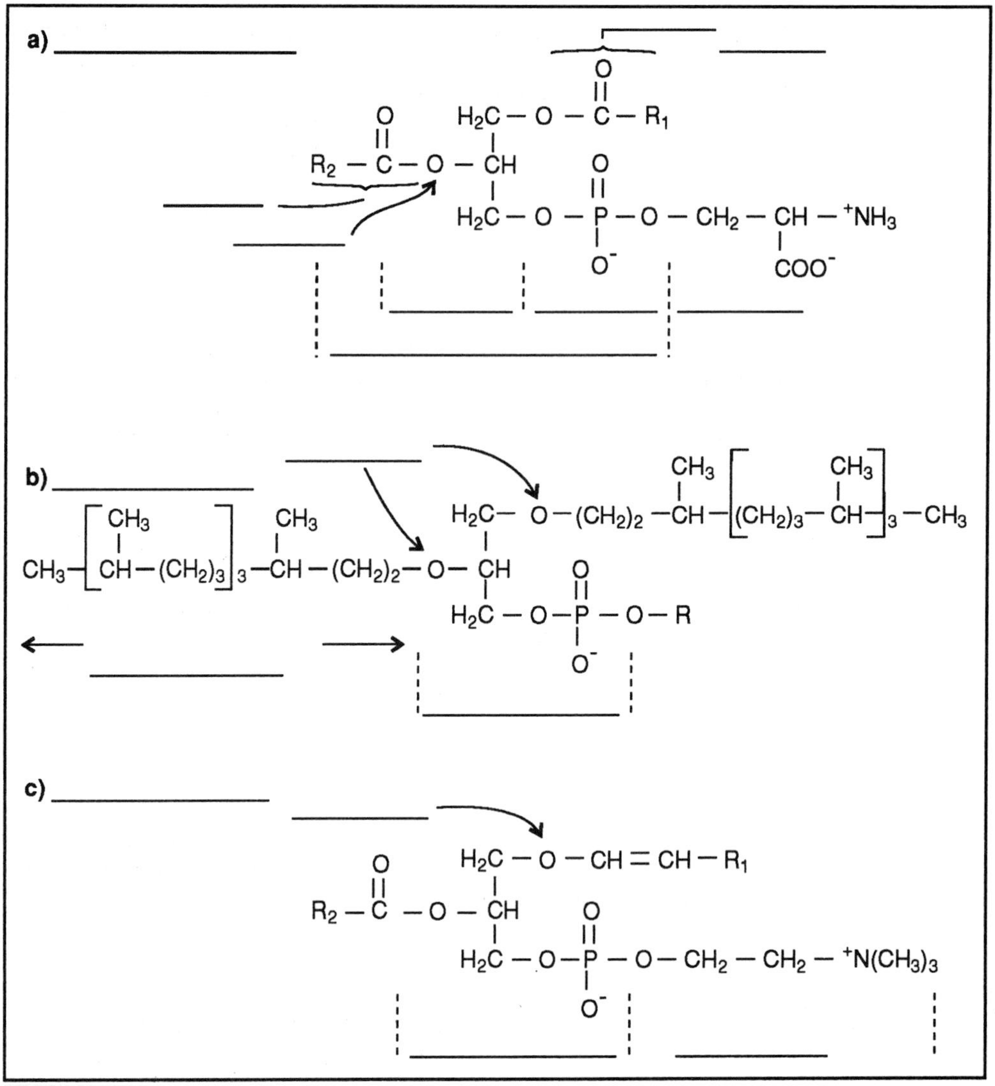

a) ________

$$H_2C-O-\overset{O}{\overset{\|}{C}}-R_1$$
$$R_2-\overset{O}{\overset{\|}{C}}-O-CH$$
$$H_2C-O-\overset{O}{\overset{\|}{P}}(O^-)-O-CH_2-CH(COO^-)-{}^+NH_3$$

b) ________

$$CH_3-[CH(CH_3)-(CH_2)_3]_3-CH(CH_3)-(CH_2)_2-O-CH$$
$$H_2C-O-(CH_2)_2-CH(CH_3)-[(CH_2)_3-CH(CH_3)]_3-CH_3$$
$$H_2C-O-\overset{O}{\overset{\|}{P}}(O^-)-O-R$$

c) ________

$$H_2C-O-CH=CH-R_1$$
$$R_2-\overset{O}{\overset{\|}{C}}-O-CH$$
$$H_2C-O-\overset{O}{\overset{\|}{P}}(O^-)-O-CH_2-CH_2-{}^+N(CH_3)_3$$

Figure 5.1 Examples of phospholipids.

glycolipids

Glycolipids, as the name implies, are compounds containing both carbohydrates and lipids. They occur in virtually all living systems and are an extremely diverse group. Those of higher animals, for instance, include cerebrosides-compounds containing fatty acid, sugar residues and sphingosine, the last unknown in prokaryotes. We will illustrate the group by simply using a few examples. The important point to remember is that they have widespread occurrence, diversity and complexity.

peptidoglycan

Cell walls of Gram positive bacteria are largely composed of peptidoglycan and they contain the carbohydrate compounds teichoic acids linked to lipids forming lipoteichoic acids. (A detailed description of the structure of bacterial cells walls is given in the Biotol text, 'The Infrastructure and Activities of Cells').

lipopoly-saccharides

Gram negative bacteria have a much more complicated cell envelope containing a relatively thin layer of peptidoglycan surrounded by an outer membrane. This outer membrane contains large quantities of lipopolysaccharides (LPS) which are complex molecules with a structure which is often unique to a genus or species. The *Salmonella*

LPS molecules for example often have a molecular weight in excess of 10,000 and are composed of three layers, an outer 'O' polysaccharide moiety, a central 'R' core inside of which is the lipid A moiety. The lipid is not a glycerol lipid but contains 6 (or 7) fatty acids, four of which are joined by ester linkage, two to each of two N-acetylglucosamines. Glycolipids consisting of mono- or digalactosyl diglycerides are also very important in photosynthetic bacteria (cyanobacteria or blue green algae) whereas only monogolactosyl diglycerides are found in green photosynthetic bacteria.

isoprene

The free five-carbon compound isoprene is rare in nature but it is the precursor of many important lipid groups as it may be joined to itself to give compounds containing from two to around twenty isoprenoid units. For example the four isoprenoid derivative product phytol, an example of which we saw in Figure 5.1 as the long-chain alcohol moiety of archaebacterial phospholipids. The six isoprenoid derivative produces squalene which is the precursor of sterols; the eight isoprenoid derivative yields the carotenoids and the eleven isoprenoid derivative yields bactoprenol. A more detailed study will follow in section 5.8.

fatty acids

Fatty acids have been mentioned several times as they are major constituents of several lipid classes. These compounds have the general formula R-COOH where R is an alkyl group. In theory the simplest fatty acid is methanoic (formic) acid where R is a hydrogen atom. Biologically speaking the simplest fatty acid is acetic acid (ethanoic) where R is a methyl group. The majority of fatty acids in biological systems are within the sixteen to twenty-carbon length group. In eukaryotes fatty acids can be fully saturated or mono-, di- or tri-unsaturated. Prokaryotes show much more diversity having fatty acids of longer than eighteen carbon chain length, branched chains, cyclopentane rings, a hydroxyl groups or odd numbers of carbon atoms. Unlike eukaryotes however they contain only saturated or mono-unsaturated examples.

It is important to remember the basic principles of fatty acid structure. The molecule looks very complicated due to its chain length but in practice it is chemically quite simple. Biologically most reactions occur within a few carbons of the -COOH end of the molecule and it is important for us to name and recognise the individual carbons relative to the carboxyl end.

SAQ 5.2

Look at the partially completed diagram of an eighteen-carbon fatty acid in Figure 5.2. The carboxyl carbon is carbon one and the opposite terminal carbon is carbon eighteen.

Can you:-

1) identify the γ carbon atom

2) identify the ω carbon atom

3) identify the α carbon atom

4) identify the β carbon atom

5) insert a double bond at Δ_{9-10}

6) complete the diagram by inserting the hydrogens on each carbon?

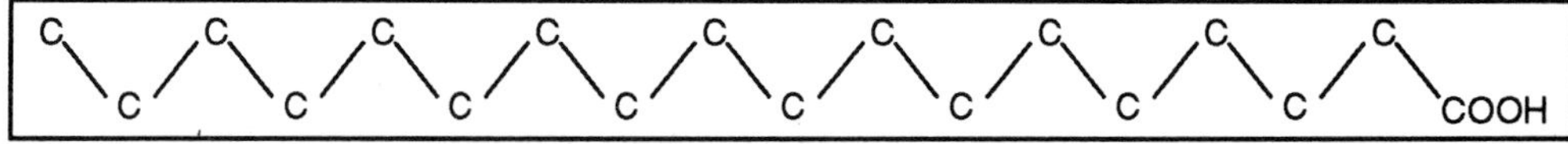

Figure 5.2 Diagram of an eighteen-carbon fatty acid.

It is important to recognise particularly the α, β and γ carbons as it is at these carbons that most reactions occur during biosynthesis and degradation.

Π What is the name of the fatty acid drawn in your completed Figure 5.2?

The answer is oleic acid, one of the most important fatty acids in eukaryotic cells.

Π Write the general formula for oleic acid

The answer is $C_{17}H_{33}COOH\Delta_{9-10}$.

Remember that the position of the double bond is very important. Bacteria have a mono-unsaturated fatty acid, vaccenic acid, with the same general formula as oleic acid but in this case the double bond is at Δ_{11-12}, making it a different compound.

5.2 The function and occurrence of lipids

Lipids have a variety of functions which may be classified into three areas. These are as fuel molecules, as concentrated energy reserve material and as structural components of the membranes of living systems. Their role as fuel molecules in aerobically-growing living systems was described in the BIOTOL text, 'Energy Sources for Cells' and will not be covered here. We will examine the biosynthesis of lipids for the production of energy reserves and also for their insertion into membranes.

storage: neutral lipids membrane: phospholipids

As indicated earlier only some eukaryotes, notably higher animals, store lipids as an energy reserve, the lipid class normally being the triglycerides. In terms of membrane lipids, phospholipids are by far the most important quantitatively. The nature of the phospholipid inevitably varies with the nature of the organism concerned. Bacteria generally contain phosphatidyl ethanolamine whereas phosphatidyl choline is the most prominent in eukaryotes. The fatty acid component of the phospholipids is very important for membrane structure and function. In the characteristic bilayer of membranes, fatty acids are orientated to concentrate the hydrophilic portions on the outside of the membrane and the hydrophobic parts towards the centre of the membrane. This type of structure limits, and therefore exercises control over, the passage of water-soluble molecules into and out of the cell.

membrane sterols

Sterols are essential components of all eukaryotic membranes. Prokaryotes, however, with the exception of *Mycoplasmas*, are almost totally devoid of sterols. Mycoplasmas are a group of bacteria which are almost unique in that they do not contain cell walls but have a special, toughened cell membrane which contains sterols.

5.3 Biosynthesis versus degradation of fatty acids

Chemically the biosynthesis and degradation of fatty acids are an almost exact reversal of each other. Biologically, however, they occur by totally independent processes involving different enzymes and different subcellular locations. Degradation in eukaryotes occurs within the mitochondrial matrix whereas biosynthesis is cytosolic in animal and micro-organisms but occurs in chloroplasts in plants. Table 5.1 summarises some of he differences between biosynthesis and degradation of the fatty acid processes.

	Degradation	**Biosynthesis**
Subcellular location	mitochondrial matrix in eukaryotes: cytosol in prokaryotes	cytosol, except in higher plants, where it occurs in chloroplasts
Enzymes involved	not associated	usually in a multi-enzyme complex
Carbon fragment released or required	acetyl CoA released	malonyl CoA and acetyl CoA
Redox carrier involved	NAD^+ and enzyme bound FAD	$NADPH + H^+$
Carrier molecule	coenzyme A	acyl carrier protein (ACP)
Other essentials		ATP, bicarbonate, citrate (in eukaryotes)

Table 5.1 A summary of the requirements for fatty acid biosynthesis and degradation.

Having two different sets of enzymes obviously makes the separate control of degradation and biosynthesis much easier. Eukaryotes but not prokaryotes have the added advantage of a membrane-separated subcellular location although this advantage is somewhat neutralised by the problems arising in providing metabolic precursors at the correct sites within the eukaryotic cell. We shall investigate this topic shortly.

activation of fatty acids

Fatty acids do not generally function in the free form: before they can enter metabolic pathways they need to be activated. Generally this activation is brought about by their condensation with coenzyme A. However, during biosynthesis they are not linked to coenzyme A but to a different molecule, the acyl carrier protein (ACP). Both coenzyme A and ACP molecules contain a phosphopantotheine prosthetic group. In the case of coenzyme A this is linked to adenosine monophosphate. In the case of ACP it is linked via serine to a polypeptide chain containing seventy-six amino acids. Thus the ACP is a very large molecule compared to coenzyme A. Figure 5.3 shows the relationship between the structures of ACP and coenzyme A. The fatty acid reacts with the -SH group of CoA and ACP, in both cases resulting in the activation of the fatty acid.

The acyl carrier protein (ACP) molecule

```
                                         CH3       O
                                         |         ||             76
HS–CH2–CH2–NH–C–CH2–CH2–NH–C–CH–C–CH2–O–P–|O – serine – amino
              ||             ||  |   |         |                 acids
              O              O  OH  CH3        O-
```

The coenzyme A molecule

```
                                         CH3       O     O
                                         |         ||    ||
HS–CH2–CH2–NH–C–CH2–CH2–NH–C–CH–C–CH2–O–P–|O–P–O–ribose–adenine
              ||             ||  |   |         |     |
              O              O  OH  CH3        O-    O-
```

The phosphopantotheine prosthetic groups of each molecule (shown in the boxes) are identical.
The -SH group on the far left of each molecule as shown above is the reactive group which combines to the fatty acid.
Thus coenzyme A is often abbreviated as CoASH.

Figure 5.3 The relationship between the structures of the acyl carrier protein (ACP) and coenzyme A.

5.4 Provision of precursors for fatty acid biosynthesis

In Table 5.1 we saw that the requirements for the biosynthesis of fatty acids were malonyl CoA (from acetyl CoA), NADPH + H^+, ATP and bicarbonate. We shall see shortly that the biosynthesis of, for example, palmitate requires 8 acetyl CoA, 14 NADPH + 14 H^+ and 7 ATP molecules. In prokaryotes these precursors can be distributed throughout the interior of the cell. In eukaryotes, however, the sites of synthesis and, therefore, availability of these precursors are not, on the face of it, ideal.

Π Can you remember where acetyl CoA is synthesised in eukaryotes and from which sources?

The answer is in the mitochondrion. Pyruvate is produced in the cytosol from the Embden Meyerhof pathway or from other quantitatively less important sources, for example, amino acids. The pyruvate enters the mitochondrion where it is degraded to acetyl CoA. Acetyl CoA is also produced within the mitochondrion from the degradation of fatty acids.

Π Can you remember from which pathway NADPH + H^+ is produced?

The NADPH + H^+ is produced largely from the oxidative half of the pentose phosphate pathway.

citrate used as source of acetyl CoA

The problem, then, for eukaryotes is that acetyl CoA is required in the cytosol but it is produced within the mitochondrion, the membrane of which is not readily permeable to it. The solution is provided by citrate, a molecule which can freely diffuse through the mitochondrial membrane. Citrate is formed in the TCA cycle (inside the mitochondrion) as follows:

$$\text{acetyl CoA} + \text{oxaloacetate} \xrightarrow{\text{citrate synthase}} \text{citrate}$$

After diffusing out of the mitochondrion the citrate is degraded by citrate lyase thus:

$$\text{citrate} + \text{ATP} + \text{CoA} \rightarrow \text{acetyl CoA} + \text{ADP} + \text{Pi} + \text{oxaloacetate}$$

The oxaloacetate is required back within the mitochondrion but the membrane is impermeable to it. Oxaloacetate is reduced to malate by cytosolic malate dehydrogenase. We have met this reaction in the reverse direction occurring during the conventional TCA cycle operation.

$$\text{oxaloacetate} + \text{NADH} + \text{H}^+ \rightarrow \text{malate} + \text{NAD}^+ \qquad \text{(E 5.1)}$$

Malate is oxidatively decarboxylated by the malic enzyme to pyruvate. This enzyme requires ATP.

$$\text{malate} + \text{NADP}^+ \rightarrow \text{pyruvate} + \text{CO}_2 + \text{NADPH} + \text{H}^+ \qquad \text{(E 5.2)}$$

The pyruvate produced can diffuse freely into the mitochondrion where it is carboxylated to oxaloacetate by pyruvate carboxylase.

$$\text{pyruvate} + \text{CO}_2 + \text{ATP} + \text{H}_2\text{O} \rightarrow \text{oxaloacetate} + \text{ADP} + \text{Pi} \qquad \text{(E 5.3)}$$

If equations 5.1-3 are drawn as a cycle as below the way in which acetyl CoA gets out of the mitochondrion can be seen.

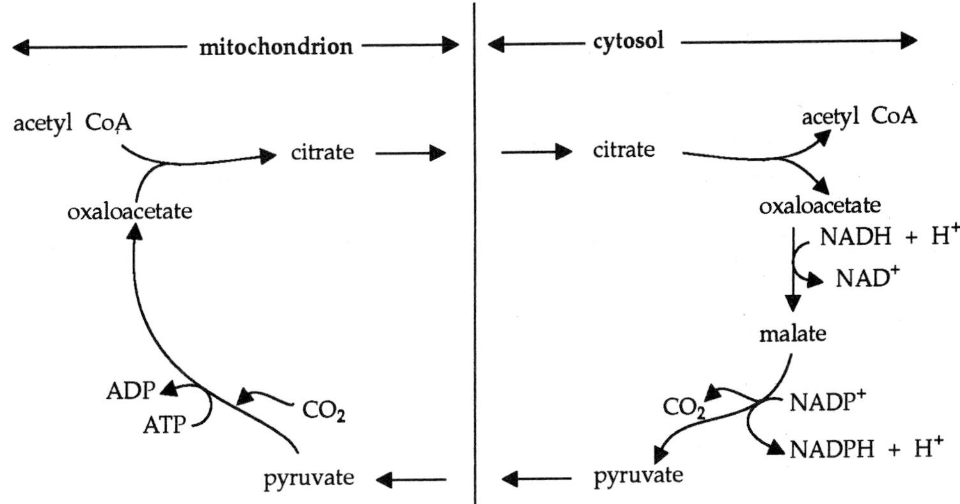

Π Can you summarise the overall reactions in equations 1) to 3)?

$$NADH + H^+ + NADP^+ + ATP + H_2O \rightarrow NADPH + H^+ + NAD^+ + ADP + Pi$$

This is a somewhat misleading overall reaction which gives no indication that the purpose is to transfer acetyl CoA out of the mitochondrion.

Π Can you see another advantage to be derived from this cyclical system in terms of fatty acid biosynthesis?

If you are not sure of the answer, think of the requirements for fatty acid biosynthesis and then look at the products of your overall reaction.

The answer is NADPH + H^+ production. Do not forget, however, that this is not 'free', because ATP has been consumed.

Now we have seen how acetyl CoA is made available we can proceed with the description of fatty acid biosynthesis.

NADPH production

At the beginning of this section we noted that 8 acetyl CoA molecules and 14 NADPH + 14 H^+ are required to synthesise palmitate. Each turn of the above cycle releases an acetyl moiety from the mitochondrion and results in the production of one NADPH + H^+. Thus 8 of the NADPH + H^+ are produced during the passage of 8 acetyl CoA molecules to the cytosol. Only six molecules of NADPH are required, therefore, from the pentose phosphate pathway. Prokaryotic cells, of course, do not contain mitochondria so they do not undergo the cyclical sequence of reactions and all of their NADPH + H^+ molecules are derived from the pentose phosphate pathway.

amino acids and carbohydrates used to synthesis fatty acids

Remember that a source of hexose sugar is required to drive the pentose phosphate pathway via glucose-6-phosphate. Acetyl CoA is largely derived from pyruvate which itself is derived largely from carbohydrate degradation. However, both acetyl CoA and pyruvate may be the products of the degradation of some amino acids so, in summary, we can say that fatty acids are synthesised from carbohydrates and/or amino acids. We shall see in the next chapter that only in the case of a small number of micro-organisms is the reverse process, amino acid and sugar synthesis from fatty acids, operative.

provision of malonyl CoA

The provision of one further precursor needs to be elucidated, that of malonyl CoA. The discovery that bicarbonate but not CO_2 was necessary for fatty acid biosynthesis quickly gave rise to the recognition of the presence of acetyl CoA carboxylase. This enzyme carboxylates acetyl CoA to malonyl CoA and has an absolute requirement for ATP and bicarbonate. Thus:

$$H_3C-\overset{\overset{O}{\|}}{C}-SCoA + ATP + HCO_3^- \longrightarrow {}^{-}O-\overset{\overset{O}{\|}}{C}-CH_2-\overset{\overset{O}{\|}}{C}-SCoA + ADP + Pi$$

requirements of biotin, bicarbonate and ATP explained

This enzyme has a biotin prosthetic group. The elucidation of the role of this enzyme explained at a single stroke the need in fatty acid synthesis for biotin, ATP and bicarbonate. As the first step proper of fatty acid biosynthesis it is not surprising that some form of control is exercised on acetyl CoA carboxylase. The enzyme is activated by the presence of slightly raised concentrations of either citrate or isocitrate.

5.5 The biosynthesis of fatty acids

5.5.1 The biosynthesis of straight chain, saturated fatty acids containing an even number of carbon atoms

production of acetyl and malonyl ACP

Biosynthesis of fatty acids is generally reckoned to start with the production of acetyl ACP and malonyl ACP from their respective CoA derivatives. The enzymes involved are specific transacylases. Thus:

$$\text{acetyl CoA + ACPSH} \xrightarrow[\text{acetyl transacylase}]{} \text{acetyl ACP + CoASH}$$

$$\text{malonyl CoA + ACPSH} \xrightarrow[\text{malonyl transacylase}]{} \text{malonylACP + CoASH}$$

In the reaction which follows the malonyl group condenses with the acetyl group to produce a four-carbon derivative and carbon dioxide is released, derived from one of the carboxyl groups of malonyl ACP.

```
CH3       O          O                        CH3        O
|         ||         ||                       |          ||
C = O +   C - CH2 - C - SACP   ——▶   C - CH2 - C - SACP + CO2 + ACPSH
|         |                           ||
SACP      O-                          O
```

One ACP molecule and one carbon atom are in bold type on each side of the equation for identification purposes.

Π Can you name the ACP derivative formed and the enzyme involved in the above reaction?

The compound formed is β-oxobutyryl ACP (or 3-oxobutyryl ACP) or, to use the name biologists use, acetoacetyl ACP. If you had difficulty with the naming of this compound, we would suggest that you re-read the response to SAQ 5.2. The enzyme, therefore, is β-oxobutyryl ACP synthase. As the enzyme will operate on compounds of varying chain length the more general name of β-oxoacyl ACP synthase is acceptable.

The reaction carried out could be shown simply as

$$C_2 + C_3 \rightarrow C_4 + CO_2.$$

The obvious question is, why not use $C_2 + C_2 \rightarrow C_4$, that is, two acetyls condensing to give the β-oxobutyryl derivative? In fact the production of butyryl from two acetyls is unfavourable energetically but malonyl plus acetyl is favourable because the decarboxylation step contributes energy.

conversion of the carbonyl to a methylene group

We have now produced a C_4 compound and to make longer ones we need to keep adding successive two-carbon fragments. Before this can proceed, however, the β-oxo group (C = O) has to be reduced to a $-CH_2$ group. We have seen biological interconversions of this type before.

Π Try to complete the diagram filling in the formulae in between the R groups and the underlined queries for the cofactors and types of reaction?

R–C(=O)–CH₂–R ⇄ (? → ?; reduction) R–C(?)–C(?)–R ⇄ (?; ?) R–C(?)=C(?)–R ⇄ (? → ?; ?) R–CH₂–CH₂–R

The answer is:

R–C(=O)–CH₂–R ⇄ (NADPH + H⁺ → NADP⁺; reduction) R–CH(OH)–CH₂–R ⇄ (H_2O; dehydration) R–CH=CH–R ⇄ (NADPH + H⁺ → NADP⁺; reduction) R–CH₂–CH₂–R

In fatty acid biosynthesis NADPH + H^+ is used. However, your answer may have used NADH + H^+ or even $FADH_2$, either of which are possible alternatives.

This sequence of reactions occurs in the TCA cycle:

oxaloacetate ↔ malate ↔ fumarate ↔ succinate

and also, in reverse, in the β-oxidation of fatty acids.

SAQ 5.3

Study Figure 5.4. From β-oxobutyryl CoA complete the three reaction sequence by inserting the missing formulae and the names of enzymes and intermediates.

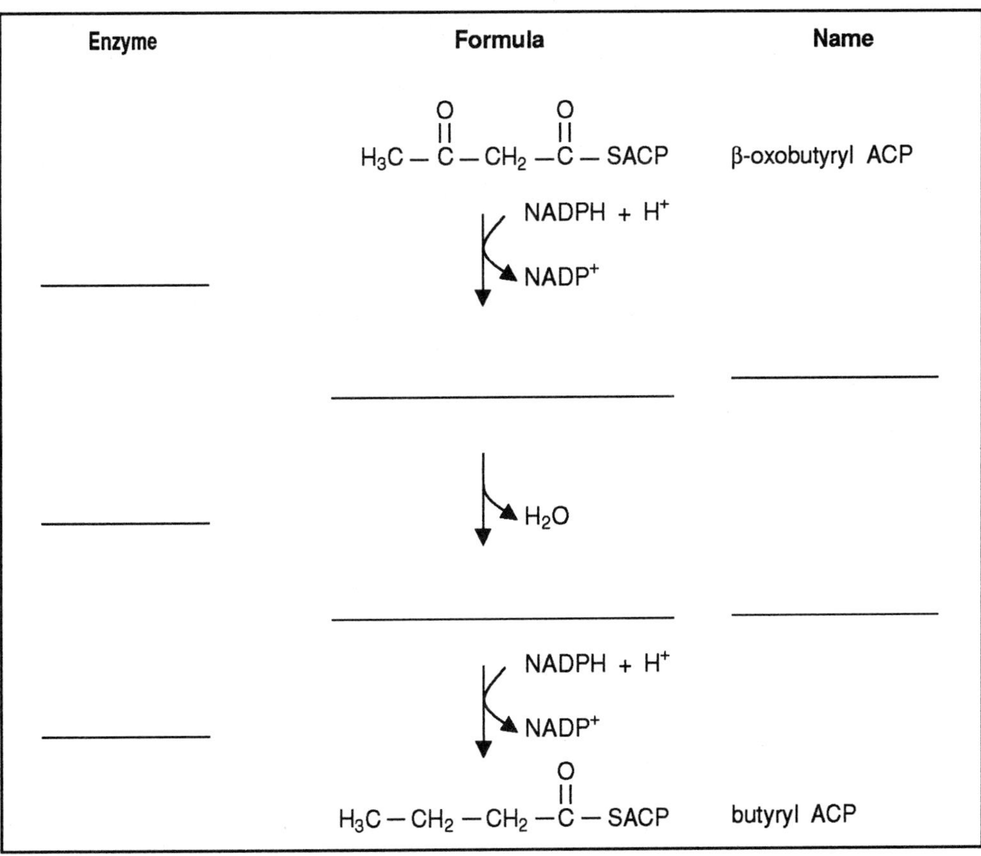

Figure 5.4 The conversion of β-oxobutyryl ACP to butyryl ACP. (See SAQ 5.3).

It is really important not only that you carry out SAQ 5.3 but that you understand it. These three reactions together with the earlier condensation reaction are repeated over and over again, each time adding on a two-carbon unit, until the correct chain length fatty acid is reached. It is much easier to learn the process at this stage when the molecular chain length is very short. The problem can look much more difficult when fourteen-carbon or sixteen-carbon compounds are being considered!

chain extension

The sequence of reactions now begins again by adding a two-carbon fragment to the four-carbon ACP unit.

Π What is required for the addition, an acetyl ACP or a malonyl ACP?

The answer is a malonyl ACP. Remember that although only a two-carbon fragment is added at each turn, the energy released during the decarboxylation of malonyl CoA is necessary for the condensation to proceed.

Thus:

Butyryl ACP + malonyl ACP → β-oxohexanoyl ACP + CO_2 + ACPSH

After an analogous series of reactions to the ones we have just studied in SAQ 5.3, the β-oxohexanoyl ACP is converted to hexanoyl ACP.

production of a sixteen-carbon compound

Let us now work our way through the production of a sixteen-carbon derivative, starting from the beginning. The first set of four reactions yielded a four-carbon derivative. Six further sets would be needed to yield a sixteen-carbon derivative. Only one acetyl ACP is required in the first condensation, all other carbons are derived from malonyl ACP derivatives, seven in total. Each set of four reactions in the cycle requires two molecules of NADPH + H^+ and releases one carbon dioxide molecule.

Putting all of this data together,

acetyl ACP + 7 malonyl ACP + 14 NADPH + 14 H^+

$\rightarrow$ C_{16} fatty acid (palmitoyl ACP) + 7 CO_2 + 14 $NADP^+$ + 7 ACPSH (E 5.4)

At the end of section 5.4 we noted that the formation of malonyl CoA from acetyl CoA required ATP according to the equation:-

acetyl CoA + ATP + HCO_3^- $\rightarrow$ malonyl CoA + ADP + Pi

Therefore production of seven malonyl CoA would require seven ATP thus:

7 acetyl CoA + 7ATP + 7HCO_3^- $\rightarrow$ 7 malonyl CoA + 7ADP + 7Pi. (E 5.5)

adding equations 4) and 5) together and assuming that 7HCO_3^- and 7CO_2 are equivalent gives:

8 acetyl CoA + 7ATP + 14NADPH + 14H^+

$\rightarrow$ palmitoyl CoA + 7CoASH + 14$NADP^+$ + 7ADP + 7Pi. (E 5.6)

The sixteen-carbon fatty acid palmitoyl ACP is the end-product of this elongation system in most organisms including all eukaryotes. One might expect that the eighteen-carbon compound, stearyl ACP, would be produced by the further addition of a two-carbon fragment from malonyl ACP. However, due to the specificity of the β-oxoacyl ACP synthase, which will not utilise anything larger than fourteen-carbon (tetradecanoyl) derivative for elongation, the process stops here. The palmitoyl group is released from the enzyme either as the CoA derivative, or it may be incorporated directly into phospatidic acid during the biosynthesis of phospholipids. The processes leading to the formation of the larger fatty acids will be dealt with later. Bacteria do not need extra systems because their enzymes are not specific for chain length.

one enzyme acts on four-carbon to sixteen-carbon derivatives

In one sense this is one of the most complicated pathways we have encountered. Each addition of a two-carbon fragment requires four reactions, seven additions therefore requiring twenty-eight reactions. If the two transacylases are included together with acetyl CoA carboxylase the list could get, theoretically, to over thirty enzymes. Fortunately cells do not require such a large number of different enzymes for this pathway because they are not chain-length specific. Thus for instance, a single enoyl ACP reductase will reduce all enoyl ACP derivatives from four-carbon to sixteen-carbon chain length.

It has been shown that all organisms contain seven components for fatty acid synthesis which are:- the acyl transacylases (acetyl and malonyl), the β-oxoacyl synthase (the condensing enzyme), β-oxoacyl reductase, β-hydroxyacyl dehydrase, enoyl reductase and an ACP molecule. As we shall see later more than one dehydrase occurs; in bacterial systems, a rather special one specific for ten-carbon chain length derivatives is involved in the production of unsaturated fatty acids.

multi-enzyme complex

Early studies of the fatty acid synthetase system were centred on yeasts, pigeon liver and *E. coli*. Lynen showed that in yeasts the seven components were present in a multi-enzyme complex of relative molecular mass around 2.3 x 10^6. This complex can be isolated from yeast as a functional, intact complex from cell-free extracts, but attempts to dissociate it result in loss of activity. The synthetase complex from pigeon liver can be isolated and separated into two major functional components. The *E. coli* system, however, was shown by Vagelos and colleagues to be separable into its seven components without loss of activity and this appears to be the case for bacteria in general.

production of eighteen-carbon compounds

Eukaryotic organisms as indicated produce a sixteen-carbon acid using the fatty acid synthetase complex though the fatty acid which is most important quantitatively is one of the eighteen-carbon chain length range. Elongation from sixteen to eighteen-carbons in eukaryotes occurs by two distinct enzyme systems, one in mitochondria and one on the endoplasmic reticulum (ER). The ER system employs malonyl CoA rather than malonyl ACP whereas the mitochondrial one employs acetyl CoA. The mitochondrial enzymes will also elongate unsaturated fatty acids. Bacteria produce a wider variety of fatty acids, including some of chain lengths exceeding twenty-carbons. Elongation occurs using the enzymes already described because, as we noted before, the bacterial β-oxoacyl ACP synthase is not particularly chain-length specific.

5.5.2 Biosynthesis of straight-chain, saturated fatty acids containing an odd number of carbon atoms

Fatty acids containing an odd number of carbon atoms are not as prevalent as even-numbered acids but are still vital to the well-being of certain organisms. In higher organisms the way to produce such compounds (notably seventeen-carbon acids) is to carry out α-oxidation of the acid containing one more carbon than is required. Hence:

$$C_{18}\text{ acid} \xrightarrow{\alpha\text{-oxidation}} C_{17}\text{ acid}$$

Bacterial systems, however, have considerably more odd-numbered acids but they produce them using the pathway described in Section 5.5.1 with one slight difference.

Π Can you think what the difference is which allows bacteria to do this?

The answer is by altering the starting, primer molecule. For even-numbered fatty acids we saw that acetyl was the primer and the sequence proceeded:

$$\text{acetyl + malonyl} \rightarrow C_4 + CO_2$$

$$C_4\text{+ malonyl} \rightarrow C_6 + CO_2 \text{ etc.}$$

use of an odd-numbered primer

In bacterial systems acetyl can be replaced by an odd-numbered primer. For example, it is often the five-carbon derivative valeryl CoA. Thus:

$$CH_3 - (CH_2)_3 - CO\ SACP + \text{malonyl SACP} \rightarrow \beta\text{-oxoheptanoyl SACP} + CO_2 + ACPS$$

and the process occurs in an identical manner to that for even-numbered acids.

5.5.3 Biosynthesis of unsaturated fatty acids

There are two quite distinct mechanisms for introducing double bonds into fatty acids. Animals, plants, eukaryotic micro-organisms and a few bacteria (for example *Corynebacterium spp., Bacillus spp., Mycobacterium spp.*) use the so-called aerobic pathway whereas the overwhelming majority of bacteria use the anaerobic pathway.

insertion of a double bond

In the aerobic pathway direct desaturation of a preformed, saturated fatty acid occurs, the process requiring molecular oxygen. The reaction is rather complex and is carried out by a particulate multienzyme system termed a mono-oxygenase or mixed function oxidase. In eukaryotes the enzyme is on the ER. One atom of molecular oxygen produces water by reacting with two hydrogens from the fatty acid. This leaves a free oxygen atom which combines with two further hydrogen atoms from NADPH + H^+. Thus, for example:

$$CH_3 - (CH_2)_7 - CH_2 - CH_2 - (CH_2)_7\ CO\ SCoA + NADPH + H^+ + O_2 \rightarrow$$

$$CH_3 - (CH_2)_7 - CH = CH - (CH_2)_7\ CO\ SCoA + NADP^+ + 2\ H_2O$$

Π Can you name the two fatty acids in the above equation?

The precursor is stearic acid and the product is oleic acid, the latter having the empirical formula:

$$C_{17}H_{33}CO\ SCoA\Delta_{9\text{-}10}.$$

Note that it is the CoA derivatives rather than the free acids which are substrates for the formation of unsaturated fatty acids. The two principal mono-unsaturated fatty acids in eukaryotes are palmitoleic acid ($C_{15}H_{29}CO\ OH\Delta_{9\text{-}10}$) and oleic acid as shown above. Both as indicated have Δ9-10 double bonds. There are two other very important unsaturated fatty acids in eukaryotes, linoleic acid and linolenic acid which respectively have the formulae:

$$C_{17}H_{31}COOH\ \Delta_{9\text{-}10}, \Delta_{12\text{-}13}, \text{ and}$$

$$C_{17}H_{29}COOH\ \Delta_{9\text{-}10}, \Delta_{12\text{-}13}, \Delta_{15\text{-}16},$$

arachidonic acid

These two are the basis for a further series of polyunsaturated acids, for example arachidonic acid but prokaryotic systems do not contain these di- and poly unsaturated fatty acids.

One problem for animals is that they cannot insert double bonds beyond C_9 of a fatty acid and therefore cannot synthesise linoleic acid, linolenic acid or their derivatives. These fatty acids are therefore dietarily essential and thus have to be provided for us from plant sources.

double bond inserted during biosynthesis

The anaerobic pathway utilised by most bacteria is so named because it does not utilise molecular oxygen even though it will proceed under aerobic conditions. In this pathway ACP rather that CoA derivatives are employed and unsaturation takes place during the actual biosynthesis of the fatty acid. The principal mono-unsaturated fatty acid of prokaryotic cells is vaccenic acid which has the general formula

$$C_{17}H_{33}COOH\Delta_{11\text{-}12}.$$

Remember from section 5.5.1 that during fatty acid biosynthesis two-carbon units are added at the end which will become the free carboxyl. Thus to arrive at a $C_{18}\ \Delta_{11-12}$ fatty acid the double bond has to be inserted at the C10 (decanoyl ACP) stage. We shall return to this point after a study of the process. Let us start then at the point where a C_8 derivative has just accepted a further two-carbon fragment to yield:

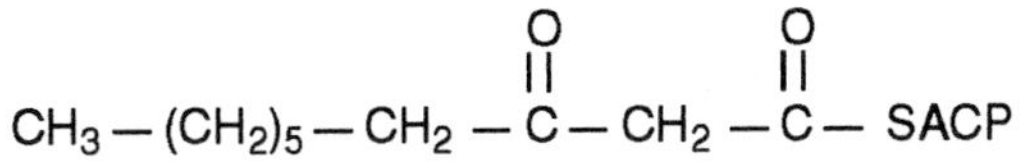

$$CH_3-(CH_2)_5-CH_2-\overset{\overset{\displaystyle O}{\|}}{C}-CH_2-\overset{\overset{\displaystyle O}{\|}}{C}-SACP$$

Π Can you name this compound?

The answer is β-oxo (or 3-oxo) decanoyl ACP.

As in conventional saturated fatty acid biosynthesis, a reduction occurs in which the β-oxo group is reduced to a β-hydroxy derivative.

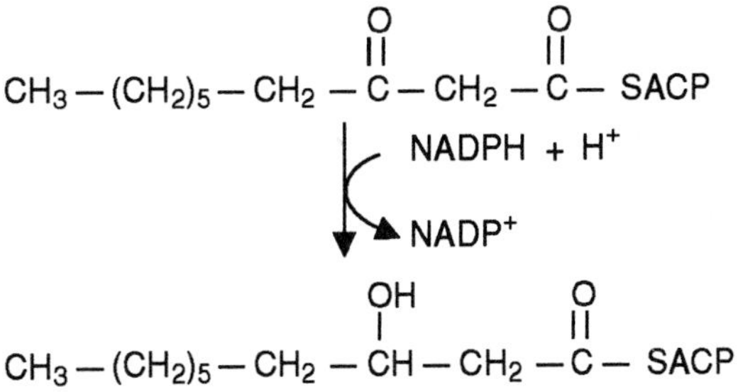

$$CH_3-(CH_2)_5-CH_2-\overset{\overset{\displaystyle O}{\|}}{C}-CH_2-\overset{\overset{\displaystyle O}{\|}}{C}-SACP$$

$$\Big\downarrow \; NADPH + H^+ \rightarrow NADP^+$$

$$CH_3-(CH_2)_5-CH_2-\overset{\overset{\displaystyle OH}{|}}{CH}-CH_2-\overset{\overset{\displaystyle O}{\|}}{C}-SACP$$

It is at this point that the difference between the two pathways occurs. Normally an α - β unsaturation occurs but in the pathway to produce vaccenic acid, a β - γ unsaturation occurs. Study Figure 5.5 carefully and familiarise yourself with the α, β and γ carbons of the β-hydroxydecanoyl SACP. The left hand branch of the pathway represents conventional fatty acids biosynthesis as studied in Section 5.5.1: following dehydration to yield an α - β unsaturated double bond a reduction occurs producing a fully saturated acyl ACP derivative which then continues by combining with a further two-carbon fragment.

On the right hand side of the pathway dehydration yields a β - γ unsaturation. Enoyl ACP reductase will not work on such intermediates, it can only recognise unsaturated α carbons, thus the double bond is left intact. In addition the β-oxoacyl SACP synthetase will readily attach another two-carbon fragment to the 3,4 decenoyl SACP which has the effect of pushing the double bond away from the carboxyl end.

in bacteria a single enzyme produces α -β and β - γ unsaturation

In bacteria, for example *E. coli*, it has been shown that it is the same enzyme which produces the α -β and the β - γ double bond. This enzyme is highly specific for C_{10} derivatives but exactly how the control is exerted to produce α - β as opposed to β -γ unsaturated products is not clear. The diagram which follows shows how vaccenic acid, a C_{18} fatty acid, is produced.

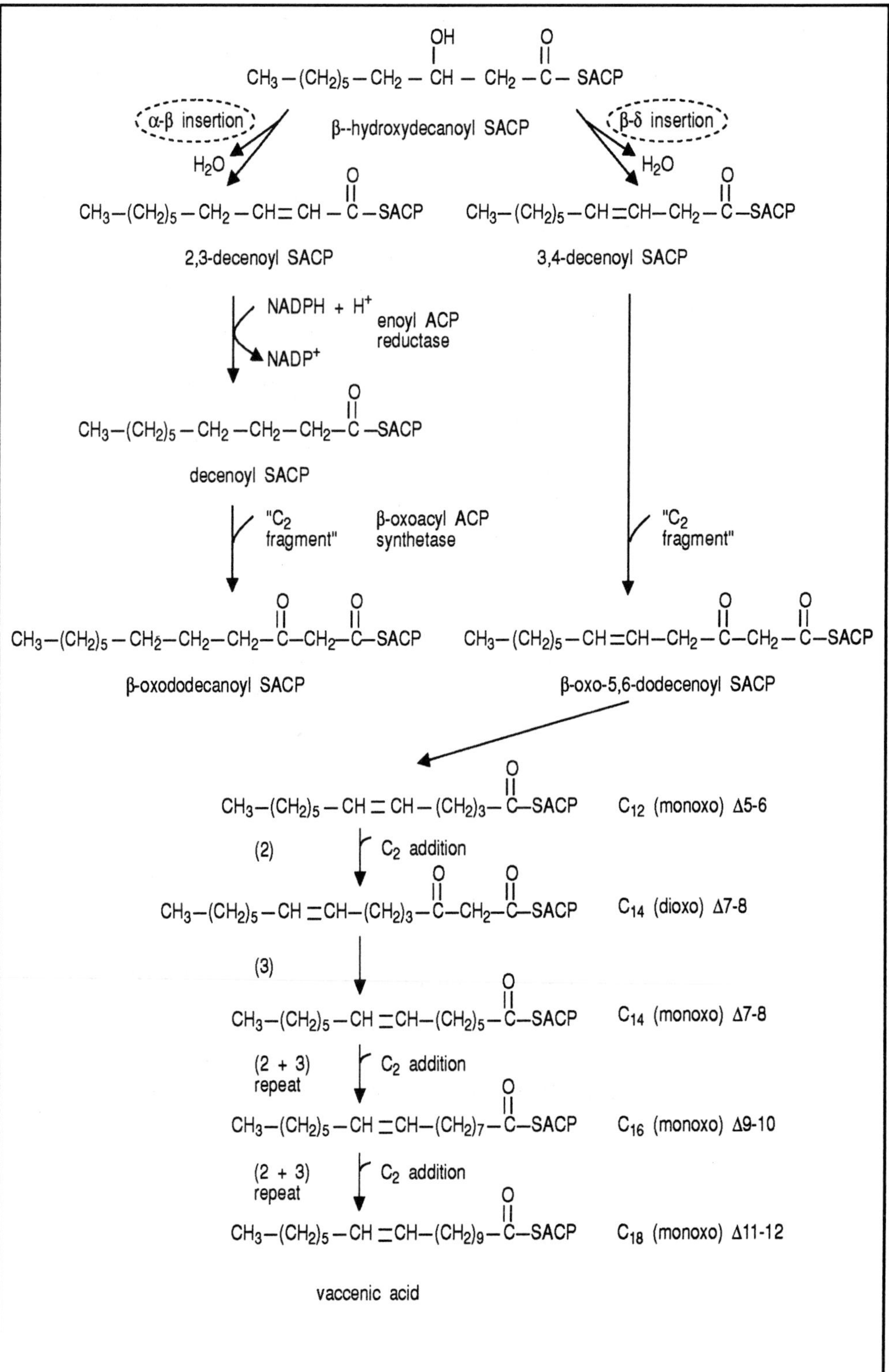

Figure 5.5 Biosynthesis of unsaturated fatty acids - the anaerobic pathway. (See text for a description).

Thus, insertion of the β - γ bond into a compound shorter than ten-carbons in length would yield the double bond between $\Delta_{13\text{-}14}$ or higher in vaccenic acid and insertion of the bond into a chain of greater than a ten-carbon derivative would yield a double bond at $\Delta_{7\text{-}8}$ or lower.

temperature affects degree of unsaturation

The ability of all living systems to produce unsaturated fatty acids is essential for several reasons. Not only are specific fatty acids required for particular structural or functional roles but the relative content of unsaturated fatty acids increases with a decrease in temperature. The melting point of lipids, which is influenced by the percentage of saturated fatty acids, affects the fluidity of membranes. The function of membranes depends on lipid fluidity and it is desirable for this property to show constancy. Thus a drop in environmental temperature (and hence a loss of fluidity) is compensated for by increasing the content of the more-fluid unsaturated fatty acids.

5.5.4 The biosynthesis of cyclopropane fatty acids

methionine donates a methyl group

Cyclopropane fatty acids are a unique feature of prokaryotic lipids occurring in significant quantities within their phospholipids. Just two are universally distributed, the nineteen-carbon acid which is derived from vaccenic acid and the seventeen-carbon derivative which is derived from palmitoleic acid. In each case the extra carbon is derived from the terminal methyl group of methionine and is added across the double bond. The reaction with vaccenic acid is as follows:

$$CH_3-(CH_2)_5-CH=CH-(CH_2)_9-\overset{\overset{O}{\|}}{C}-R$$

vaccenic acid

↓ CH_3 from methionine

$$CH_3-(CH_2)_5-\underset{\diagdown}{CH}-\underset{\diagup}{CH}-(CH_2)_9-\overset{\overset{O}{\|}}{C}-R$$
$$CH_2$$

11,12- methylene octadecanoic acid (lactobacillic acid)

The R prefix is employed because the reaction occurs when the fatty acid is bound to glycerol as part of a phospholipid molecule. Phospholipid biosynthesis will be discussed in a following section.

Members of the genus *Bacillus*, particularly those which prefer to grow at high temperatures, have more complex cyclic groups; for example the presence of a cyclohexyl fatty acid and another containing a peptidolipid ring are well documented.

5.5.5 Biosynthesis of branched-chain fatty acids

Branched-chain fatty acids are another characteristic feature of prokaryotic lipids not found in eukaryotes. The number of derivatives is somewhat limited and the term branched is somewhat misleading as these fatty acids have a single methyl group joined to the penultimate or second from last carbon. The primer molecule for these acids would be isovalerate or methyl butyrate respectively. For example:

$$CH_3-\overset{\overset{CH_3}{|}}{CH}-CH_2-\overset{\overset{O}{||}}{C}-SACP \xrightarrow{+\ 6\ \text{"2-carbon fragments"}} CH_3-\overset{\overset{CH_3}{|}}{CH}-(CH_2)_{13}-\overset{\overset{O}{||}}{C}-SACP$$

isovaleryl ACP → 15-methyl stearyl ACP

$$CH_3-CH_2-\overset{\overset{CH_3}{|}}{CH}-\overset{\overset{O}{||}}{C}-SACP \xrightarrow{+\ 6\ \text{"2-carbon fragments"}} CH_3-CH_2-\overset{\overset{CH_3}{|}}{CH}-(CH_2)_{12}-\overset{\overset{O}{||}}{C}-SACP$$

2-methylbutyryl SACP → 14-methyl stearyl ACP

The mycobacteria are exceptional in that they have two other examples of long-chain fatty acids with several branches, for example mycolipenic acid which has a twenty four-carbon backbone and methyl groups at C_2, C_4 and C_6.

5.6 Biosynthesis of triglycerides

glycerol-3-phosphate is the precursor of triglycerides

Triglycerides or triacylglycerols are, as the name implies, a combination of glycerol esterified to three fatty acids. Such compounds are the main energy storage compounds in animals and are therefore common within cells. They are, however, rare or absent in bacterial systems. Their biosynthesis requires the presence of glycerol-3-phosphate.

Π Can you remember from where glycerol-3-phosphate is derived?

The main supply is via reduction of dihydroxyacetone phosphate (DHAP), an intermediate of glycolysis. It may also arise from phosphorylation of free glycerol using ATP and glycerol kinase.

The biosynthesis of triglycerides proceeds by a simple four-step reaction, two steps of which are shared with the biosynthesis of phospholipids (Figure 5.6). This process is independent of ATP, the energy for esterification being supplied by cleavage of the fatty acyl - CoA bond.

The first two reactions, which are carried out by all living systems as a means of producing phospholipids, would, in prokaryotes, employ ACP instead of CoA derivatives.

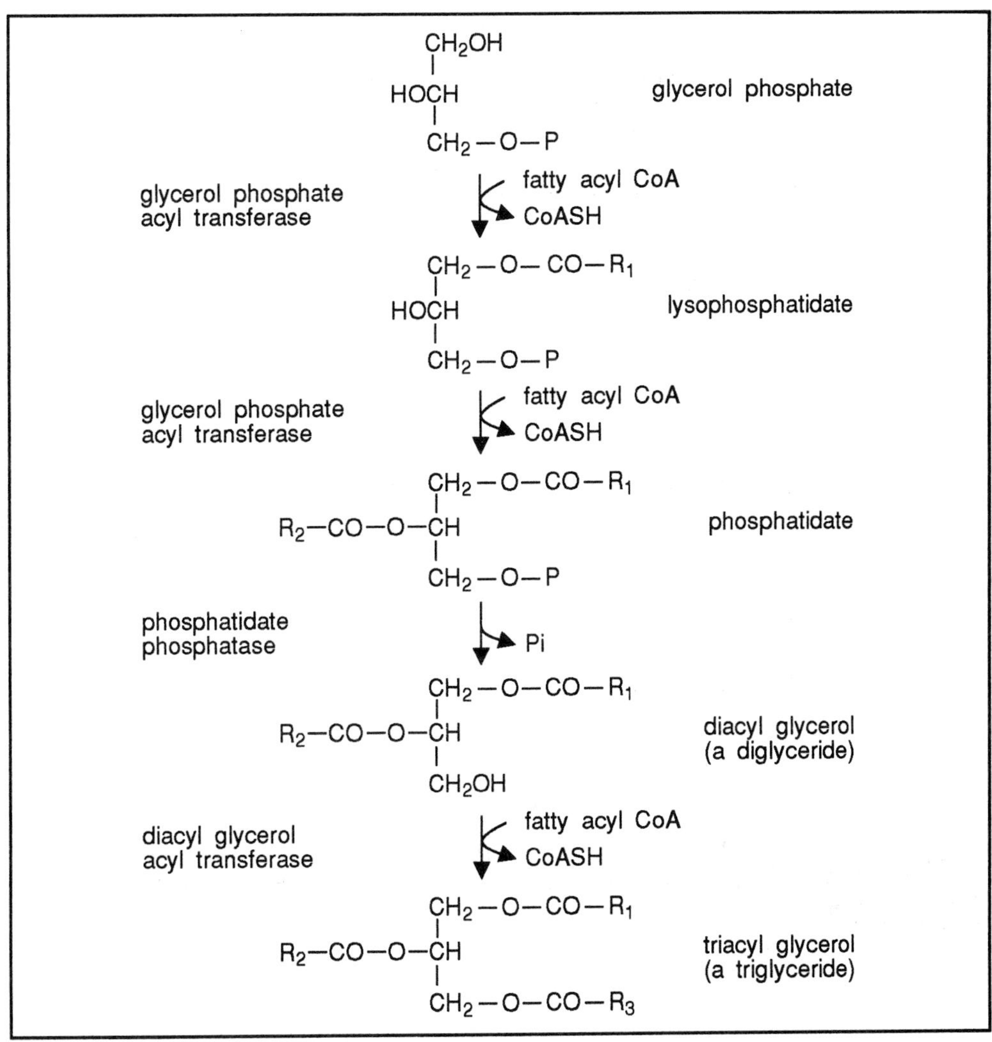

Figure 5.6 A simplified diagram to show the biosynthesis of triglycerides.

5.7 Biosynthesis of phospholipids

phosphatidic acid is the precursor of phospholipids

Let us first recap briefly on the overall structure and occurrence of phospholipids. They are present in all biological membranes and, therefore, are required and synthesised by all living systems. All arise from the precursor metabolite phosphatidic acid (phosphatidate; see Figure 5.6) which contains glycerol esterified to two fatty acids and an inorganic phosphate. Several types of phospholipid are synthesised by adding to the phosphate one of a variety of compounds, such as choline, serine and ethanolamine. All are present in all systems but the relevant proportions of each vary considerably, for example phosphatidyl choline is predominant in eukaryotes but generally phosphatidyl ethanolamine is predominant in prokaryotes. One can generalise and say that the biosynthetic processes are largely similar when considering phospholipid biosynthesis in eukaryotes and prokaryotes and both occur on substrates embedded in membranes, prokaryotes on the plasma membrane and eukaryotes or the ER. We shall now look in some detail at biosynthesis in a named prokaryote, *Escherichia coli*.

The first two reactions are as discussed in Figure 5.6, namely:

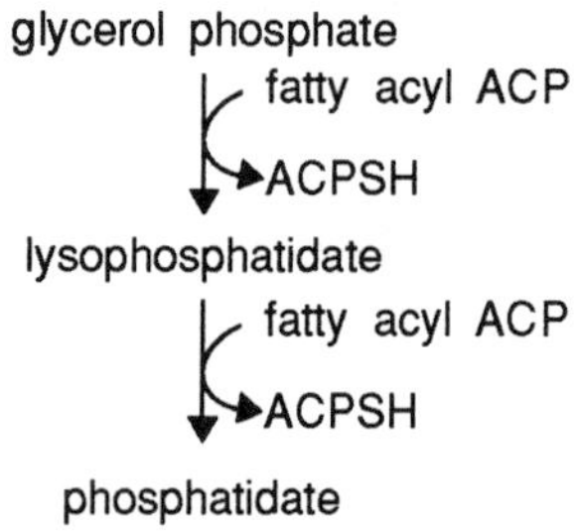

glycerol phosphate
↓ fatty acyl ACP → ACPSH
lysophosphatidate
↓ fatty acyl ACP → ACPSH
phosphatidate

The next stage in *E. coli* is to add a compound to the phosphatidate but before this happens activation of the phosphatidate is needed and this requires the presence of a cytidine carrier. Cytidine is a pyridine base, which, like the purine adenosine, for example, can produce mono-, di- and triphosphate derivatives. Cytidine triphosphate is employed in all living systems in phospholipid biosynthesis and here reacts with the phosphatidate to give a cytidine diphosphate diglyceride (CDP diacyl glycerol).

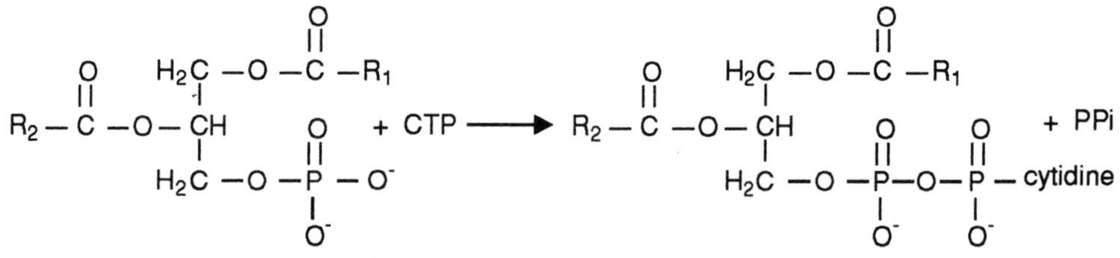

$$\text{phosphatidate} + \text{CTP} \longrightarrow \text{CDP diacyl glycerol} + \text{PPi}$$

phosphatidate: $H_2C{-}O{-}C(=O){-}R_1$ / $R_2{-}C(=O){-}O{-}CH$ / $H_2C{-}O{-}P(=O)(O^-){-}O^-$

CDP diacyl glycerol: $H_2C{-}O{-}C(=O){-}R_1$ / $R_2{-}C(=O){-}O{-}CH$ / $H_2C{-}O{-}P(=O)(O^-){-}O{-}P(=O)(O^-){-}\text{cytidine}$

The enzyme is phosphatidate cytidyl transferase.

This molecule then reacts with serine to produce phosphatidyl serine and in doing so releases cytidine monophosphate (CMP).

Π Can you remember anything about serine, its chemical groups or formula for example?

Serine is one of the α-amino acids which were discussed in Chapter 4. It is a three-carbon compound and is one of only two amino acids to contain a hydroxyl group.

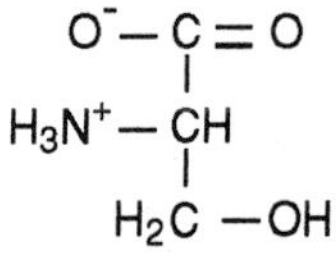

$$O^-{-}C{=}O$$
$$H_3N^+{-}CH$$
$$H_2C{-}OH$$

Phosphatidyl ethanolamine is produced from phosphatidyl serine by a decarboxylation reaction, removing the COO^- group from the serine residue.

SAQ 5.4

Examine Figure 5.7. Using the information above complete the diagram of the reactions producing phosphatidyl serine and then phosphatidyl ethanolamine. Write down the formulae and attempt to name the enzymes on the lines indicated on the left of the figure.

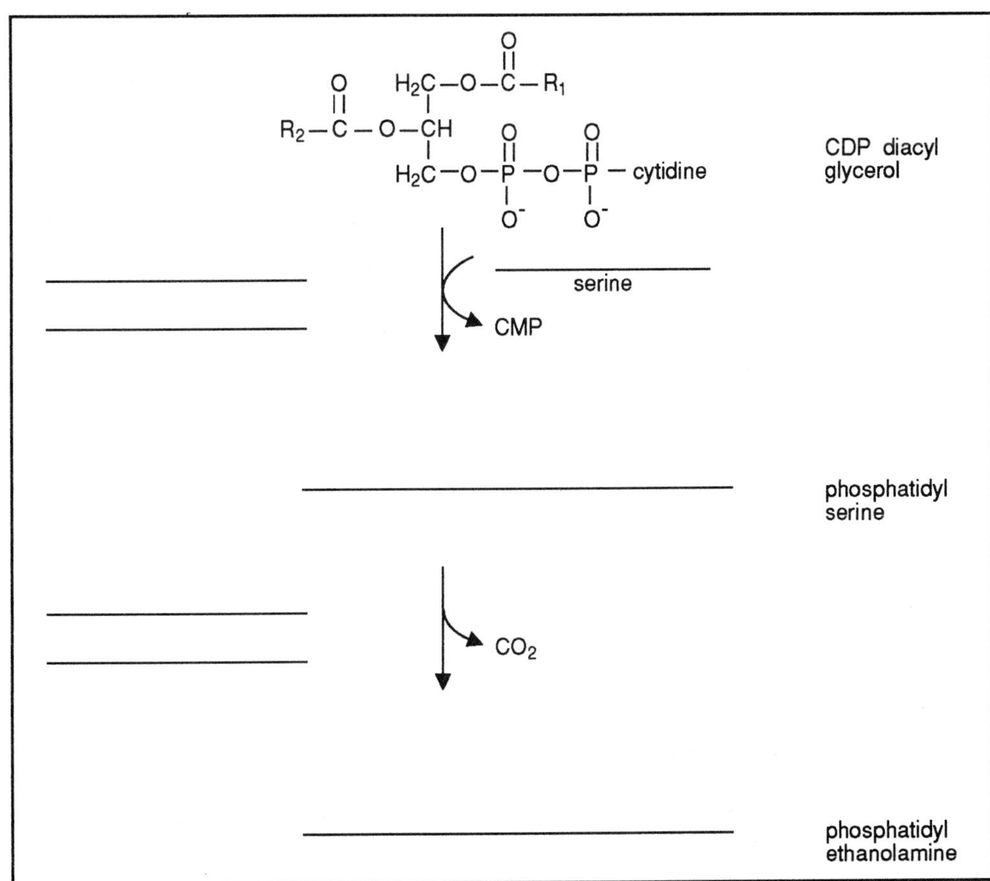

Figure 5.7 The biosynthesis of phosphatidyl serine and phosphatidyl ethanolamine in prokaryotes. (See SAQ 5.4).

In addition to phosphatidyl serine and phosphatidyl ethanolamine two further phospholipids are very important in prokaryotes. These are phosphatidyl glycerol and diphosphatidyl glycerol also known as cardiolipin. Figure 5.8 shows a simplified diagram of the biosynthesis of these two compounds.

Cardiolipin represents a significant proportion of bacterial lipids and around 20% of the lipids of the inner mitochondrial membrane. The biosynthetic pathway of this and phosphatidyl glycerol are the same in all living systems, as indicated in Figure 5.8. Note that we have adapted a short-hand nomenclature to represent the structure of the lipids.

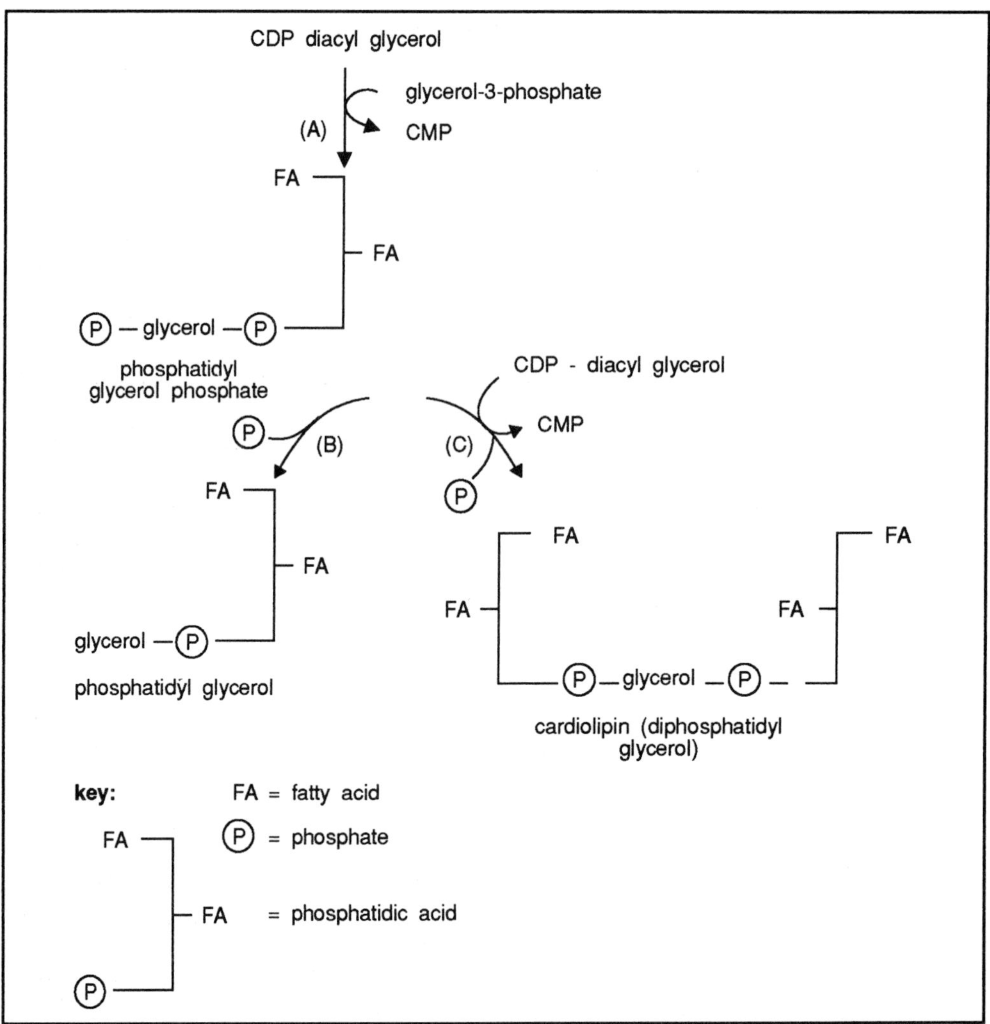

Figure 5.8 The biosynthesis of phosphatidyl glycerol and cardiolipin by prokaryotes. For (A), (B) and (C) see text.

Π Can you name the enzymes A, B and C in Figure 5.8?

Enzyme A is glycerol phosphate phosphatidyl transferase; enzyme B is phosphatidyl glycerophosphatase; enzyme C is cardiolipin synthetase.

Π Can you draw the structural formula of cardiolipin (using R for the fatty acid alkyl groups)? The molecule is quite complicated in terms of the number of carbons but when considered in its component parts should not present any real difficulty. This task should be seen as one to help remove the fear of complex molecules and, on completion of the correct formula, should generate a sense of well being for the next section.

The structural formula for cardiolipin

```
                   O                                  O
                   ||                                 ||
  O     H2C -O - C - R1                    R4 -C-O-CH2     O
  ||      |                                          |     ||
R2-C-O-CH        O                     O        HC-O-C-R3
          |       ||                    ||         |
        H2C -O - P -O-CH2- CH- CH2-O- P-O-CH2
                  |              |          |
                  O-             OH         O-

   glycerol   phosphate   glycerol      phosphate   glycerol
```

R_1, R_2, R_3 and R_4 represent the alkyl groups of the four fatty acids.

archaebacterial phospholipids have branched side chains

To complete the section on phospholipids we must now look briefly at the archaebacterial phospholipids. We looked at an example earlier (Section 5.1, Figure 5.1 of this chapter) and noted that these lipids were unique in having two very unusual properties. Firstly they have an ether rather than an ester link between the side chain and the glycerol, a property shared only by plasmalogens. Secondly the side chain is a branched structure which is made up of multiples of the five-carbon compound isoprene. The hydrocarbons are derived from the C_{20} compound - phytane - or the C_{40} biphytane. One group of archaebacteria, the thermoacidophiles, have cyclopentane rings along the biphytane chain. Biosynthesis of the hydrocarbon chains will be studied in Section 5.8.2 of this chapter. The formation of an ether rather than an ester bond occurs because glycerol reacts with an alcohol, phytanol, rather than with a fatty acid. Formation of ether bonds is by an enzyme unique to archaebacteria.

5.8 Biosynthesis of lipids from isoprene derivatives

isoprene used to add five carbons at a time

We have seen in Section 5.1 that the compound isoprene, though rare in nature in its free form, is the precursor for several important lipid groups. Isoprene can dimerise to produce a 10-carbon derivative, terpene and successive additions of either a further isoprene unit or condensation of terpenes will give longer molecules. The simple sketch following shows the complexity of our task in studying this group of compounds.

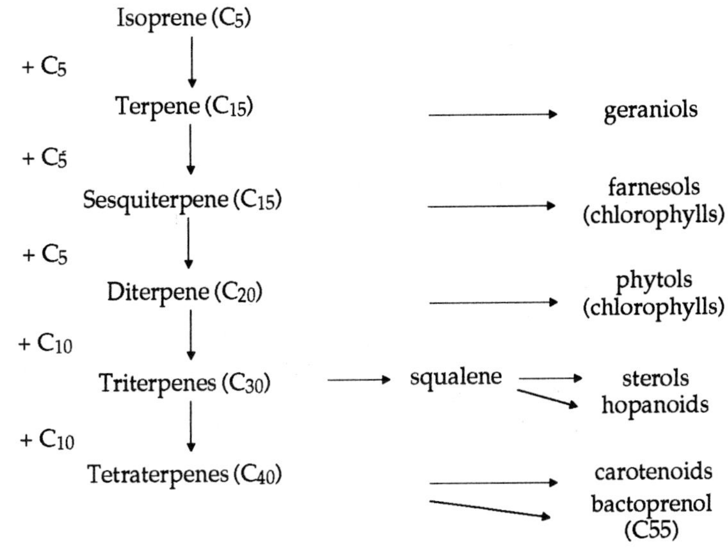

As we have already noted not all of these derivatives are present in all organisms, for example sterols are reckoned to be virtually absent from all bacterial cells except *Mycoplasmas*.

5.8.1 Production of the active five-carbon precursor molecule

The heading uses the phrase 'active five-carbon precursor molecule' rather than isoprene because, as we shall see shortly, the activated five-carbon compound for all biological systems is isopentanyl pyrophosphate.

all carbons of cholesterol are derived from acetate

The story of the elucidation of this pathway began in the early 1940s when elegant work by Block and his colleagues showed that all of the carbons of cholesterol were derived from acetate molecules. Further experiments using acetate labelled only at carbon one or at carbon two showed that the labelling was fairly evenly divided, 12 of the 27 carbons of cholesterol arising from the carboxyl of acetate and the other 15 from the methyl group.

The next step forward was the conclusive evidence that squalene, a thirty-carbon compound, is an intermediate in cholesterol biosynthesis. These two factors still left scientists a long way from the elucidation of the pathway until a third piece of the jigsaw, totally unrelated to the previous work, was discovered.

Using bacterial systems it was found that certain acetate-requiring mutants would use mevalonic acid (six-carbon compound) if acetate was unavailable. Furthermore: 1) acetate was shown to be a natural precursor of mevalonic acid; 2) mevalonic acid was readily decarboxylated to a five-carbon unit; 3) this five-carbon unit was converted by bacteria to squalene; and 4) the five-carbon unit could be used by liver extracts to produce cholesterol.

Thus the overall outline was established.

The biosynthesis of mevalonate is accomplished in all living systems by the condensation of three molecules of acetyl CoA; the chemistry is the same though the details vary slightly in different organisms.

$$\begin{matrix} COSCoA \\ | \\ CH_3 \end{matrix} + \begin{matrix} COSCoA \\ | \\ CH_3 \end{matrix} \longrightarrow \begin{matrix} COSCoA \\ | \\ CH_2 \\ | \\ C{=}O \\ | \\ CH_3 \end{matrix} + CoASH$$

Π Can you name the four-carbon product? Where have we met this compound before?

The compound is usually called acetoacetyl CoA in biological texts; you could have used the chemical name β-oxo (or 3-oxo) butyryl CoA. We studied this compound in Section 5.5.1 because it is the first step in fatty acid biosynthesis although derivatives of ACP rather than CoA were used. From this point the pathways of fatty acid and cholesterol synthesis diverge and are totally independent of each other.

In the next reaction the acetoacetyl CoA condenses with another molecule of acetyl CoA to yield 3-hydroxy-3-methyl glutaryl CoA.

```
acetoacetyl CoA
                                                  COSCoA
COSCoA                                              |
|                                                  CH2
CH2                                                 |
|         ------------------------------>    HO — C — CH3   +   CoASH
C = O                                               |
|             hydroxymethyl                        CH2
CH3           glutaryl CoA                          |
       +      synthetase                           COOH

CH3
|
COSCoA
acetyl CoA
```

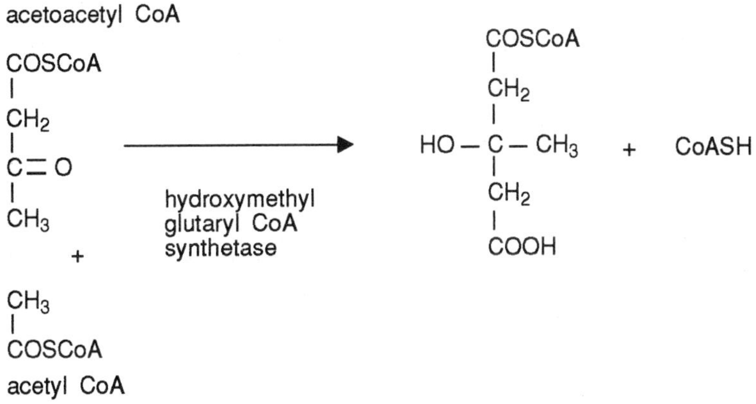

As in many earlier cases, do not be put off by these long names: we have had experience of 2-oxoglutarate as an intermediate of the TCA cycle. The six-carbon derivative above does not have the 2-oxo group, instead it has a methyl and a hydroxyl on carbon 3.

The 3-hydroxy-3-methyl glutaryl CoA can, if the cell requires, be cleaved to acetoacetate plus acetyl CoA or it may be converted by a two-stage reaction to mevalonic acid. The reaction to produce mevalonic acid is important because it is irreversible and commits the organism to produce sterols or other squalene derivatives. In this reaction the CoASH is removed and the carboxyl group to which it was attached is reduced using NADPH + H^+ to an alcohol group (effectively a double reduction via an aldehyde).

Π Can you 1) draw the reaction sequence and, 2) name the enzyme catalysing the reaction?

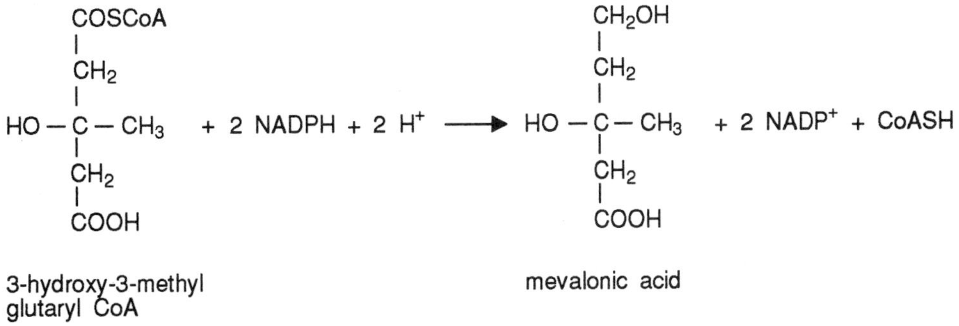

```
   COSCoA                                          CH2OH
   |                                               |
   CH2                                             CH2
   |                                               |
HO — C — CH3   + 2 NADPH + 2 H+  ——>   HO — C — CH3   + 2 NADP+ + CoASH
   |                                               |
   CH2                                             CH2
   |                                               |
   COOH                                            COOH

3-hydroxy-3-methyl                              mevalonic acid
glutaryl CoA
```

Note that 2 NADPH + 2 H^+ are required - four hydrogens in total - to reduce the acid via an aldehyde to an alcohol.

The enzyme is simply 3-hydroxy-3-methyl glutaryl CoA reductase. You could not in this case call it a dehydrogenase because it is a one way reaction.

The final stage in the production of the active five-carbon unit, isopentanyl pyrophosphate, is a three step sequence of successive additions of phosphate from ATP, the first two giving a pyrophosphate moiety on the alcohol group and the third adding to the hydroxyl on carbon 3. The final derivative produced is unstable and it loses phosphate and decarboxylates to give isopentenyl pyrophosphate.

Π The reaction sequence is shown below, can you name the intermediates A to C? Note the appearance of the double bond in the product.

mevalonate —(1) ATP → ADP→ A —(2) ATP → ADP→ B —(3) ATP → ADP→ C —(4) → Pi + CO_2 → isopentenyl pyrophosphate

Compound A is 5-phosphomevalonate
Compound B is 5-pyrophosphomevalonate
Compound C is 3-phospho-5-pyrophosphomevalonate.

Π From our earlier studies can you name the types of enzymes in the sequence?

Enzymes 1 and 2 are both kinases; note that ATP is required. Reactions 3 and 4 occur as a single reaction. In nature they are catalysed by a single enzyme called pyrophosphomevalonate decarboxylase.

SAQ 5.5

Can you devise an overall equation for the sequence of reactions from 3 acetyl CoA molecules to the isopentenyl pyrophosphate? Is the sequence demanding in terms of energy equivalents in that if all of the starting material had been used by an aerobic organism for the sole purpose of producing ATP, how much has been 'lost'?

Remember that for a complicated pathway it is safer to write down the reactions individually and then cancel out to get the overall reaction.

5.8.2 Biosynthesis of polyisoprenoid compounds

active five-carbon compound is the basis for many large molecules

Having arrived at an active five-carbon unit the next step is to produce polymers of these compounds, namely 10, 20 and 30 carbons etc. In many ways these compounds are quite simple and recognisable as repeating units of the isoprenoid nucleus. However, complications arise because the molecules are long and can be twisted into a variety of shapes. We shall investigate this latter problem shortly with respect to squalene.

The first step in this section is obviously the combination of two five-carbon compounds to yield a ten-carbon derivative. Before this happens the following occurs:

$$CH_2{=}C(CH_3){-}CH_2{-}CH_2{-}O{-}\overset{O}{\overset{\|}{P}}(O^-){-}O{-}\overset{O}{\overset{\|}{P}}(O^-){-}O^- \longleftrightarrow CH_3{-}C(CH_3){=}CH{-}CH_2{-}O{-}\overset{O}{\overset{\|}{P}}(O^-){-}O{-}\overset{O}{\overset{\|}{P}}(O^-){-}O^-$$

isopentenyl pyrophosphate   dimethylallyl pyrophosphate

Π Can you name the enzyme?

The reaction is an isomerisation, the enzyme isopentenyl pyrophosphate isomerase.

It is a condensation of isopentenyl pyrophosphate and dimethylallyl pyrophosphate which give rise to the ten-carbon geranyl pyrophosphate.

Further additions require isopentenyl pyrophosphate so the dimethylallyl pyrophosphate is acting only as a primer. Figure 5.9 shows the biosynthesis of the twenty-carbon phytyl pyrophosphate. Before looking at Figure 5.9, examine the flow diagram below to understand clearly what is happening.

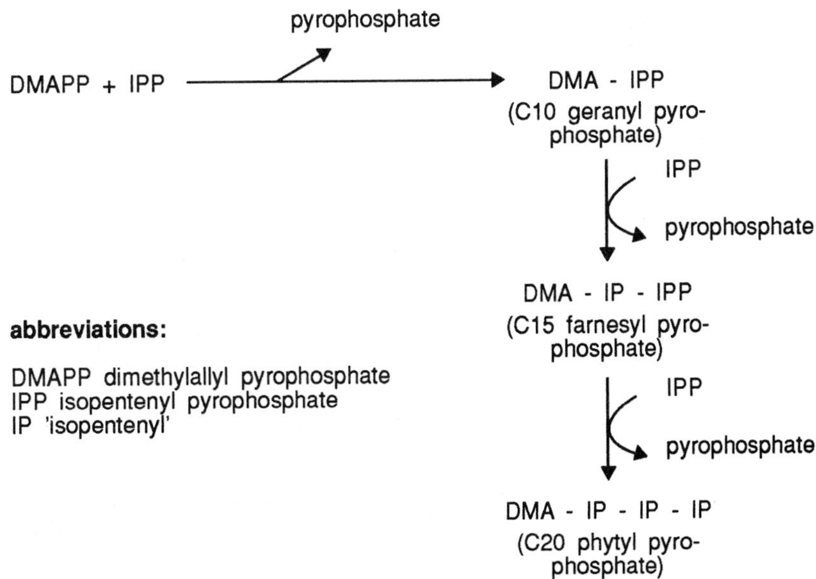

migration of double bond

Now examine Figure 5.9. Try looking at the first condensation, the production of geranyl pyrophosphate, and then try the other two reactions for yourselves. It is not as difficult as it looks! One thing to be aware of is the migration of the double bond when isopentenyl pyrophosphate condenses each time.

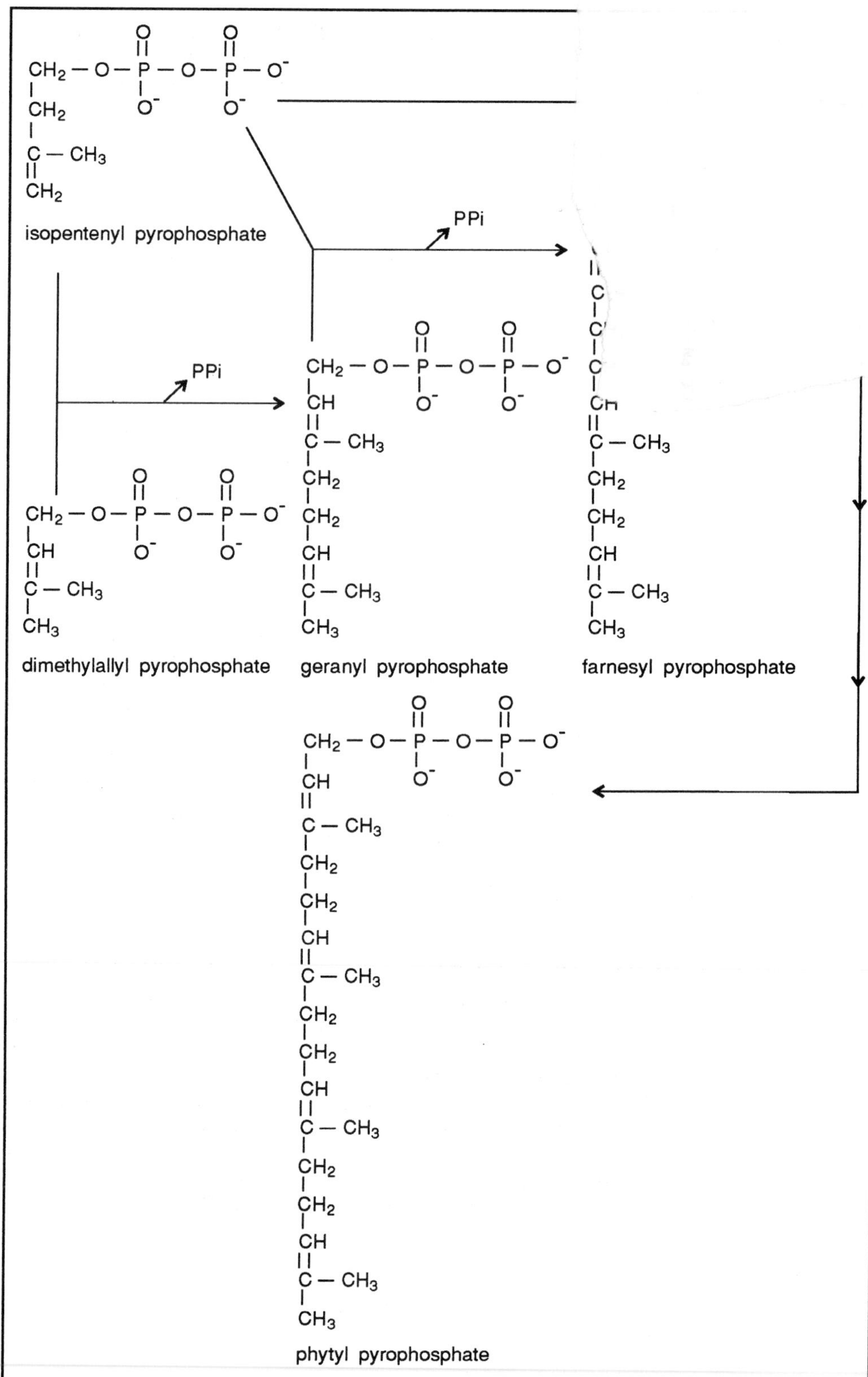

Figure 5.9 The biosynthesis of the twenty-carbon compound phytyl pyrophosphate.

It is about the C_{15} or C_{20} level that different mechanisms for elongation are employed. Organisms requiring squalene for any purpose produce it not by addition of six five-carbon groups but by condensation of two fifteen-carbon farnesyl pyrophosphates to give presqualene pyrophosphate which is then reduced to squalene. The reaction proceeds:

$$CH_3-\underset{\displaystyle CH_3}{\underset{|}{C}}=CH-CH_2-CH_2-\underset{\displaystyle CH_3}{\underset{|}{C}}=CH-CH_2-CH_2-\underset{\displaystyle CH_3}{\underset{|}{C}}=CH-CH_2-O-\underset{\displaystyle O^-}{\underset{|}{\overset{\displaystyle O}{\overset{||}{P}}}}-O-\underset{\displaystyle O^-}{\underset{|}{\overset{\displaystyle O}{\overset{||}{P}}}}-O^-$$

farnesyl pyrophosphate

NADPH + H^+ ; farnesyl pyrophosphate

$NADP^+$; 2 pyrophosphates

$$CH_3-\underset{\displaystyle CH_3}{\underset{|}{C}}=CH-CH_2-(CH_2-\underset{\displaystyle CH_3}{\underset{|}{C}}=CH-CH_2)_2-(CH_2-CH=\underset{\displaystyle CH_3}{\underset{|}{C}}-CH_2)_2-CH_2-CH=\underset{\displaystyle CH_3}{\underset{|}{C}}-CH_3$$

squalene

squalene forms a ring structure

It is at this point that the molecules get very complicated particularly in that their structure can be represented in several ways. We have seen the linear structure of squalene above, a molecule containing six double bonds and no rings. Figure 5.10 shows the molecule rearranged prior to ring formation before producing sterols or, the bacterial 'equivalent' hopanoids together with cholesterol and a common bacterial hopanoid. It is really beyond the scope of this chapter to go into greater detail regarding sterols but it is worth spending a few minutes following the repeating five-carbon isoprenoid units in the squalene model in Figure 5.10.

One or two other compounds derived from polyisoprenoid units should be mentioned due to their distribution and importance.

Firstly bactoprenol (undecaprenyl alcohol) is a fifty five-carbon compound derived from eleven isoprenoid units and is found in the cell membrane of prokaryotes. It lies transversely across the membrane with the long, non-polar hydrocarbon chain anchored in the membrane and the polar end pointing out of the membrane structure. It serves as a flexible means of transporting out of the cell disaccharide pentapeptide units for peptidoglycan biosynthesis as well as the sugar residue for Gram negative outer membranes. These structures will be dealt with in more detail in the next chapter.

Secondly there are a variety of C_{40} derivatives, the carotenoids, which are associated with the photosynthetic apparatus of all relevant living systems.

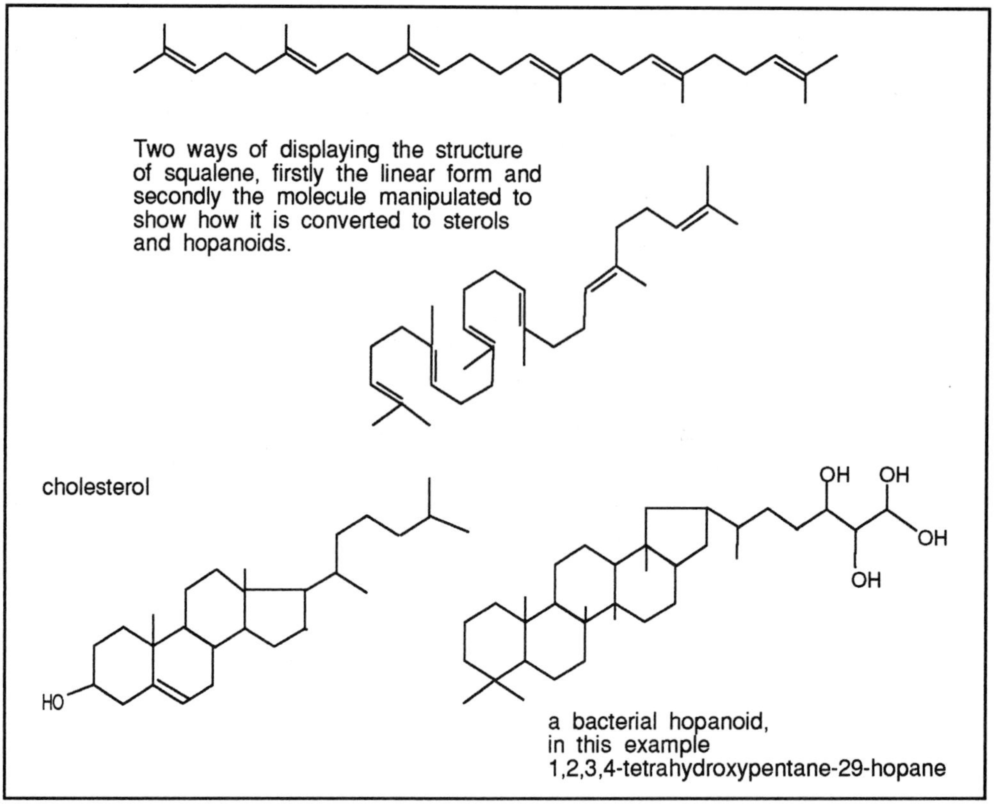

Two ways of displaying the structure of squalene, firstly the linear form and secondly the molecule manipulated to show how it is converted to sterols and hopanoids.

cholesterol

HO

OH OH

OH

OH

a bacterial hopanoid, in this example 1,2,3,4-tetrahydroxypentane-29-hopane

Figure 5.10 The structure of squalene and two compounds derived from it, cholesterol and a bacterial hopanoid.

Carotenoids differ from organism to organism being either linear or having ring structures at either end. The ring structure may be aromatic. The group is very complex, the purple bacteria alone displaying over thirty different carotenoids.

5.9 Microbial biotransformations (bioconversions) of steroids

steroid hormones

We have looked at one example of a sterol, cholesterol, which is a compound absent from bacterial systems but either cholesterol or a close relative are components of all eukaryotic membranes. In addition, higher organisms may modify these compounds to produce steroid hormones.

biotransformation of steroids

Cholesterol (a twenty seven-carbon compound) is metabolised in the adrenal gland to produce a variety of twenty one-carbon compounds, for example cortisone. In addition the ovary and testis produce respectively the female and male sex hormones which are eighteen-carbon and nineteen-carbon compounds. Over the last forty or fifty years chemotherapeutic uses have been found for hormones; for example cortisone treatment relieves inflammation and can control arthritis and various allergies. In addition their use in oral contraception has now become widespread. This created a need for such steroids but the only major source of naturally-occurring steroids is plants. Animals only produce small quantities for their own use and bacteria do not produce them at all. The problem with plant sterols is that they have minor differences from the highly specialised human steroids and chemical conversion is either difficult or impossible. It was found that some eukaryotic micro-organisms and, surprisingly, some bacteria were capable of carrying out certain specific modifications of the steroid nucleus. Once the correct microbe has been found for a particular task, conversion may occur rapidly and with high specificity, giving high yield.

A good example for us to look at is the oxidation of carbon 11 atom of the sterol nucleus. All adrenocortical hormones have an oxygen atom (either C=O or C-OH) at carbon 11 but cholesterol and plant sterols such as ergosterol do not. Specific oxidation of only carbon 11 proved virtually impossible chemically but it can be done by *Rhizopus nigricans* which is a fungus. In the pathway cholesterol is converted to progesterone and then comes the key step, oxidation of progesterone to 11-hydroxyprogesterone by *Rhizopus*. Two simple chemical reactions then convert this to cortisone. Several other hydroxylations, isomerisations and formation of double bonds are well documented in the literature and the main positions at which the biotransformation occurs are shown in Figure 5.11. Some of the products of hydroxylation are extremely valuable.

The structure of the common sterol cholesterol.

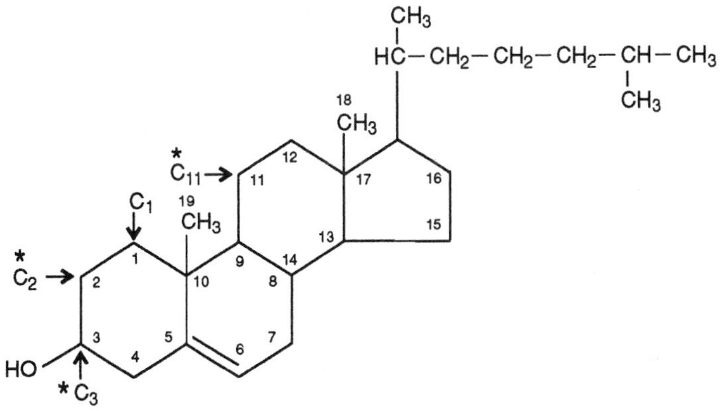

Figure 5.11 Diagram to show examples of where microbial biotransformations of the steroid nucleus. The positions in which specific biotransformations are carried out are marked *.

It is now possible to modify many if not all of the individual carbons selectively using micro-organisms. The most important industrial biotransformation involves the insertion by *Rhizopus nigricans* of a hydroxyl group at C_{11} at the progesterone to 11-hydroxyprogesterone stage during the conversion of sterols to cortisone.

Introduction of a double bond between carbons 1 and 2 by an *Arthrobacter* species has also proved useful in assisting in a number of transformations, both chemical and biological.

Several of the substituent groups on the steroid nucleus may be in either the α (below the plane) or β position (above the plane), Their position is important giving rise to a variety of different compounds. Many transformations have been used or attempted commercially, for example the conversion of the 3-α-OH group to the 3-β-OH group using *Nocardia restrictus*.

Summary and objectives

In this chapter we have examined the biosynthesis of the structually and metabolically important lipids. We have learnt that even with ubiquitous compounds such as fatty acids, some differences occur in their biosynthetic pathways. We have examined the biosynthesis of acyl fatty acids, triglycerides and phospholipids and briefly explored the biosynthesis of the main isoprenyls.

Now that you have completed this chapter you should be able to:

- correctly name and label drawn structures of the common lipid classes found in eukaryotic cells;
- distinguish the role of the acyl carrier protein from that of coenzyme A;
- list the reactions involved in the biosynthesis of the different types of fatty acids in prokaryotes and eukaryotes;
- distinguish between the apparatus for fatty acid biosynthesis and that for degradation;
- describe the biosynthesis of and assess the importance of plasmalogens;
- list the reactions involved in the biosynthesis of the major phospholipids;
- discuss the importance of glycolipids in bacteria;
- list the reactions involved in the synthesis of polyisoprenoid compounds such as bactoprenol and squalene;
- brieflý describe the reactions in the biosynthesis of selected sterols, carotenoids, quinones and chlorophylls;
- give industrially useful-examples of the ways in which micro-organisms can be used to perform sterol interconversions.

The biosynthesis of carbohydrates

Introduction 144

6.1 The occurrence of carbohydrates in living systems 144

6.2 The biosynthesis of glucose 146

6.3 The biosynthesis of hexoses, pentoses and tetroses 157

6.4 The biosynthesis of disaccharides, oligosaccharides and polysaccharides 159

6.5 The commercial importance of polysaccharides of micro-organisms 168

Summary and objectives 170

The biosynthesis of carbohydrates

Introduction

pyruvate is the starting compound

The production of carbohydrates from simple precursors generally starts from pyruvate or the metabolically closely related phosphoenol pyruvate and proceeds via a single pathway, gluconeogenesis, to yield glucose. The processes of degradation of glucose by the Embden Meyerhof pathway (glycolysis) and its synthesis (gluconeogenesis) have several enzymes in common and thus effective control of the two pathways is required.

In this chapter we shall briefly examine the requirement for both simple and complex carbohydrates by living cells. In this section we shall discover that the quantitative demand will depend on the nature of the carbon source used by particular organisms. Higher animals for example can satisfy their carbohydrate requirements from their diet: at the other extreme many bacteria will grow successfully on a range of non-carbohydrate carbon sources and could therefore need to synthesise all of their carbohydrates.

The biosynthesis of glucose from pyruvate or alternatives will then be studied in detail. Inevitably as seven of the enzymes are in common with those of the Embden Meyerhof pathway we shall first need to revise the latter, compare the two pathways closely and then study the mechanisms controlling them.

The biosynthesis of other hexoses and pentoses can be dealt with fairly simply and quickly followed by an examination of the biosynthesis of selected polysaccharides by both eukaryotes and prokaryotes.

Finally a short section will demonstrate, by the use of selected examples, the industrial importance of bacterial polysaccharides.

6.1 The occurrence of carbohydrates in living systems

one tenth of dry weight is carbohydrates

All living systems contain carbohydrates, in simple form such as hexoses and pentoses, through to high molecular weight complex structures - the homopolysaccharides or heteropolysaccharides. As a rough generalisation, cells contain 70-90% by weight of water. Approximately one tenth of the dry weight of the cell is made up of polysaccharide and a twentieth occurs as sugars and simple sugar derivatives. Thus there is less carbohydrate material than protein though both materials are absolutely vital to the well-being of the cell.

The free sugars within cells are largely present as glucose or its phosphorylated derivatives, glucose-1-phosphate and glucose-6-phosphate. Interconversion of glucose to several other common hexoses can be carried out by all living systems and is generally not energy requiring. We shall consider this point later in section 6.3.

The interconversion of hexoses, pentoses and tetroses, particularly the production of ribose and deoxyribose, generally occurs via the pentose phosphate pathway: again we shall briefly review the requirements and biosynthetic processes for pentose and tetrose sugars in section 6.3.

glucose is central

Glucose is quite clearly the central monosaccharide compound as it has so many functions and is at so many branch points in living cells. Not only is it the compound to which the vast majority of carbohydrates are degraded, it also acts as a starting point for the biosynthesis of most polysaccharides. In addition it is the principal energy source for the majority of living systems, usually being degraded to pyruvate via the Embden Meyerhof pathway or, to a lesser extent, via the pentose phosphate pathway. These two pathways are responsible for providing eight of the twelve fundamental precursor metabolites, the TCA cycle, which is fed by pyruvate, providing the other four.

Π Can you remember what is meant by the term 'precursor metabolite'?

They are the twelve simple metabolites which are vital to all living cells in that it is from these twelve compounds that all biosyntheses proceed.

homopolymers and heteropolymers

Polysaccharides are chemically divided into several groups. In simple terms there are the homopolymers - compounds containing many molecules of a single monosaccharide unit; the heteropolymers - compounds containing many molecules of two or more monosaccharides; and complex polysaccharides - compounds containing polymers of carbohydrate material linked to other, non-carbohydrate compounds.

The nature of the polysaccharides varies from organism to organism and there are enormous numbers of different polysaccharides already well-characterised. Relatively few different types are found in higher animals through to larger numbers from eukaryotic micro-organisms but the overwhelming majority comes from bacterial sources.

Polysaccharides are either α or β linked and rarely, if at all, do combinations of α and β linkages occur in the same molecule.

Π Can you remember the generalisation for the function of α-linked compared to β-linked polysaccharides?

Broadly speaking α-linked polysaccharides, for example starch and glycogen, are compounds for energy storage whereas β-linked polysaccharides, for example cellulose, chitin and peptidoglycan, are for structural purposes.

starch and glycogen

The two principal energy storage polysaccharides are the starches, usually produced by plants, and glycogen, usually produced by animals. Starch is quantitatively the second most important plant product (after cellulose) and its production is exploited by man as his principal food source. Glycogen is the major energy reserve in higher animals, its biosynthesis and concentration being largely under the control of the liver in mammalian systems. Some bacterial cells produce starch (for example photosynthetic bacteria) and many others produce glycogen. Glycogen production is generally reckoned to occur in cells as a result of limiting growth conditions (for example excess carbon, little or no nitrogen). This glycogen may then be used for basal metabolism, that is to keep the cell alive during periods of carbon starvation. The intracellular concentration and turnover of glycogen are relatively low in prokaryotes and its production is regarded as a consequence of environmental conditions rather than a process occurring by choice. One exception is that *Clostridium spp.* accumulate glycogen just before sporulation and this glycogen is metabolised rapidly during spore formation.

An almost infinite number of β-linked structural polysaccharides are known. Plants produce several polysaccharides including cellulose, the single most abundant polysaccharide on earth. Man produces connective tissue and related polymers, insects produce several polymers, notably chitin - a homopolymer of N-acetylglucosamine. Bacterial polysaccharides exhibit an almost unlimited structural variation and are the principal location of unusual sugars. We have met several of these in preceding chapters. One polysaccharide which is present in all bacterial cell walls is peptidoglycan. In addition, Gram negative bacteria have an outer membrane containing polysaccharides of the 'O' chains of the lipopolysaccharides. Capsules and slime layers produced by certain bacteria are, almost without exception, polysaccharide compounds.

The majority of bacterial polysaccharides show antigenic properties and are highly specific for a given genus or even given species. We shall look at the biosynthesis of individual important polysaccharides and also the industrial implications of these compounds later in the chapter.

6.2 The biosynthesis of glucose

This section will be divided into three parts, firstly a brief study of the precursor metabolites required for glucose biosynthesis followed by a detailed study of the pathway by which glucose is synthesised and finally a study of the control mechanisms involved.

6.2.1 The production of precursor metabolites for glucose biosynthesis

precursor metabolites

The two most important precursor metabolites utilised for glucose synthesis in most living systems are pyruvate and oxaloacetate. Many organisms are heterotrophic, they require an organic carbon source for growth, and are aerobic. They will, therefore, generally use the Embden Meyerhof pathway and the TCA cycle for energy and precursor production.

SAQ 6.1

Study Figure 6.1. It is a summary diagram of the Embden Meyerhof pathway and the TCA cycle. Imagine that an organism no longer has access to glucose from its environment, in fact it has a requirement for glucose. Can you from your previous studies indicate at least one source for the intermediates marked with a question mark?

It is worth spending a little time comparing your answer to the one at the end of the book.

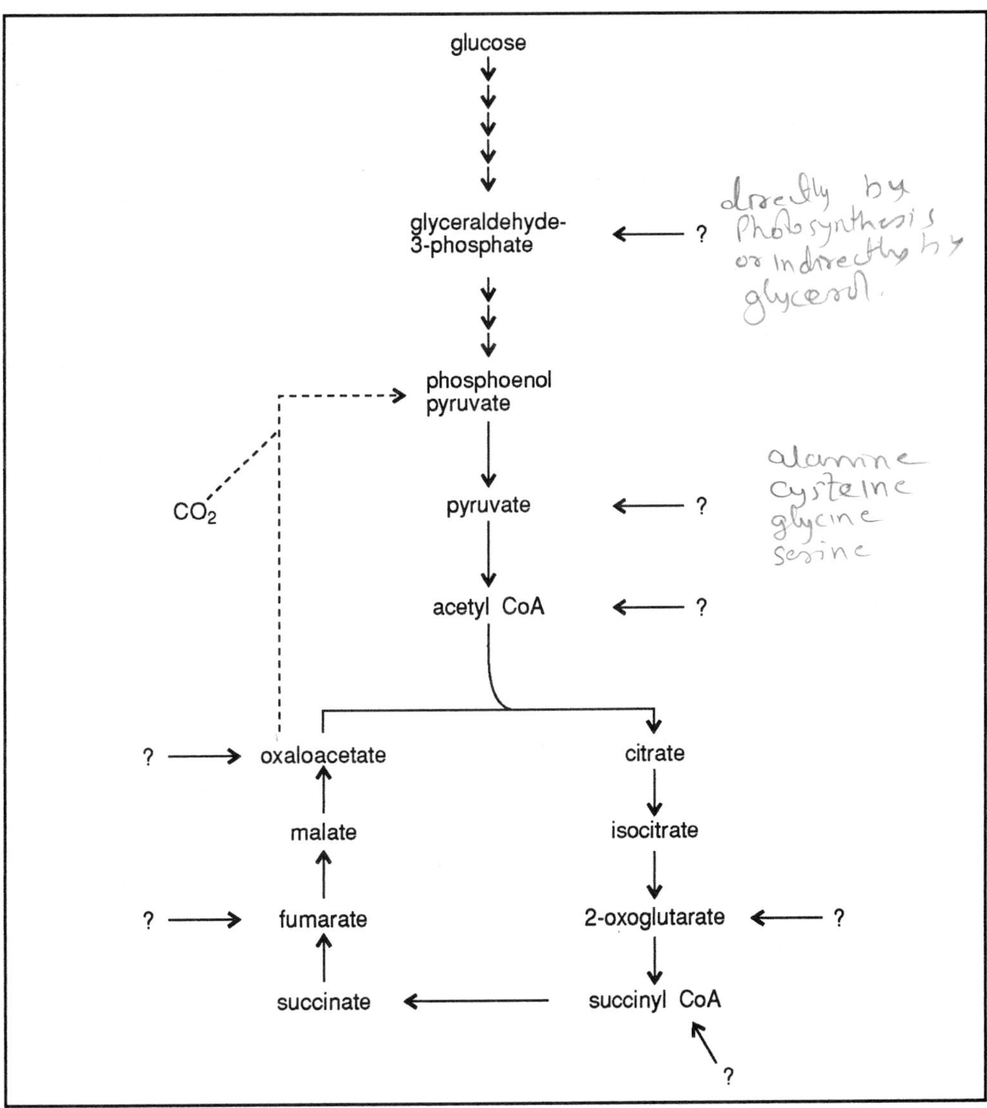

Figure 6.1 The Embden Meyerhof pathway and TCA cycle (see SAQ 6.1).

Π What is the source of glycerol in the answer to SAQ 6.1?

Remember that glycerol is present in many lipids, for instance triglycerides and phospholipids. Higher animals store and utilise triglycerides as a principal energy source and thus their turnover of such compounds is high.

SAQ 6.2

Which of the compounds listed in the answer to SAQ 6.1 could be used for glucose biosynthesis in higher animals and which could not? Explain your answer.

Compare your answer carefully with that at the end of the book. If you do not recall the reasons why acetyl CoA cannot be used as a carbon source in higher animals, you should refer to Chapter 7 in Biotol text, 'Principles of Cell Energetics'. Remember that plants and many bacteria can utilise fatty acids or acetate as a source of glucose because they contain the glyoxylate cycle enzymes, isocitrate lyase and malate synthase, which effectively allows them to produce a four-carbon succinyl derivative from two acetyl molecules.

Several bacteria will grow on one-carbon organic compounds as sole source of carbon and energy and after an initial rather rare sequence of reactions in which they produce a three-carbon derivative from three one-carbon compounds, their glucose production is quite conventional. The three-carbon compound is produced in a cyclical manner.

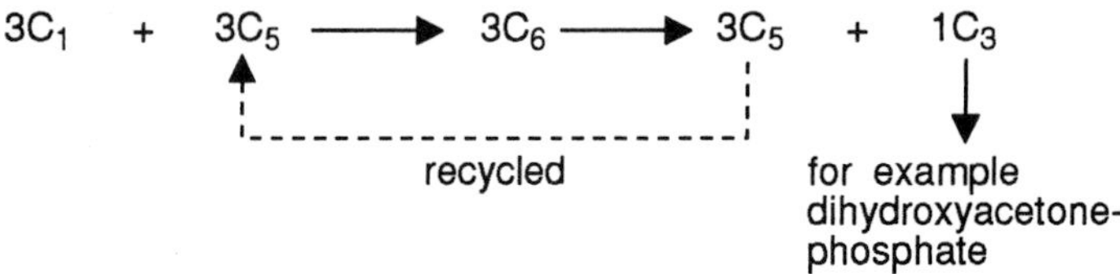

In summary for this section:

- virtually all organisms can utilise for glucose biosynthesis any compound which on degradation yields an intermediate of either the TCA cycle or the Embden Meyerhof pathway;

- compounds producing acetyl CoA can only be used by organisms which have the glyoxylate cycle enzymes, for example plants, most fungi, many protozoa and bacteria. Thus higher animals can produce fatty acids from glucose but not glucose from fatty acids;

- organisms growing on organic one-carbon compounds initially produce a three-carbon intermediate of the Embden Meyerhof pathway by a process virtually unique to their metabolism;

- although we have not stressed it in this chapter, you should remember that photosynthetic organisms utilise carbon dioxide and convert three molecules of this compound to a three-carbon Embden Meyerhof pathway intermediate.

6.2.2 The biosynthesis of glucose from pyruvate and oxaloacetate

Embden Meyerhof revisited

We noted earlier that seven of the ten reactions of the Embden Meyerhof pathway are common to the process of gluconeogenesis. Thus for us to study gluconeogenesis adequately we should first revise the Embden Meyerhof pathway.

SAQ 6.3

Study Figure 6.2. It is a partially completed diagram of the Embden Meyerhof pathway. Can you:

1) fill in the missing intermediates.

2) add the cofactors involved.

3) derive an overall reaction for the process.

4) have a try at naming the enzymes, at least the type of enzyme.

5) carefully label the arrows as either one way or bidirectional?

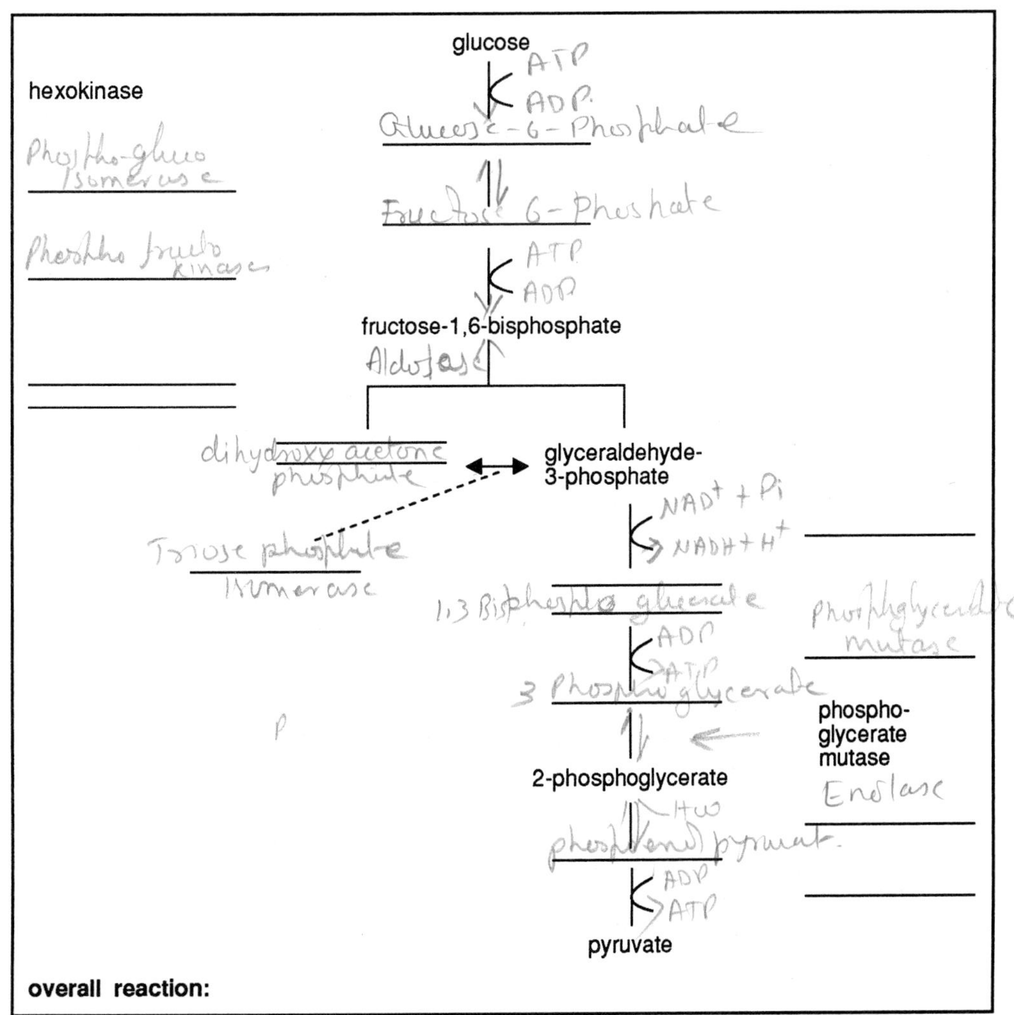

Figure 6.2 The Embden Meyerhof pathway (see SAQ 6.3).

Most of the intermediates of the pathway should now be almost second nature to you and recognition of enzyme types, if not precise names, relatively easy. Note carefully that three reactions, the first, third and the last were not readily reversible. Finally, if the overall equation is causing concern, remember that the dihydroxyacetone phosphate is converted to a second molecule of glyceraldehyde-3-phosphate which is then converted to a second pyruvate. This has the effect of producing the extra NADH + H^+ and two ATP molecules.

Let us now turn our attention to gluconeogenesis. Figure 6.3 shows the pathway and the three asterisks show the stages where gluconeogenesis is at variance with the Embden Meyerhof pathway. Let us examine these in more detail.

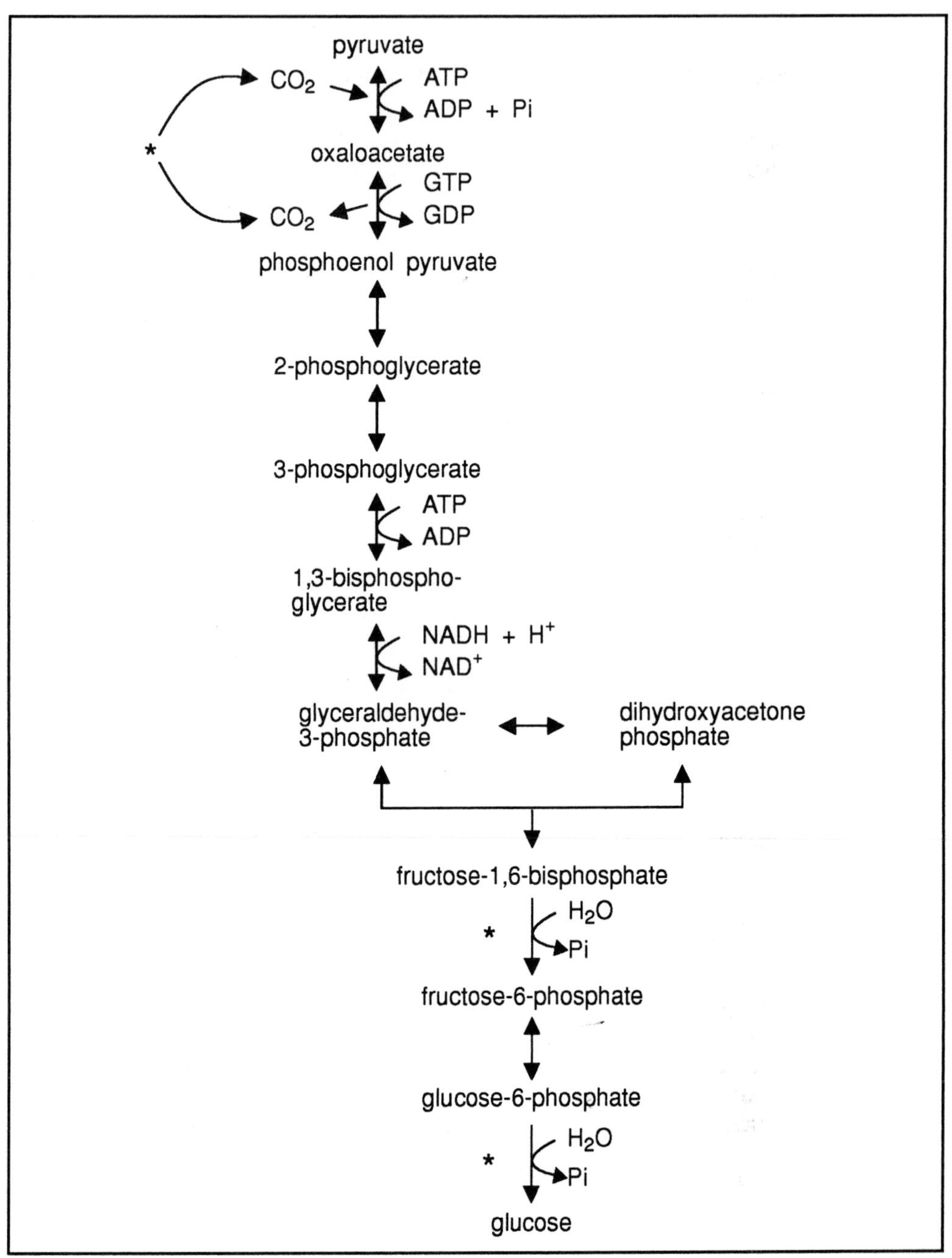

Figure 6.3 Gluconeogenesis.

Pyruvate is converted to phosphoenol pyruvate by a two-stage reaction via oxaloacetate. Direct phosphorylation of pyruvate to yield phosphoenol pyruvate (the reversal of the Embden Meyerhof reaction) is not feasible thermodynamically so the more favourable two-stage reaction is required, thus:

$$CO_2 + CH_3{-}\overset{\displaystyle O}{\overset{\|}{C}}{-}COO^- \;\underset{(1)}{\overset{ATP \quad ADP + Pi}{\longleftrightarrow}}\; {}^-OOC{-}CH_2{-}\overset{\displaystyle O}{\overset{\|}{C}}{-}COO^- \;\underset{(2)}{\overset{GTP \quad GDP}{\longleftrightarrow}}\; CO_2 + CH_2{=}C(COO^-){-}O{-}\overset{\displaystyle O}{\overset{\|}{P}}(O^-){-}O^-$$

GTP, remember, is guanosine triphosphate, an alternative purine triphosphate to adenosine triphosphate.

Π Can you name the two enzymes above?

Enzyme (1) is pyruvate carboxylase and enzyme (2), more difficult to name, is phosphoenol pyruvate carboxykinase. Both reactions are reversible. Putting the two reactions together gives:

$$\text{pyruvate} + ATP + GTP \quad \leftrightarrow \quad \text{phosphoenol pyruvate} + ADP + GDP + Pi$$

In some plants, eukaryotic micro-organisms and bacteria an alternative, direct reaction can occur catalysed by the enzyme pyruvate orthophosphate dikinase. Here two high energy phosphate bonds are required simultaneously hence ATP is converted to AMP + PPi.

$$CH_3{-}\overset{\displaystyle O}{\overset{\|}{C}}{-}COO^- + Pi \;\xrightarrow{ATP \quad AMP + PPi}\; CH_2{=}C(COO^-){-}O{-}\overset{\displaystyle O}{\overset{\|}{P}}(O^-){-}O^-$$

The second of the three specific reactions of gluconoegenesis is the conversion of fructose-1, 6-bisphosphate to fructose-6-phosphate. The enzyme in gluconeogenesis is the cytosolic fructose bisphosphatase, or hexose bisphosphatase, as it is not absolutely specific for fructose, and it catalyses the irreversible reaction:

$$\text{fructose-1, 6-bisphosphate} + H_2O \rightarrow \text{fructose-6-phosphate} + Pi$$

Note that this essentially represents a loss of a high energy phosphate bond as inorganic phosphate is produced.

Π Can you remember the enzyme carrying out the reverse reaction in the Embden Meyerhof pathway and what was special about it?

phospho-fructokinase

The reaction was carried out by phosphofructokinase which is the only enzyme unique to the Embden Meyerhof pathway and is a major metabolic control point.

Finally, the conversion by glucose-6-phosphate to glucose is not carried out by hexokinase in reverse but by another phosphatase, glucose-6-phosphatase.

$$\text{glucose-6-phosphate} + H_2O \rightarrow \text{glucose} + \text{Pi}$$

We have included this final reaction because it does sometimes occur. However, much of the glucose-6-phosphate is used directly to make other monosaccharides or polysaccharides.

Π Can you now derive an overall reaction for gluconeogenesis from two pyruvate to glucose?

Your answer should be:

$$2 \text{ pyruvate} + 4 \text{ ATP} + 2 \text{ GTP} + 2 \text{ NADH} + 2 H^+ + 2 H_2O$$

$$\rightarrow \text{glucose} + 4 \text{ ADP} + 2 \text{ GDP} + 6 \text{ Pi} + 2 \text{ NAD}^+$$

The six high energy phosphate bonds are required together with the loss of 2NADH + $2H^+$ molecules which could yield six further ATP molecules by the electron transport chain and oxidative phosphorylation. The requirement for six high energy phosphate bonds compared to a net gain of two in the Embden Meyerhof pathway is the price paid for having to circumvent the energetically unfavourable pyruvate to phosphoenol pyruvate stage.

6.2.3 The subcellular locations of the enzymes of gluconeogenesis and the control of gluconeogenesis versus glycolysis

location of pyruvate carboxylase

Gluconeogenesis, like the Embden Meyerhof pathway, is known to occur in the cytosol of all living systems. However, a problem arises in eukaryotes because pyruvate carboxylase is mitochondrial. This enzyme catalyses the major anaplerotic reaction of the TCA cycle which is the production of oxaloacetate from pyruvate.

Pyruvate, if produced in the cytosol, will readily diffuse into the mitochondrion, in this case for conversion to oxaloacetate. Oxaloacetate, however, is transported out of the mitochondrion as malate. A mitochondrial malate dehydrogenase reduces oxaloacetate to malate which passes out of the mitochondrion and is oxidised back to oxaloacetate by a malate dehydrogenase situated on the ER. The process can be depicted schematically as follows:

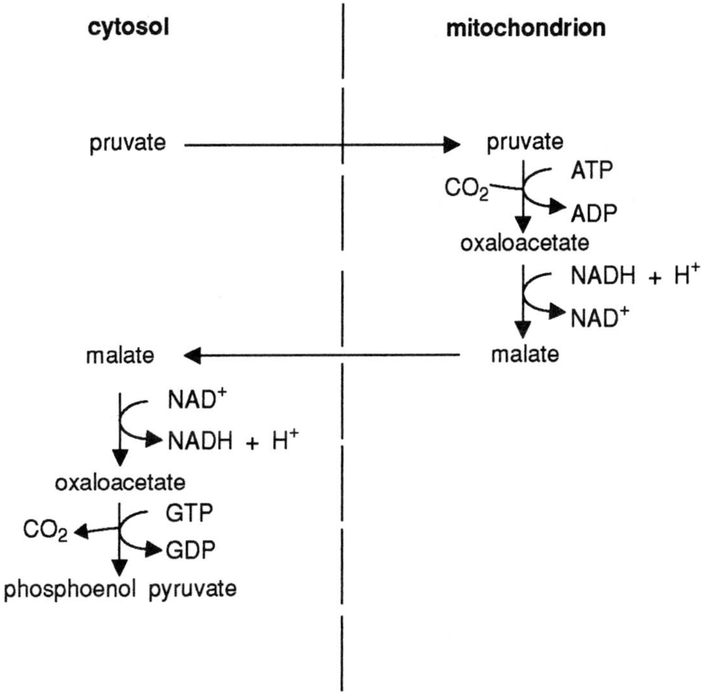

We are now in a position to carry out a detailed look at the problem of control of gluconeogenesis versus the Embden Meyerhof pathway of glycolysis.

Π At which stages of each pathway could you expect the two pathways to be controlled and why?

The answer is the first and the last stage of each pathway (thus the first and the last in one direction is the last and the first respectively of the reverse direction). The first stage is the obvious one to control as it avoids unnecessary build up of unwanted intermediates. In these two pathways phosphofructokinase is an obvious control point. It is the only enzyme unique to one of the pathways so if this enzyme is affected then only that pathway in which it occurs is directly controlled.

Π Can you suggest which compounds might affect control?

In broad terms gluconeogenesis occurs when the ATP concentration is high, pyruvate levels are high and perhaps the glucose levels are low. Alternatively the Embden Meyerhof would be required when energy or intermediates are required. Thus the prime factors which may be expected to influence the enzymes are the concentrations of ATP (therefore of ADP and AMP), glucose and acetyl CoA.

energy charge

There is a relatively constant amount of ATP + ADP + AMP in each cell. It is the relative proportions of these which alter dramatically. If, for instance, the ATP concentration increases, one or both of the other two together must show a concomitant decrease. The extent to which all the adenosine groups are 'filled' with high energy phosphates, that is all the adenosine is present as ATP (an unlikely occurrence) gives rise to the energy charge or adenylate charge of the cell: when all is present as ATP then the charge is 1.0, which is maximal. If AMP were the only molecule present then the energy charge would be zero. The value of the energy charge can be calculated using the simple equation:

$$\text{energy charge} = \frac{[\text{ATP}] + 0.5\ [\text{ADP}]}{[\text{ATP}] + [\text{ADP}] + [\text{AMP}]}$$

As a generalisation energy charges of 0.75 to 0.9 are normal. It has long been known that the energy charge is an important controller of certain allosteric enzymes including some of those we are considering here. This topic is covered in more detail in the following chapter.

Let us check that you understand when the two pathways will be used.

SAQ 6.4

Below are a number of cellular situations. Which of them would tend to lead to the occurrence of the Embden Meyerhof pathway and which to gluconeogenesis?

1) Energy charge of the cell is high.

2) Energy charge of the cell is low.

3) Acetyl CoA concentration is high.

4) TCA cycle intermediates present at low concentration.

5) Glucose is readily available.

At this point study Figure 6.4. This figure, perhaps a little bit complicated, tries to show firstly the differences between the two processes, secondly the key enzymes involved in these differences and finally the control influences exerted on the key enzymes. Study the diagram for a few minutes before proceeding to the following explanation.

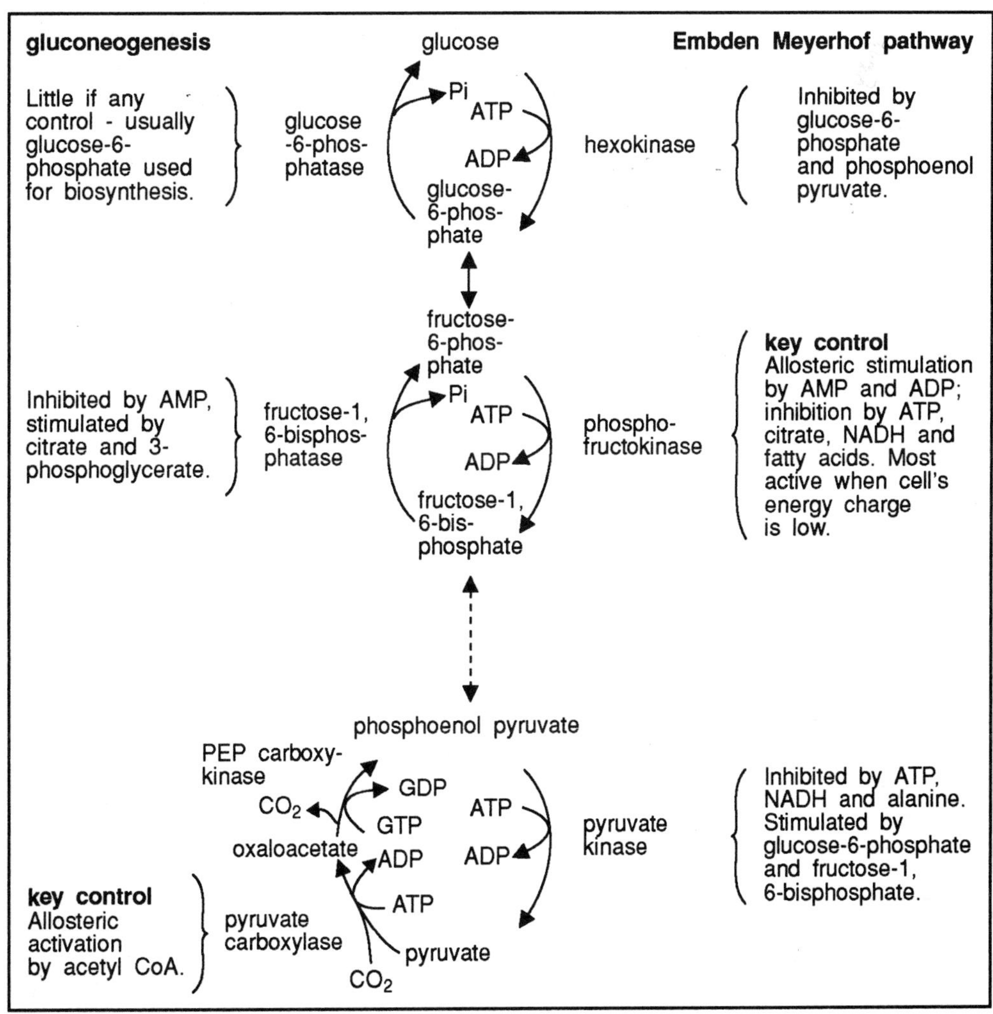

Figure 6.4 Diagram to indicate control points and mechanisms for gluconeogenesis and the Embden Meyerhof pathway.

With the exception of glucose-6-phosphatase all of the enzymes which occur in only one of the two processes take part in control. The key controlling factor of the Embden Meyerhof pathway is phosphofructokinase. This is a good choice because it occurs early on in the pathway and is unique to the Embden Meyerhof pathway. Phosphofructokinase is stimulated by the presence of raised concentrations of AMP and/or ADP but inhibited by increased concentrations of ATP. It is also inhibited by increased concentrations of citrate, fatty acids and NADH + H^+, all compounds which in high concentration would indicate that a cell is rich in intermediates and has a high energy charge. In summary this enzyme is most active when the energy charge is low, least active when the energy charge is high. Note that the somewhat unusual properties of substrate inhibition (ATP) and product stimulation (ADP) are exhibited by this enzyme.

role of phospho-fructokinase

The key controlling effect of gluconeogenesis is exerted by the first enzyme of the pathway, pyruvate carboxylase. This enzyme is allosterically activated by acetyl CoA. This is a complicated but very logical control method.

Effectively the acetyl CoA concentration controls the level of oxaloacetate in the cell: high acetyl CoA encourages oxaloacetate production. Oxaloacetate is required for gluconeogenesis but is also a key intermediate of the TCA cycle.

Π What do you think will decide which of the two pathways oxaloacetate enters?

The answer is mainly the adenylate charge. If ATP concentration is high the need for energy from the TCA cycle is lessened thus gluconeogenesis will be favoured. If ATP levels are low then the TCA cycle will be favoured.

The control pressures exerted on the other enzymes in Figure 6.4 follow the general trend and it will pay dividends to mentally follow the inhibitions and check that you understand the logic behind each process.

summary of control

In summary, when the cell has ample ATP and fuel molecules such as acetyl CoA, citrate and NADH + H^+ available, gluconeogenesis is promoted and the Embden Meyerhof pathway is inhibited. The converse is true when the energy charge is low and the concentration of fuel molecules is low.

futile cycles

One interesting metabolic problem you may have heard of is the presence of so called 'futile metabolic cycles'. Take as an example one pair of reactions from Figure 6.4:

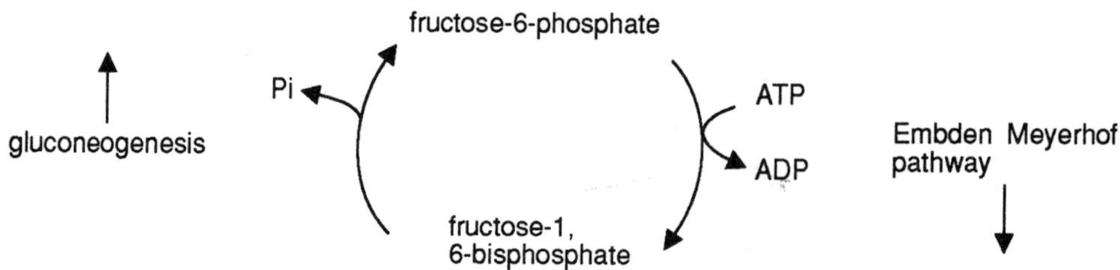

The two equations are:

$$\text{fructose-6-phosphate} + \text{ATP} \longrightarrow \text{fructose-1,6-bisphosphate} + \text{ADP}$$
$$\text{fructose-1,6-bisphosphate} + H_2O \longrightarrow \text{fructose-6-phosphate} + \text{Pi}$$

summed:

$$\text{ATP} + H_2O \longrightarrow \text{ADP} + \text{Pi} \text{ !}$$

In mammalian systems it has been shown that gluconeogenesis and the Embden Meyerhof pathway can occur simultaneously. This leads to the question - could this process - a 'futile cycle' be happening? The answer is probably yes: the reasons suggested include:

- a biological imperfection leading to inefficiency;
- a device to 'burn-off' excess ATP;
- a means (in mammals) of generating heat.

6.3 The biosynthesis of hexoses, pentoses and tetroses

The most important hexose sugar is glucose though several others, notably fructose, galactose and mannose are very widespread. In addition other sugars, for example rhamnose, have a more restricted distribution and others are possibly unique to a given species of bacterium.

sugar interconversions

Sometimes sugars require attachment to a carrier molecule before interconversion can occur, the usual carrier being uridine diphosphate (UDP) or occasionally thymidine diphosphate (TDP). Figure 6.5 shows a simplified diagram detailing the production of hexoses from gluconeogenic intermediates. Many more intermediates are found in living systems particularly as the diagram includes examples of oxidised sugars (glucuronic acid) and substituted sugars (glucosamine). The purpose of the diagram is to indicate that living systems generally can produce hexose derivatives with relative ease from the central glucose and fructose precursors. Note the key ('Central') position of fructose-6-phosphate.

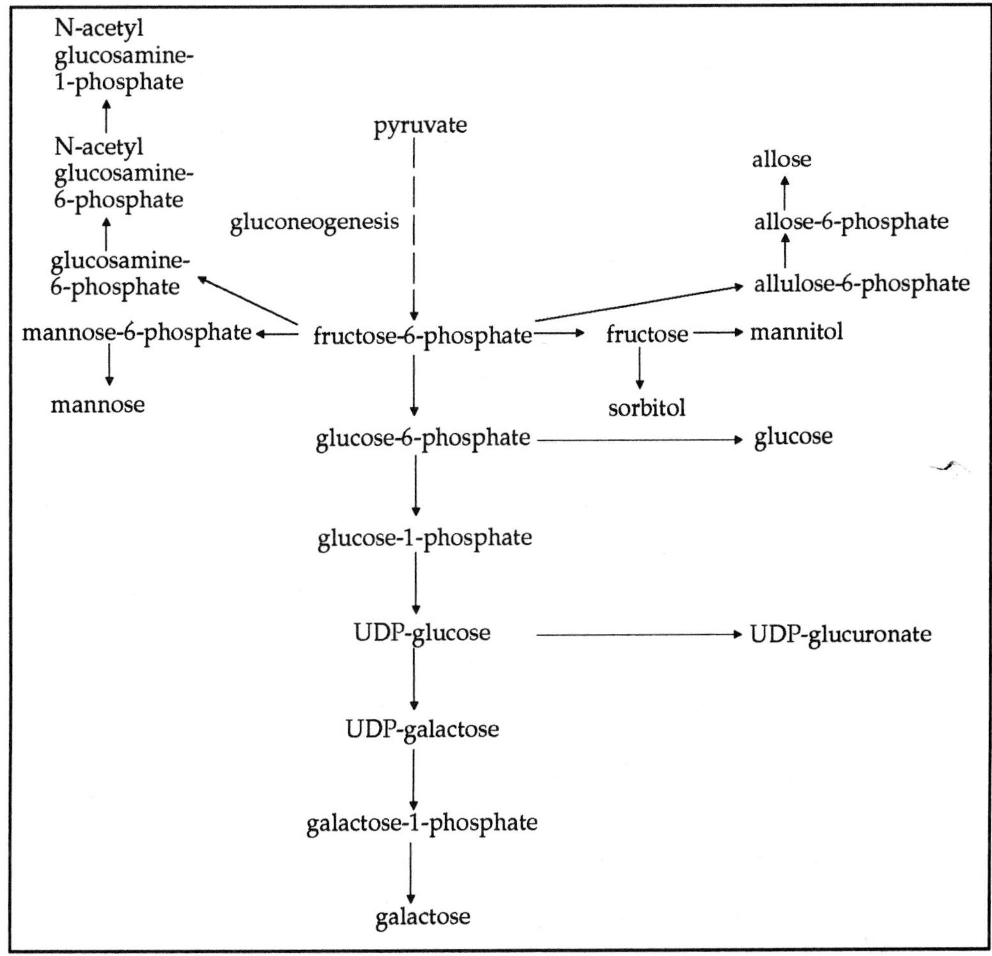

Figure 6.5 The biosynthesis of hexoses, hexitols and other carbohydrate derivatives from gluconeogenic precursors.

Pentose sugars are produced generally by the pentose phosphate pathway (as are tetroses) which yields ribose-5-phosphate, ribulose-5-phosphate and xylulose-5-phosphate following oxidative decarboxylation of glucose. As with hexoses, interconversion of pentoses is readily achieved with the central intermediate appearing to be xylulose-5-phosphate.

The production of deoxysugars, particularly deoxyribose, is probably via one of two mechanisms. The first, demonstrated in *E. coli,* reduces the ribose of ADP, UDP, GDP or CDP, the hydrogens being donated by NADPH + H^+ via an intermediate carrier - thioredoxin. Other organisms are known to use the ribonucleoside triphosphates for reduction, a process in which vitamin B_{12} is involved.

N-acetyl muramic acid

As a last example we can examine the production of a rather special hexose derivative, N-acetyl muramic acid. This compound is a component of the peptidoglycan moiety of bacterial cell walls. All bacteria, with the exception of *Mycoplasmas,* contain this peptidoglycan wall component which is unique to prokaryotes. The uniqueness is bestowed by the monosaccharide N-acetylmuramic acid which is synthesised from the ubiquitous N-acetylglucosamine as follows:

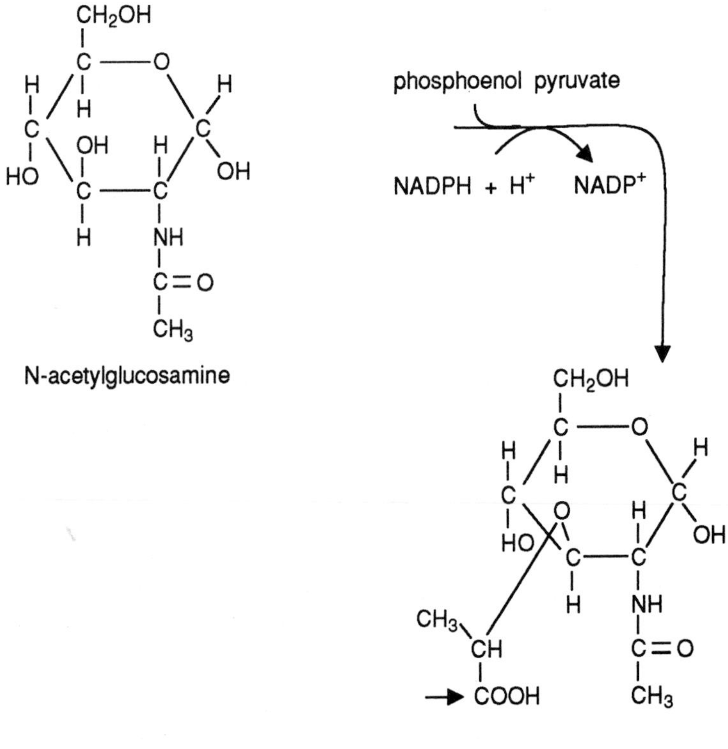

N-acetylglucosamine

phosphoenol pyruvate

NADPH + H^+ → $NADP^+$

N-acetylmuramic acid

It is the insertion of the three-carbon lactyl group onto the carbon three of N-acetyl glucosamine which gives this molecule its singular role and location in nature - as a structural component of bacterial cell walls. This is because the binding of the peptide to the glycan occurs through the lactyl carboxyl group. Without it there would be no link between these two parts of the wall.

6.4 The biosynthesis of disaccharides, oligosaccharides and polysaccharides

Due to the complexity and diversity of polysaccharides it would be quite easy to produce an extremely long list of different types of polysaccharide biosyntheses without really doing this topic justice or achieving an overall understanding. Thus the section will start with an explanation of the basic problems involved in linking monosaccharides to each other followed by a general statement incorporating a few important, selected examples.

Firstly let us remind ourselves of the basic structure of glucose in its pyranose ring form.

Π Can you draw the pyranose ring form of α glucose indicating the number of each carbon atom?

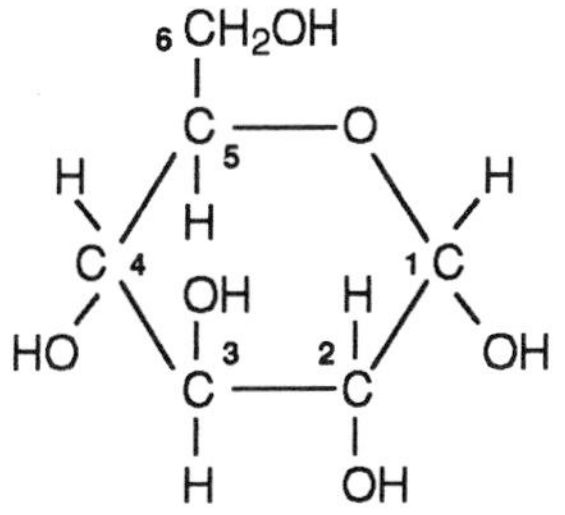

Remember that the position of the OH groups, whether above or below the plane of the ring, is important. For example alter the position of the OH on carbon four and the molecule becomes galactose.

Substituents may be added at the hydroxyl groups of carbon one, carbon two, carbon three, carbon four and carbon six and the sugar is capable of binding to one or more monosaccharide groups. Do not forget that the OH of carbon one can be below the plane or above the plane; the compound above is α-glucose, if the OH was above the plane of the ring it would be β-glucose.

When two monosaccharides link together it is by a condensation reaction and usually involves the carbon one of one monosaccharide with the carbon two, carbon three, carbon four or carbon six of a second residue. This link is a glycoside bond. Most polysaccharides contain C_1-C_4 linkages allowing the formation of long, straight chain polymers. Let us look at the structure of a disaccharide using as our example two glucose residues linked together. The two glucose molecules may be linked by either an α or a β link between carbon one and carbon two, three, four or six of the second molecule. Thus there are many possibilities.

cellobiose The following example shows the disaccharide unit found in cellulose. This is called cellobiose and its formula is:

β-D-glucopyranosyl-(1-4)-β-D-glucopyranose.

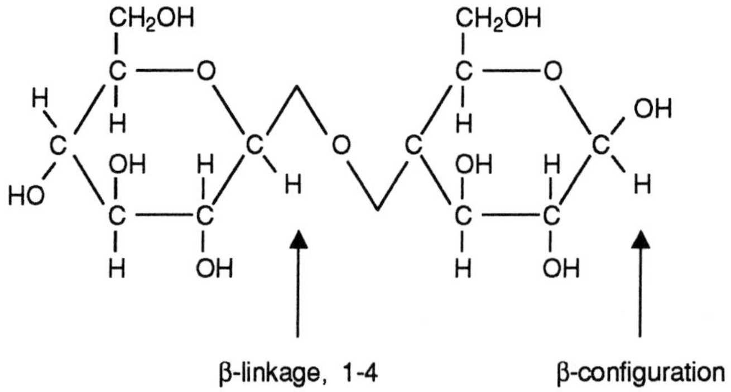

β-linkage, 1-4

β-configuration

maltose

Maltose, the disaccharide precursor of starches and glycogen, has the formula:

α-D-glucopyranosyl-(1-4)-α-D-glucopyranose

branch points

'Branch point' disaccharides within glycogen and starches may have 1-6 or even 1-2 or 1-3 bonds. To finish this section draw the disaccharide having the formula:

α-D-glucopyranosyl-(1-6)-α-D-glucopyranose

without looking at the structure shown below

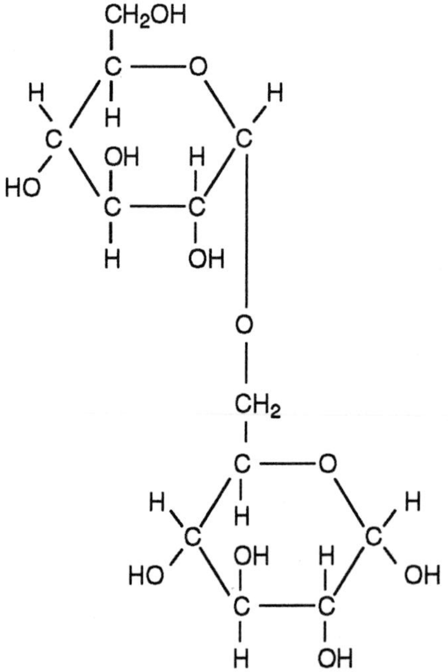

We now need to study the way in which individual monosaccharides are put together. The condensation reactions obviously require energy in some form but as we shall see it is not simply a matter of the following occurring:

2 glucose + ATP $\rightarrow$ a disaccharide + ADP + Pi

Generally speaking the sugar needs first to be activated - this is by phosphorylation to produce glucose-1-phosphate or glucose-6-phosphate. Remember the formation of glucose-6-phosphate occurs during gluconeogenesis. During polysaccharide biosynthesis this is usually converted to the glucose-1-phosphate derivative.

Π Why do you think that this latter conversion occurs?

The reason is because the phosphate group needs to be on the carbon which is going to participate in a reaction and virtually all linkages are of the type α or β (1-X) where X represents carbon 1, 2, 3, 4 or 6 of the second molecule.

The next step is to transfer the activated sugar to a carrier molecule - usually, but not exclusively, uridine diphosphate. Let us look at the production of polymers of α-D-glucose units as our examples.

glycogen structure

In mammalian systems glycogen is the main storage polysaccharide. It is a polymer composed of mainly α-(1-4) linkages but is highly branched having α-(1-6) linkages every eight to twelve glucose residues. Glycogen production proceeds as follows:

$$\text{glucose-6-phosphate} \leftrightarrow \text{glucose-1-phosphate}$$

$$\text{glucose-1-phosphate} + \text{UTP} \rightarrow \text{UDP-glucose} + \text{PPi}$$

$$\text{UDP-glucose} + (\text{glucose})_n \rightarrow \text{UDP} + (\text{glucose})_{n+1}$$

$$\text{where } (\text{glucose})_n = \text{glycogen}$$

The primer chain $(\text{glucose})_n$ needs to be four or more glucose units long but the enzyme is more active with much longer chains. A branching enzyme is required to insert an α-(1-6) branch point when required. Bacteria also synthesise glycogen, as mentioned earlier in this chapter, but utilise ADP-glucose rather than UDP-glucose.

starch structure

Plants generally produce one of two kinds of starch, the linear α(1-4) linked amylose or the branched amylopectin - α-(1-4) with α-(1-6) branches. Plants, like bacteria, tend to use the ADP sugar derivatives.

dextrans

Yet another α-linked glucose polymer group are the dextrans which have an α-(1-6) backbone and branch points, generally α-(1-3) but they may be α-(1-2) or α-(1-4). Some of these compounds are important industrially (next section), for example the dextrans produced by *Leuconostoc dextranicum*.

Finally some *Streptococcus spp.* produce α-glucans which are largely α-(1-3) linked glucose polymers.

In summary, this is a very complex area in which we have only mentioned one type of linkage (alpha) and only a single monosaccharide (glucose). From these examples it should be apparent that the linkage is very important, that is, 1-2, 1-3, 1-4 or 1-6 because variation here gives rise to very different polymers with highly characteristic properties. Remember that all of those discussed were α-linked. This is also important in determining polysaccharide properties in that α-linked polymers are generally energy storage compounds and most organisms including humans can degrade them to yield glucose for energy and carbon production.

cellulose

A simple change to β-linkages, however, gives rise to a very different compound. The polymer having β-(1-4) linked glucose is cellulose which is a structural polysaccharide synthesised not only by plants but also by bacteria and fungi. The significance of the β-link is enormous. It results in the formation of a straight polymer, many of which line up in parallel fibres. These form the basis of the strength of wood. The biosynthesis of cellulose is of similar overall type to starch biosynthesis with UDP-glucose derivatives being used though some contributions by GDP-glucose may occur. The significance is that the β-linkage renders this molecule resistant to any digestive enzymes of mammals or most other organisms. Thus any cellulose taken in within our diet passes through the body in virtually unchanged form; it constitutes the fibre of our diet. We rely largely on bacteria and fungi for cellulose degradation and turnover of its carbon.

By looking at the way in which glucose can polymerise with other glucose units we have gained an insight into the variety and complexity of the products, and the importance of the α or β bonds.

One of the problems faced by unicellular organisms which have a cell wall or a membrane external to their plasma membrane eg Gram-negative bacteria, is the location of polysaccharide biosynthesis and the functional site of the polymer product.

Π What do you think the problem is when considering site of biosynthesis and functional location?

cell wall formation

Biosynthesis occurs within the cytosol but the functional location is on the outside of the cell wall membrane. Biosynthesis requires a plentiful supply of energy and intermediates, neither of which would be available on the outside of the cell. The theoretical answer is to synthesise the polymer on the inside of the cell and use it on the outside. There are problems here because cell membranes are delicate, continuous structures which do not posses large 'holes' for the passage of enormous molecules. Cells solve this by transporting outwards the largest possible piece of cell wall and then incorporating it into the wall at specific sites.

peptidoglycan

To study the ingenious way in which bacteria solve the location problems let us study the biosynthesis and incorporation of the peptidoglycan component of bacterial cell walls. We will first look at the overall structure before following the biosynthesis in detail. At the end of Section 6.3 we studied the production of N-acetyl muramic acid from the N-acetyl glucosamine. Both of these two sugar residues are incorporated into the peptidoglycan in a very regular alternating chain all linked β-(1-4). This is rather like producing long pieces of string but obviously more is required to give some rigidity to the overall structure. Small tetrapeptide chains are linked to each N-acetyl muramic acid and these tetrapeptide chains cross link to each other to yield the complete structure which can be regarded as net-like. Gram positive and Gram negative cells do have profound differences in overall structure but the principle of peptidoglycan synthesis is similar. Much of the early work was carried out on *Staphylococcus aureus* which has a repeating disaccharide-tetrapeptide unit in its peptidoglycan which has the structure shown in Figure 6.6.

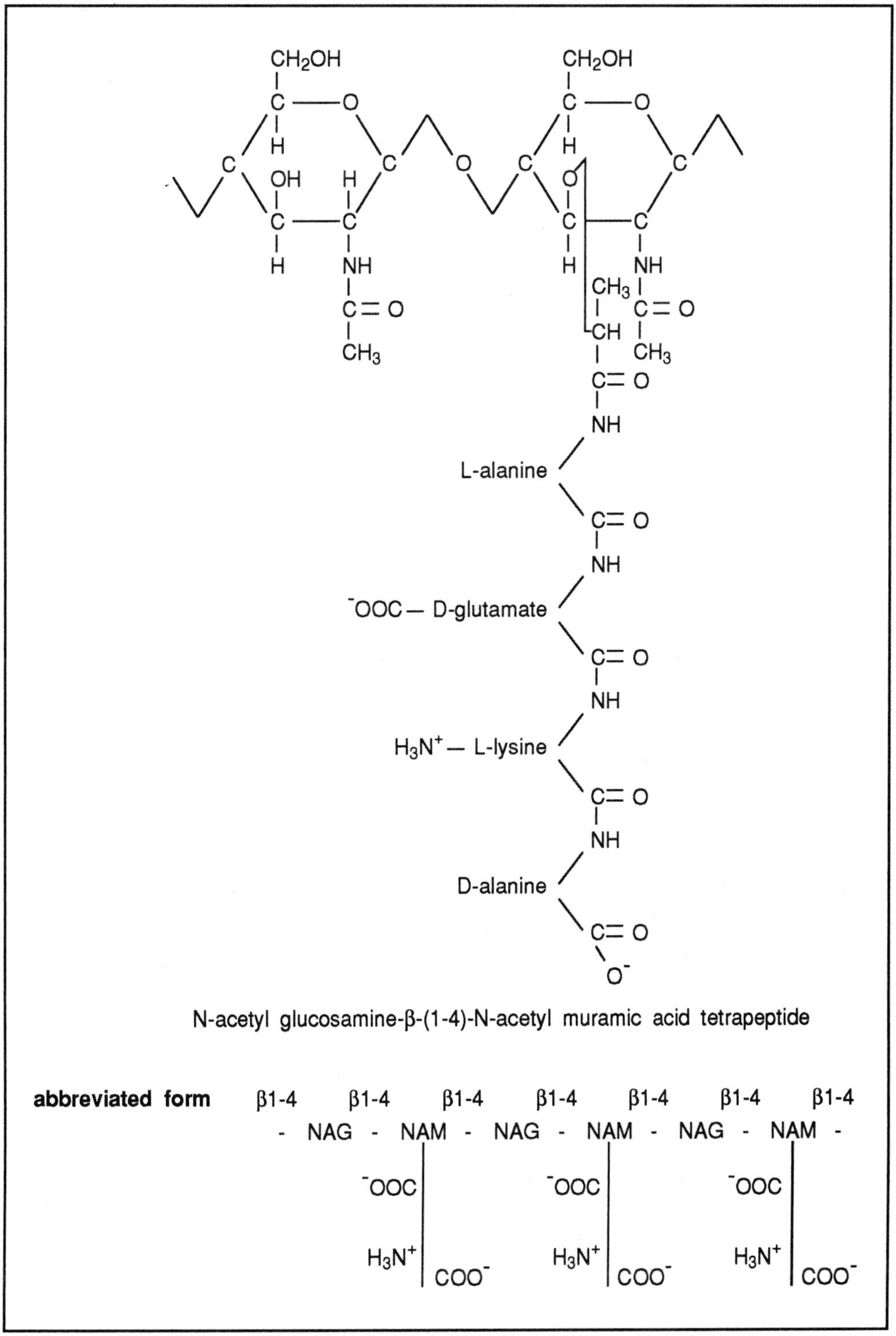

CH_2OH CH_2OH

NH NH

$C=O$ $C=O$

CH_3 CH_3

CH_3 — CH — $C=O$

NH

L-alanine

$C=O$

NH

^-OOC — D-glutamate

$C=O$

NH

H_3N^+ — L-lysine

$C=O$

NH

D-alanine

$C=O$

O^-

N-acetyl glucosamine-β-(1-4)-N-acetyl muramic acid tetrapeptide

abbreviated form β1-4 β1-4 β1-4 β1-4 β1-4 β1-4 β1-4

\- NAG - NAM - NAG - NAM - NAG - NAM -

^-OOC ^-OOC ^-OOC

H_3N^+ COO^- H_3N^+ COO^- H_3N^+ COO^-

Figure 6.6 The repeating disaccharide tetrapeptide unit of *Staphylococcus aureus* peptidoglycan. NAG = N-acetylglucosamine ; NAM =N-acetylmuramic acid.

Each of the four amino acids has an amino and a carboxyl group and these enter into a series of peptide bonds joining to the -COOH of the N-acetyl muramic acid. There is, therefore, a spare carboxyl group at the end of the chain (D-alanine) and in addition an extra amino group on the L-lysine (a diamino acid). The abbreviated form will help us to examine how these long chains containing regular repeating disaccharide tetrapeptide units are incorporated into the net-like structure of the completed wall.

In a completed cell wall the terminal carboxyl of the tetrapeptide is linked (via a special bridge, see below) with the spare amino group of the lysine of an adjacent peptidoglycan chain - a process called cross-linking. Each tetrapeptide, therefore, cross links twice in Gram positive organisms, once to its neighbour on the 'left' and once to its neighbour on the 'right'. Cross linking is achieved by the insertion of a bridging peptide. In the case of *S. aureus* the bridge is a pentaglycine molecule, also formed by peptide bonds. Study the diagram in Figure 6.7 which shows these linkages in simplified form.

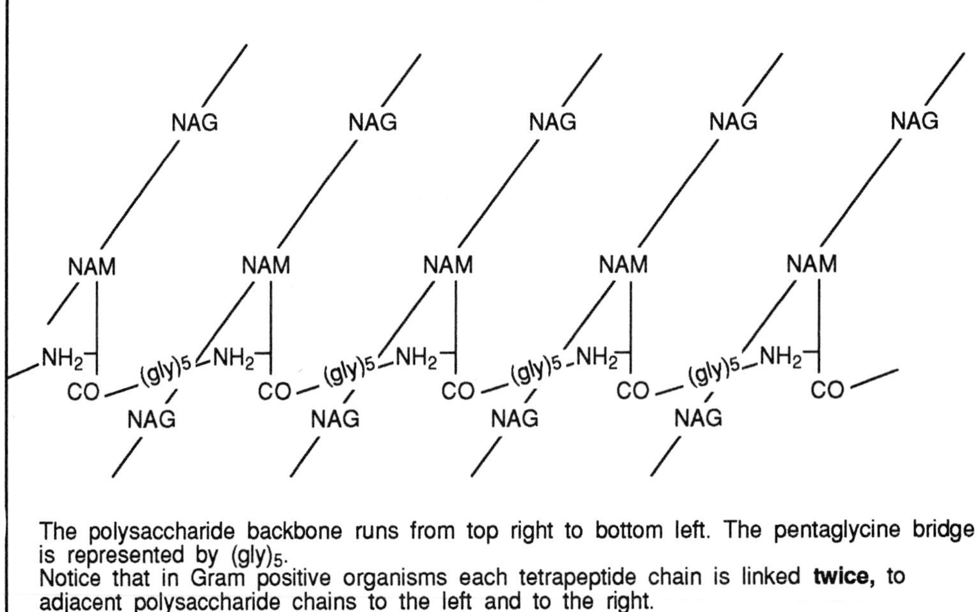

The polysaccharide backbone runs from top right to bottom left. The pentaglycine bridge is represented by $(gly)_5$.
Notice that in Gram positive organisms each tetrapeptide chain is linked **twice,** to adjacent polysaccharide chains to the left and to the right.

Figure 6.7 The peptidoglycan structure of *Staphylococcus aureus.*

Let us now consider the biosynthesis of this massive structure which may represent up to 50% of the dry weight of the cell.

The process starts with N-acetyl glucosamine which is activated by phosphorylation to form N-acetyl glucosamine-1-phosphate. Half of this is used to synthesis N-acetyl muramic acid, an outline of which was given at the end of section 6.3. Thus:

N-acetyl glucosamine-1-phosphate
↓ UTP → PPi
UDP-N-acetyl glucosamine-1-phosphate
↓ phosphoenol-pyruvate → Pi; NADPH + H^+ → $NADP^+$
UDP-N-acetyl muramic acid

The amino acids L-alanine, D-glutamate and L-lysine are added one by one and then D-alanine dipeptide is added to give a pentapeptide chain. An explanation for the extra alanine will be given shortly. Thus:

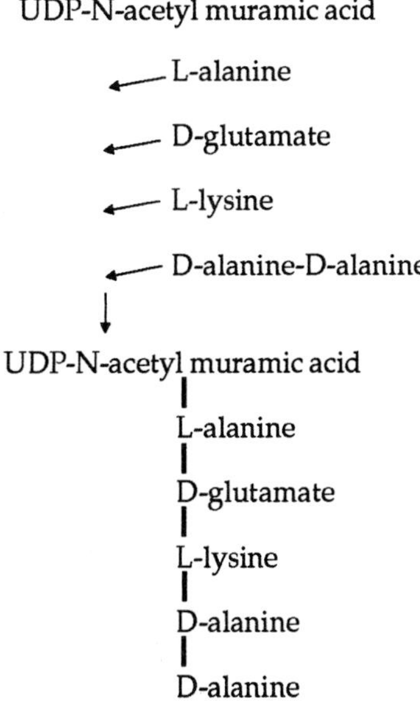

bactoprenol revisited

Before continuing we must digress slightly and comment on a compound we discussed in the last chapter, bactoprenol.

Π What can you remember about the above compound?

Firstly, it is a lipid. More specifically it is one of the isoprenoid derivatives, and in particular it is the fifty five-carbon compound with the chemical name of undecaprenyl alcohol. This compound acts as a specific transporter of the cell wall precursor molecule across the cell membrane.

The reaction proceeds as follows with the bactoprenol positioned initially pointing to the inside of the membrane and in a phosphorylated form.

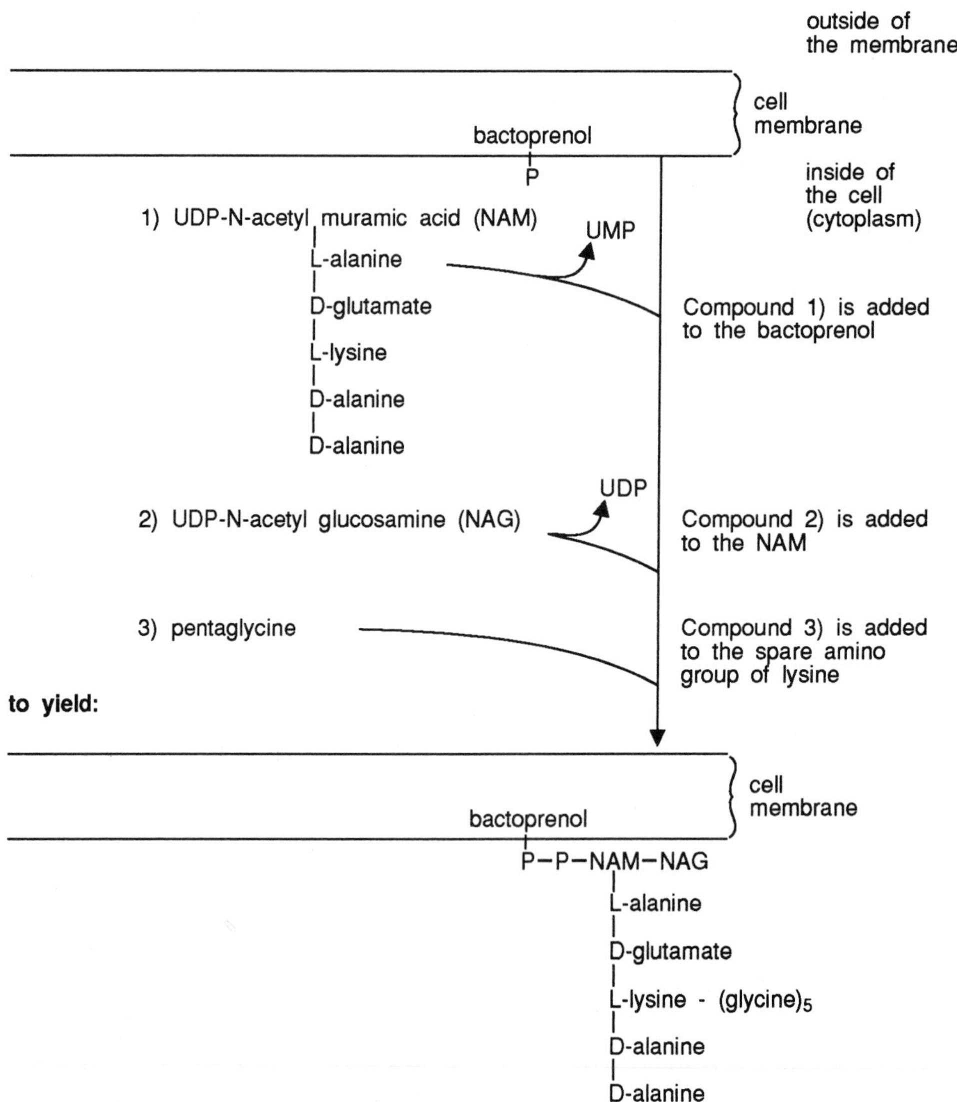

The joining of the repeating disaccharide to bactoprenol renders the whole molecule hydrophobic and it can now move through the lipid bilayer of the membrane and protrude on the outside of the membrane.

Π What two processes are required to fit the disaccharide hexapeptide into the cell wall?

The first is to link the NAM to a NAG at the end of a NAG-NAM-NAG chain and the second is to cross link the peptide chains to produce the net structure.

Π Can you think where the energy of these energy-requiring steps comes from?

bactoprenol phosphate-NAM bond and D-alanine link broken to release energy

The energy for incorporation of the NAM-NAG hexapeptide to another NAG is derived from the breaking of the bactoprenol phosphate-NAM bond. The energy to cross link the peptide chains is obtained in a novel way. The cross linking reaction is between the glycine of the pentaglycine bridge and the D-alanine in position four of the chain. The D-alanine at position five has been attached to provide the energy source for binding the pentaglycine bridge. The terminal peptide bond is hydrolysed and the energy liberated is conserved perfectly to drive the cross linking reaction. It would be useful at this point to refer back to Figure 6.7 which shows the cross links in place. The D-alanine is not lost as it can be absorbed back into the cell and re-used.

SAQ 6.5

The following eight steps summarise the synthesis and incorporation of the repeating unit of peptidoglycan. Put them into sequence.

1) Addition of pentaglycine, for bridging purposes.

2) Addition of a second sugar residue.

3) Biological activation of the monosaccharide (phosphorylation).

4) Incorporation by a two-stage process into the growing cell wall.

5) Selective step-wise additions of five amino acids - four necessary for cell wall structure and one to provide extracellular 'energy'.

6) Transfer of the monosaccharide-pentapeptide to the specific transporter bactoprenol prior to transport across the membrane.

7) Transfer to a UDP carrier molecule.

8) Transport across the membrane.

The mechanism of biosynthesis of peptidoglycan is extremely important to us medically because the fact that peptidoglycan structures are unique to bacterial cell wall allows us to make use of antibiotics which specifically interfere with cell wall biosynthesis. For example bacitracin interferes with the action of the bactoprenol; ristocetin and vancomycin inhibit incorporation of the transported molecule into the growing cell wall and finally penicillins and cephalosporins inhibit cross linking of the peptide chain.

teichoic acids

Teichoic acids, compounds containing glycerol molecules, N-acetyl glucosamine and polyribitol phosphate, are synthesised and transported out across the membrane in similar fashion via the isoprenoid phosphate carrier system. Teichoic acids are highly charged molecules largely conferring the overall negative charge to the cell.

The lipopolysaccharide (LPS) molecules of Gram negative cells are extremely diverse and complicated, often being characteristic of or unique to a bacterial species or genus. Their overall biosynthesis and trans-membrane relocation occurs in similar manner to the methods described for peptidoglycan.

SAQ 6.6

Decide whether the following statements are true or false.

1) Plants but not animals can convert fats to glucose.

2) β-linked polysaccharides are used mainly for the storage of energy while α-linked compounds have structural functions.

3) Certain bacteria can utilise three one-carbon compounds to make a three-carbon compound from which glucose can be synthesised.

4) Phosphofructokinase is the only enzyme unique to the pathway of gluconeogenesis and it is a major metabolic control point in that pathway.

5) Cellobiose is the disaccharide building unit of cellulose.

6.5 The commercial importance of polysaccharides of micro-organisms

The compounds belonging to the starches and glycogen group will not be considered further than this early mention: their importance, largely as food material for carbon and energy supply, is well known and has been discussed frequently during earlier chapters of this book.

vaccines

Many of the remaining polysaccharides, particularly β-linked structural polysaccharides, occur at or near to the outside of the cell, generally external to the cell membrane. Thus, in addition to any chemical properties they have which we may exploit, we have to consider them from an immunological point of view. Most bacterial polysaccharides are antigenic and some are used to help us identify bacteria. Others are actually produced commercially for their ability to thicken or gel aqueous solutions. In addition they may act as emulsifiers and stabilisers and, more recently, their use to enhance or alter the flavour and colour of foods has increased.

There are of course many polysaccharides from plant origin such as the various gums, cellulosic derivatives and agars but we shall concentrate on compounds of microbial origin.

uses of Xanthan gum

Xanthan gum is produced as an extracellular polysaccharide by *Xanthomonas campestris*. It is composed largely of glucose units linked β-(1-4) but contains mannose (two units) and glucuronic acid in a three residue side chain linked β-(1-3) to occasional glucose molecules. Its potential uses are incredibly diverse; it acts to thicken aqueous solutions or suspensions and it is important in stabilising oil-water emulsions. It is now registered for use as a food additive and is used in many foods, for example salad dressings, cheese products, syrups, dehydrated soups and gravies. Production was several thousand tons per annum in the late 1970s in the USA, approximately one quarter of which was used in the food industry.

dextran gels

Bacterial dextrans are also very important in industry. These compounds are produced almost entirely from a single strain of *Leuconostoc mesenteroides* and contain α-(1-6) linked glucose polymers with a variable number of α-(1-3), α-(1-4) and α-(1-2) linked branch points. Products of controllably different molecular weights can be produced by varying the growth conditions. One use for these dextrans is to produce cross-linked dextran gels (eg Sephadex) of differing molecular weights; these products act as molecular sieves and are used for the fractionation of compounds of different molecular weight, for example the separation of proteins. Dextran of molecular weight around 70,000 is used as a plasma extender - a suitable way of restoring blood volume or pressure to people following accident or shock.

alginates

Alginates have been extracted commercially from seaweed for over fifty years and are used in a wide variety of applications in the food and heavy industries, for example in meringues, gels, icing and frozen foods together with adhesives and the printing and dyeing of textiles. Its production by *Psedomonas aeruginosa* and more recently the more acceptable *Azotobacter vinelandii* have caused several industrial concerns to investigate the feasibility of bacterial production of alginate.

Π Can you think of an advantage in the bacterial production of alginate?

You may suggest a number of possibilities but an important one is that production would no longer be seasonal but on a year round basis.

Other less well established polysaccharides of microbial origin are available as gelling agents or for increasing viscosity though not all are passed for use in foods. For example curdlan -a β-(1-3) glucan from *Alcaligenes faecalis;* pullulan - a largely α-(1-6) linked glucan from the fungus Aureobasidium pullulan; baker's yeast glycan - cell wall material from yeasts containing glucan and mannan. The last is promising in that yeast and its products are more acceptable than bacteria to the general public in human foods. The glycan imparts a taste of fat or edible oils to the mouth and could satisfy some taste requirements in low calorie food.

Summary and objectives

This chapter has examined the biosynthesis of carbohydrates. We have shown that carbohydrates may be metabolically produced from a wide variety of the biochemicals including the amino acids. Particularly important is the process of gluconeogenesis. We have examined the control of carbohydrate biosynthesis and shown that it largely depends upon the energy charge of the cell. We have also examined the biosynthesis of various polysaccharides and discussed the importance of these biomolecules.

Now that you have completed this chapter you should be able to:

- show that quantitatively the most important biosynthetic route in carbohydrate biosynthesis is pyruvate to glucose;
- list the sources of pyruvate for monosaccharide biosynthesis;
- list the reactions involved in gluconeogenesis;
- critically discuss the metabolic processes which control the Embden Meyerhof pathway and gluconeogenesis;
- list the reactions by which glucose may be converted to other hexoses, pentoses, tetroses and substituted monosaccharides;
- discuss the biosynthesis of homopolysaccharides, heteropolysaccharides and complex polysaccharides;
- discuss the problems involved when the biosynthesis of polysaccharides occurs within the membrane and polysaccharides are required outside the membrane;
- by use of selected examples show the medical and industrial importance of bacterial polysaccharides.

The integration and regulation of metabolism

7.1 The strategy of metabolism 172

7.2 Coupling agents in metabolism 174

7.3 The maintenance of metabolic homeostasis through energy coupling 177

Summary and objectives 190

The integration and regulation of metabolism

You will have realised by now that cellular metabolism is an extremely complex process. On the one hand we have a whole series of catabolic pathways in which complex biochemicals (nutrients) are broken down to produce energy (ATP), reducing power and simple organic molecules. On the other hand, there is another complex series of pathways in which the products of catabolism are used to produce the new biochemicals needed for the synthesis of new cell material. We are therefore dealing with many different compounds which are being converted into new compounds using a wide variety of enzymes.

To be efficient, the cell needs to be able to co-ordinate this very complex milieu of reactions in such a way that the molecules needed for cell synthesis are produced in the correct proportions and at a rate needed for balanced cell synthesis. Furthermore, the cell also needs a strategy to co-ordinate the intricate network of reactions even when faced with changing circumstances, for example when the availability of nutrients changes. This chapter deals with this aspect of cell metabolism. We will examine the strategy adopted by cells to integrate metabolism through the use of coupling agents, especially upon the role of energy change. In the next chapter we will explore how cells control the flux of materials through particular pathways.

7.1 The strategy of metabolism

coupling of pathways

We know already that the basic strategy of metabolism is to form ATP, reducing power and building blocks for biosynthesis. This requires the coupling of the catabolic pathways, which generate ATP, reducing power and the fundamental carbon skeletons with the anabolic pathways (biosynthesis), that convert the carbon skeletons into the molecules and macromolecules necessary for growth and replication. The pathways are coupled through chemical coupling agents. These are compounds that participate in both catabolic and anabolic pathways and thus provide a link between them.

metabolism controlled for efficiency and economy

Coupling agents are found at all levels of metabolic correlation and show a wide range of sophistication in design. They give rise to metabolic sequences that have a functional interdependence. In addition to the requirement for the coupling of metabolic pathways, the cell must control the flux through metabolic pathways so that supply and demand for materials, energy and reducing power are balanced. To achieve this, the rate of catabolism is controlled not by the concentration of nutrients available in the environment of the cell but by the cell's second-to-second needs for energy in the form of ATP. Similarly the rate of biosynthesis of cell components is adjusted to immediate needs. For example, cells synthesise amino acids just fast enough to keep pace with the rate of their utilisation. Indeed, an important biological principle is that cells control their metabolism to achieve maximum efficiency and economy.

integration and control provided by coupling agents

So, it seems that the cell requires ways to control the rate of reaction sequences as well as a means of integrating (linking) reaction sequences. Chemical coupling agents provide a means of both integration and control of metabolic pathways (Figure 7.1).

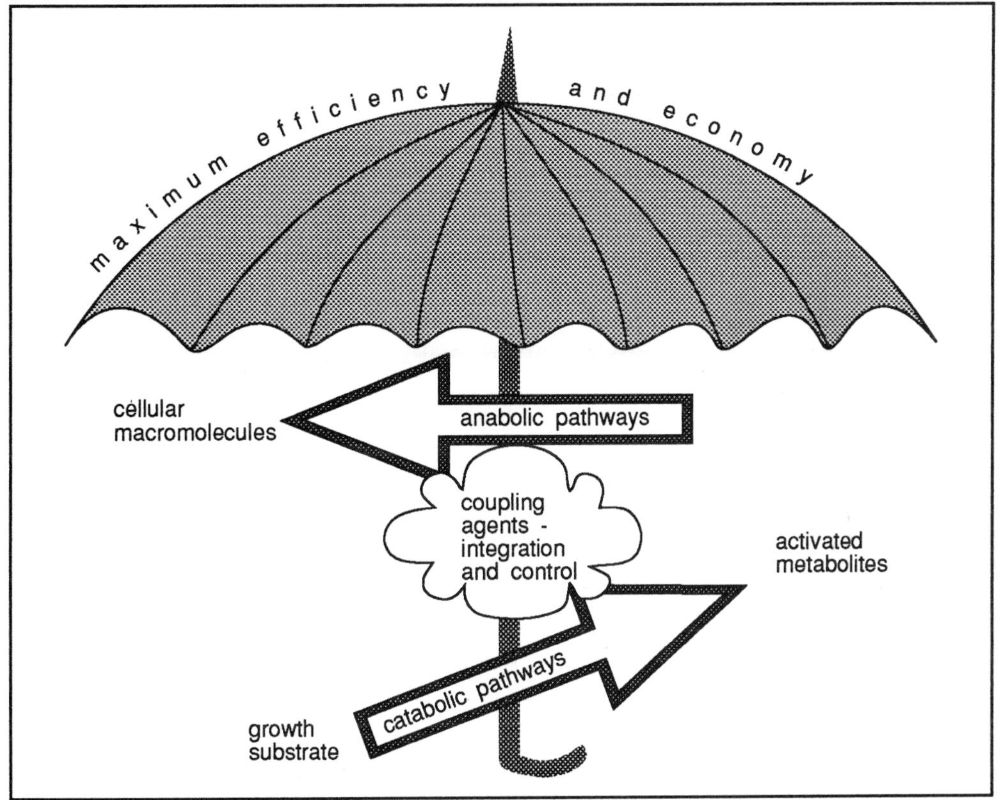

Figure 7.1 The position of coupling agents in the integration and control of metabolism.

Π Complete the following statement.

Maximum efficiency and economy by cells is achieved by .... and .... of metabolic pathways.

The words we hoped you would insert are 'integration' and 'control'.

It is easy to see why the moving parts of a mechanical machine much be integrated and controlled so perhaps the co-ordination of metabolism should be expected. However, the cell is far more sophisticated than mechanical machines and it is reasonable to ask, how is metabolic co-ordination achieved in the cell? To gain an appreciation of the magnitude of this task we will now consider the metabolic activities of a simple bacterial cell.

80% of energy consumed is used in protein synthesis

Many bacteria can simultaneously synthesise all their thousands of molecular components from just three simple precursors - glucose, ammonia and water. Considering protein synthesis alone, a bacterial cell synthesises simultaneously around 3000 different kinds of protein molecules. Each of these protein molecules contains a minimum of 100 amino acid units in a chain. Yet bacterial cells, under near optimum conditions for growth, require only a few seconds to complete the synthesis of any single protein molecule. Further, the different proteins are produced in the required proportions and since proteins are informational biopolymers the amino acids must be connected in a defined, predetermined sequence. The living cell is able to achieve such high rates of protein synthesis at the expense of a large portion of its metabolically-derived energy. Indeed, bacteria devote almost all of their energy to biosynthesis and of this, roughly 80% is channelled into protein synthesis. In the bacterium, *Escherichia coli*, around 2.5 million ATP molecules per second are invested in biosynthesis. Since the total ATP content in this organism is approximately five million, the cell carries sufficient energy for only two seconds work!

Consideration of protein synthesis in bacteria thus gives an indication of the large rates of ATP regeneration which must be maintained in order to match ATP requirements. The example also emphasises the need for metabolic regulation to meet the second-to-second need of the cell. Similar comments apply to such primary metabolic precursors as NADH and NADPH.

SAQ 7.1

Identify each of the following statements as True or False. If false give a reason for your response.

1) Chemical coupling agents are common intermediates of catabolic and anabolic pathways.

2) Functional interdependence of metabolic sequences arises from the coupling of catabolic and anabolic pathways.

3) The rate of catabolism is controlled by the concentration of available substrate.

4) Bacteria have a store of ATP that can be used during growth in the absence of nutrients in the environment.

7.2 Coupling agents in metabolism

The largest-scale coupling in a living cell is between 'energy-yielding' (catabolic) sequences and 'energy-requiring' (anabolic) sequences. From the comments made earlier concerning ATP in *E. coli* it should not be surprising to learn that ATP is a coupling agent in energy metabolism. Indeed, the use of ATP as a coupling agent is central to metabolic integration, at this level, for all cells.

carrier molecules

We know that ATP is a carrier molecule that carries a phosphoryl group in an activated form. You will also have encountered other carrier molecules in your study of metabolism - several of these are listed in Table 7.1.

	metabolite	group carried in activated form	equivalent ATP generation value
carrier molecules	ATP (to ADP)	phosphoryl	1
	ATP (to AMP)		2
	NADPH	electrons	3 or 4
	NADH	electrons	3
	$FADH_2$	electrons	2
	nucleoside triphosphates (GTP, CTP, UTP)	nucleotides	1 or 2 (as for ATP)
	coenzyme - A	Acyl	
	biotin	CO_2	
	tetrahydrofolate	one-carbon units	
	S - adenosyl-methionine	methyl	
	thiamine pyro-phosphate	aldehyde (ie C_2 units)	
activated metabolites	glucose-6-phosphate		1 (1)
	phosphoenol pyruvate		1 (2)
	acetyl - CoA		1 (2 or 1)

Table 7.1 ATP equivalent values for some carrier molecules and activated metabolites. Values are for aerobic metabolism in a typical eukaryotic cell. Values in parentheses are ATP equivalents used in the production of the activated metabolites. (See text).

Carrier molecules stoichiometrically 'link' metabolic pathways together in the cell.

Π Should carrier molecules be regarded as coupling agents?

Yes, because they participate stoichiometrically in metabolic pathways and provide a link between them.

activated metabolites

Other metabolites besides carrier molecules participate in coupling, and they must be taken into account in a quantitative consideration of metabolic interrelations. These compounds are activated metabolites and may be assigned a value in terms of ATP equivalents, on the basis of the number of ATPs that would be required for their formation or, in some cases, the number of ATPs that could be obtained from their utilisation. Table 7.1 presents values of ATP equivalents for some carrier molecules and activated metabolites that participate in coupling.

You will see from Table 7.1 that ATP equivalent values are based on processes rather than compounds. This is most evident in the case of ATP itself; two phosphorylations are required to regenerate ATP from AMP, but only one for regeneration from ADP.

You are already familiar with the assignment of three ATPs per NADH oxidised and two ATPs per $FADH_2$; these are the P/O ratios which are well established for typical eukaryotic cells. However, you may wonder why a value of three or four ATPs have been assigned to the oxidation of NADPH. In some cells the NADPH system is always far more reduced than the NADH system. This difference in reduction is often so large that the free energy change for the reaction $NADPH + NAD^+ \rightarrow NADP^+ + NADH$ is of the order of the free-energy of hydrolysis of ATP. In some cells, it is thus appropriate to consider the oxidation of NADPH to be metabolically-equivalent to NADH plus an additional ATP ie to four ATP molecules.

amphibolic pathways

Some of the carrier molecules and activated metabolites shown in Table 7.1 are intermediates of central metabolic pathways. This reflects the fact that these pathways have anabolic as well as catabolic roles - they are amphibolic pathways. Figure 7.2 summarises the amphibolic nature of glycolysis and the TCA cycle.

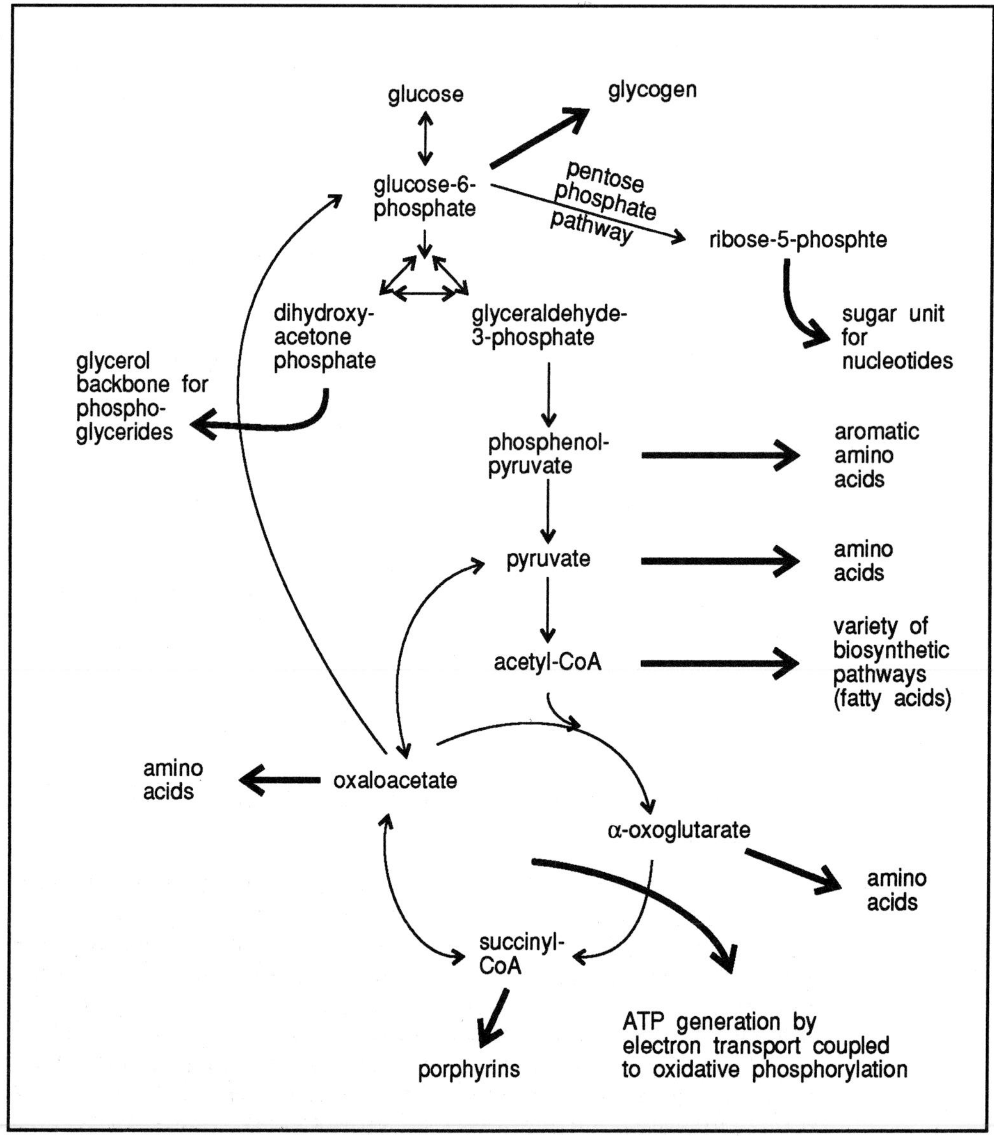

Figure 7.2 Summary of key crossroads of heterotrophic carbon metabolism.

SAQ 7.2

Match the carrier molecule with the group carried in activated form.

Carrier molecule	Group carried in activated form
S-adenosylmethionine;	electrons;
guanosine triphosphate;	acyl;
biotin;	methyl;
NADPH;	CO_2;
pyruvate;	one-carbon unit;
coenzyme-A;	nucleotide.
tetrahydrofolate.	

SAQ 7.3

Complete the following statements, which emphasise the anabolic and catabolic roles of central metabolic pathways using the words provided below:

phosphoenol pyruvate; glycerol; acetyl CoA; TCA; pentose phosphate; porphyrins; glycolysis; nucleotides; tetrahydrofolate; phosphoglycerides.

1) Dihydroxyacetone phosphate formed during [ ] gives rise to the glycerol backbone of [ ].

2) [ ], a glycolytic intermediate, provides part of the carbon skeleton of the aromatic amino acids.

3) [ ], the common intermediate in the breakdown of most growth substrates, provides a two-carbon unit for use in a wide variety of biosyntheses.

4) Succinyl-CoA, formed in the [ ] cycle, is one of the precursors of [ ].

5) Ribose-5-phosphate, which is formed by the [ ] pathway, is the source of the sugar unit of [ ].

6) [ ] is a carrier of one-carbon units for biosynthesis.

For the remainder of this chapter we will consider metabolic regulation mainly in terms of energy coupling. It is appropriate to do this because energy coupling plays the largest part in coupling in living cells and ATP participates in every extended metabolic sequence.

7.3 The maintenance of metabolic homeostasis through energy coupling

anabolic and catabolic pathways tend to be separate

From pathways considered previously we know that few pathways are amphibolic and catabolic and anabolic pathways are almost always distinct. For example, the pathway for the synthesis of fatty acids is different from that of their degradation. Similarly, glycogen is synthesised and degraded by different sets of reactions. This separation enables both catabolic and anabolic pathways to be thermodynamically favourable at all times. A biosynthetic pathway is made exergonic by coupling it to the hydrolysis of sufficient ATP; for example, four more high energy phosphate bonds are spent in converting pyruvate into glucose in gluconeogenesis than are gained in converting glucose into pyruvate in glycolysis. The additional requirement for energy assures that gluconeogenesis is highly exergonic under all cellular conditions.

There are many other examples where more ATP is consumed in the synthesis of a metabolite than can be obtained from its degradation. You are already familiar with the two further examples given below:

- Complete aerobic respiration of glucose to $CO_2 + H_2O$ generates 38 molecules of ATP per glucose molecule; photosynthetic synthesis of glucose from CO_2 requires an equivalent of 66 molecules of ATP (assuming 4 ATP could be produced from each NADPH consumed). When these pathways are coupled there is a net loss of ATP: $-66 + 38 = -28$.

- Similarly, for the conversion of 8 acetyl CoA into palmityl CoA and its breakdown to 8 acetyl CoA there is a net loss of ATP: $-63 + 35 = -28$.

Since each of the two oppositely directed sequences is thermodynamically favourable and a complete set of enzymes for each is present, you might expect that they would be continuously active and operate as a cycle. However, this would introduce a problem.

Π What problem do you envisage would occur if the two oppositely directed metabolic pathways operated as a cycle?

futile cycle

More ATP would be used in the synthetic half of the cycle than could be regained in the degradative half and the cell would be poorer in ATP after each turn on the cycle. This would result in the draining of the cell's metabolically-available energy with no return. Such a cycle is termed a futile metabolic cycle (see Figure 7.3).

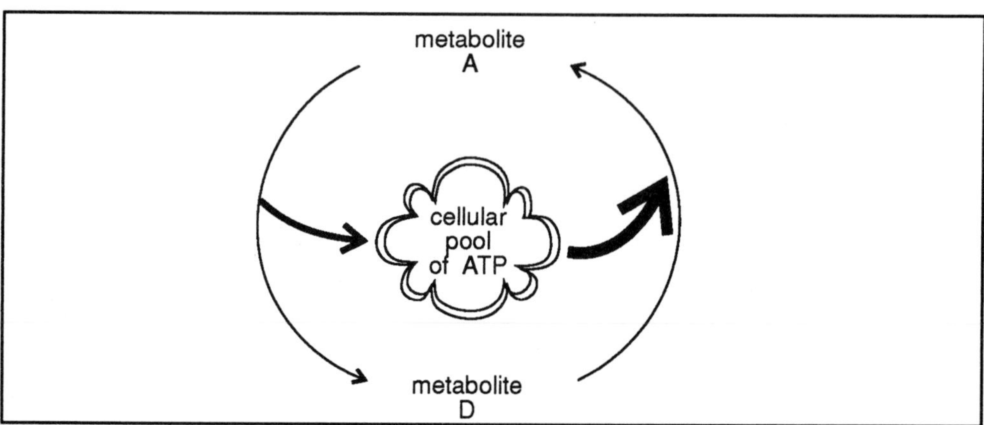

Figure 7.3 The operation of a futile cycle in which the production of A from D consumes more ATP than is generated by the conversion of A to D.

Clearly, strict controls to prevent the operation of futile cycles are essential. Since ATP is the primary metabolic coupling agent in these sequences, it is not surprising, that the adenylate system (ATP-ADP-AMP) plays a major role in the prevention of futile cycles. We will now examine how it does this.

7.3.1 The adenylate energy charge

mass action affect

Most metabolic sequences operate most of the time at considerably less than their maximum rate. A low concentration of substrate is one factor that might limit the rate of a reaction. This is a simple mass action effect but is not an adequate explanation of the control of futile cycles. If such potential cycles functioned cyclically, the net reaction would be the hydrolysis of ATP, and they would not be limited by mass action until the cell's ATP supply had been virtually exhausted. Clearly more effective controls are necessary.

An essential point concerning metabolic regulation is that the rates of metabolic pathways are governed more by the activities of key enzymes than by mass action effects.

Π Suggest two fundamentally different ways in which a cell might control the rate of an enzyme-catalysed reaction.

We anticipate that you would suggest that control may be exerted by:

- changing the concentration of the enzyme in the cell;
- changing the rate at which the enzyme functions ie changing its activity.

protein synthesis is expensive

change of enzyme activity

Let us think for a moment about controlling the rate of an enzyme-catalysed reaction by decreasing its concentration in the cell. Seen from an energetic point of view, the production of protein is an 'expensive' process and, moreover, it proceeds relatively slowly. Hence, a cell cannot respond rapidly to a change in its environment by adapting its metabolism through a change in enzyme concentration. Short-term regulation of cellular metabolism does not involve modulation of the amount of the enzyme but the rate at which a key enzyme functions.

allosteric interaction

The actual mechanism of controlling the activities of enzymes is the subject of the next chapter, and will not be considered in detail in this chapter. Reactions which are essentially irreversible are potential control sites and the first irreversible reaction in a pathway is usually an important control element. By irreversible reaction we mean a reaction whose equilibrium heavily favours the formation of products (ie Keq = $\frac{[\text{Products}]}{[\text{Reactants}]}$ is large). One of the most important control mechanisms is allosteric interaction. Allosteric literally means 'other site'.

allosteric inhibition

allosteric activation

signal metabolite

Allosteric interaction can take several forms. In one, the binding of a substrate itself has an effect on binding of other substrate molecules by the enzyme. Such enzymes are usually composed of several subunits. Increases in substrate concentration usually have a greater effect on the rate of enzyme reaction. In the case of allosteric inhibition, the enzyme interacts with an effector molecule as a site other than the active site. This leads to a reduction in the rate at which the enzyme-catalysed reaction takes place. In allosteric activation, the binding of the effector leads to an increase in the rate of the enzyme-catalysed reaction. The compound which modulates the activity of the enzyme is sometimes called the signal metabolite. (Allostery is described in greater detail in the Biotol text, 'The Molecular Fabric of Cells.'

Allosteric interaction allows for rapid (almost instantaneous) changes in enzyme activity, which are necessary to ensure efficient and effective coupling of metabolic processes. We can visualise an allosteric enzyme regulating the flow path through a pathway in the following way. Imagine the allosteric enzyme acts like a tap which controls the flow in a pipe (Figure 7.4a). If the enzyme binds with an inhibitor, this effectively closes the tap and the flow in the pipe is reduced (Figure 7.4b). If it binds with an activator, this effectively opens the tap an the flow in the pipe increases (Figure 7.4c). The binding of both activator and inhibitor results in an intermediate flow rate (Figure 7.4d). Note that we can regard activators as positive modulators, while inhibitors act as negative modulators of enzymes.

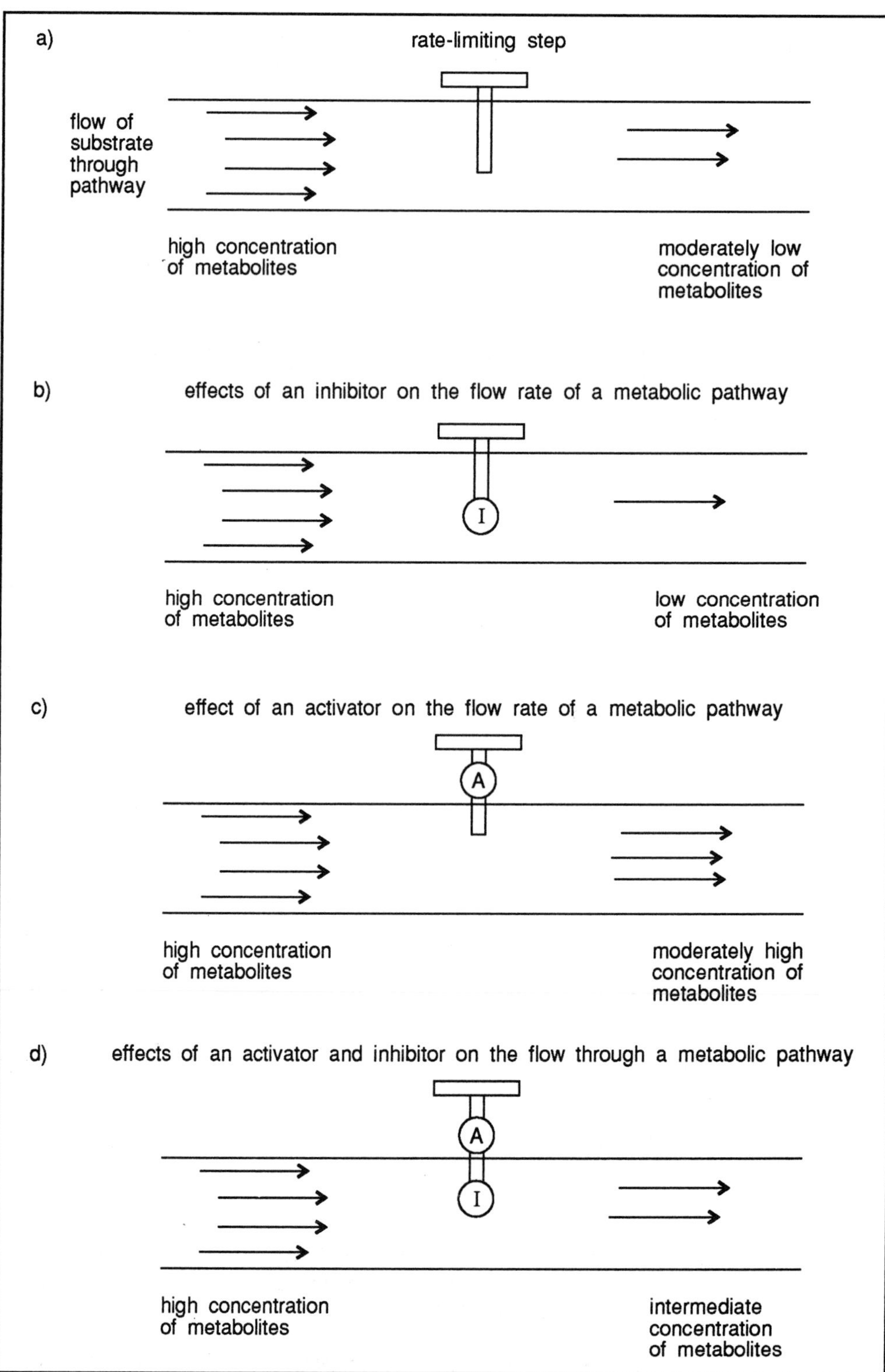

Figure 7.4 Generalised interactions of the rate-limiting enzyme and its effect on metabolic flux (see text).

SAQ 7.4

Using the principles of regulation of enzyme activity we have just learnt, see if you can answer the following question:

Phosphofructokinase (PFK), a key regulatory enzyme of glycolysis, catalyses the following reaction:

$$\text{Fructose-6-phosphate} + \text{ATP} \xrightarrow{\text{PFK}} \text{Fructose-1, 6-diphosphate} + \text{ADP} + \text{Pi.}$$

Glycolysis is a pathway whose function is to generate ATP and biosynthetic precursors. Thus it is not surprising to find that citrate and ATP are negative modulators of phosphofructokinase and AMP and Pi are positive modulators.

Will the concentration of fructose-6-phosphate and fructose-1, 6-diphosphate be Low or High under each of the following conditions?

a) High citrate.

b) High AMP.

ATP, ADP, AMP, NADH, NAD^+, NADPH and $NADP^+$ are all important allosteric modulators in cells but from an energetic point of view the primary metabolic coupling agent, the ATP-ADP-AMP (adenylate) system, is the most important. A number of metabolically-important enzymes are allosterically-regulated by ADP and ATP. AMP is also an important modulator of certain allosteric enzymes; it is formed by the removal of pyrophosphate from ATP which occurs in certain biosynthetic reactions:

$$\text{ATP} \rightarrow \text{AMP} + \text{pyrophosphate}$$

AMP can be rephosphorylated to ADP by the action of adenylate kinase:

$$\text{ATP} + \text{AMP} \rightarrow 2\text{ADP}.$$

adenine nucleotides are like a storage battery

The ATP-ADP-AMP system resembles a storage battery in its ability to accept, store and supply chemical energy. The energy status of the cell and hence the 'poise' of its allosteric modulators may be expressed by the energy charge of the cell. This is the extent to which the ATP-ADP-AMP system is 'filled' with high-energy phosphate groups. If all the adenine nucleotides in the cell are as ATP, the adenylate system is completely filled and is considered to have an energy charge of 1.0. At the other extreme, if all the adenine nucleotides are present as AMP, the system is empty of high-energy phosphate groups and has an energy charge of 0.

Another way to describe energy charge is a linear measure of the amount of metabolic energy held in the adenine nucleotide pool.

Π What is the value of energy charge in each of the following cases? 1) All the adenine nucleotide is present as ADP; 2) the adenine nucleotide is present as an equimolar mixture of ATP and AMP.

In both cases the ATP-ADP-AMP system is only half-full of high-energy groups, so the energy charge is 0.5.

calculations of energy charge

The energy charge of the ATP-ADP-AMP system can easily be calculated for any given set of concentrations of ATP, ADP and AMP by the equation:

$$\text{Energy charge} = \frac{\text{ATP} + 0.5\text{ADP}}{\text{ATP} + \text{ADP} + \text{AMP}}$$

The ATP-ADP-AMP system can be charged by stoichiometric coupling to catabolic sequences, and discharged by stoichiometric coupling to biosynthetic or other ATP-requiring processes. If no rapid energy-requiring processes are going on in the cell, oxidative phosphorylation will increase the ATP concentration while at the same time the concentration of ADP and AMP will be reduced. Conversely, a high energy consumption will lead to a low ATP concentration and high concentrations of ADP and AMP.

metabolic homeostasis

From the above argument you may think that the energy charge fluctuates widely in relation to the energy needs of the cell. In practice, the reverse is true - the energy charge is stabilised. This is a necessary prerequisite for metabolic homeostasis. Stabilisation of the energy charge results from having the rate of metabolic sequences regulated by the energy charge.

The effect of energy charge on ATP-generating and ATP-utilising sequences are shown in Figure 7.5. We can see that these curves have steep portions at relatively high energy charge values. The energy charge value is stabilised at the response midpoints of the two curves. The steepness of the curves at this point allows for effective and sensitive control of ATP-generating and ATP-utilising sequences by small changes in energy charge. For most cells the energy charge value is near 0.85.

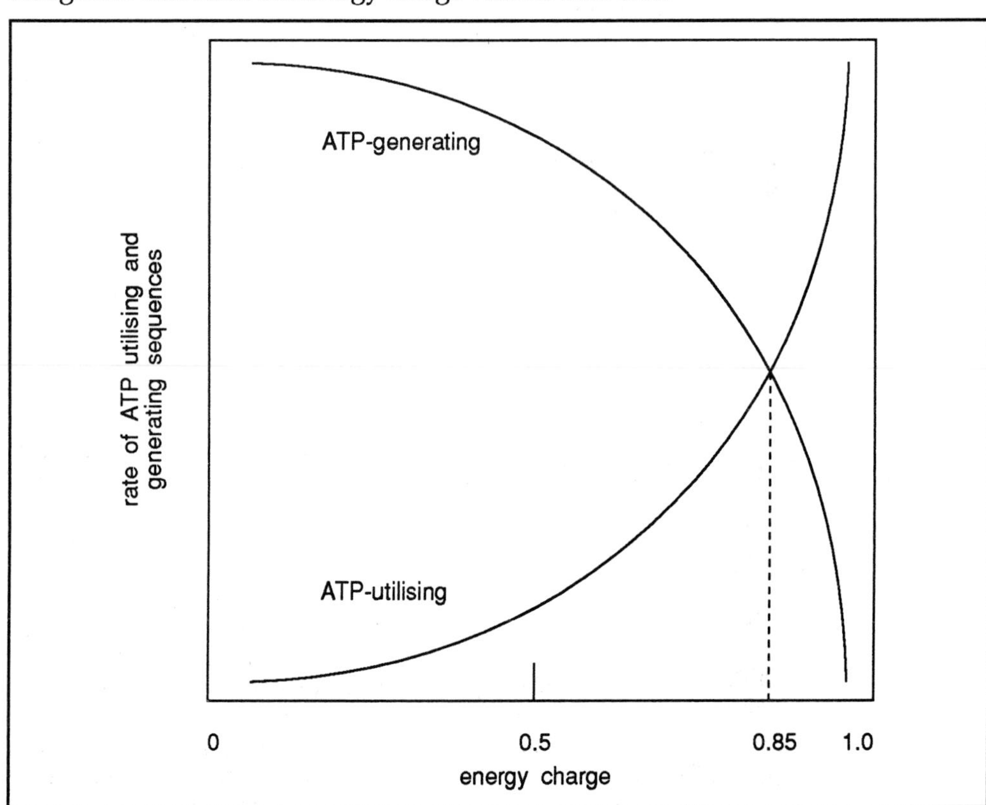

Figure 7.5 Effect of energy charge on ATP-generating and ATP-utilising sequences.

Π Complete the following statements: 1) Rates of sequences that generate ATP should [ ]with increasing energy charge. 2) Rates of sequences that use ATP should decrease with [ ] energy charge.

1) Rates of sequences that generate ATP should **decrease** with increasing energy charge.

2) Rates of sequences that use ATP should decrease with **decreasing** energy charge.

SAQ 7.5

1) We estimate that ATP, ADP and AMP are present in a culture in the ratios of 3:2:1. What is the energy charge in the culture?

2) If the AMP is all converted to ATP what will be the energy charge?

SAQ 7.6

Identify each of the following statements as True or False. If false give a reason for your response.

1) Most cells maintain an energy charge of around 0.85.

2) The energy charge is proportional to the size of the adenine nucleotide pool.

3) The energy charge is halved if the concentration of ATP is halved.

4) The more sensitive the metabolic activities respond to variation in the energy charge, the less the charge will vary.

5) A reduction in energy charge will increase the rate of ATP-generating pathways.

6) At an energy charge of 0.99 the rate of ATP-utilising pathways will be high.

7) During periods of carbon substrate starvation the energy charge is stabilised around 0.85.

7.3.2 How energy charge interacts with pathway intermediates

overlapping controls

To obtain adequate metabolic control the cell has to consider its metabolic requirements, as well at its requirement for energy. This gives rise to overlapping controls.

To illustrate the need for overlapping controls, we will consider an aerobic cell metabolising carbohydrate. We know that glycolysis and the TCA cycle are amphibolic pathways; that is, they serve both catabolic and anabolic roles. Not only are these sequences needed as a source of NADH to drive electron transport phosphorylation, but they also supply the starting materials for the biosynthesis of amino acids and other cellular constituents that serve as building blocks for macromolecular synthesis. Thus if glycolysis was controlled solely by energy charge, the production of biosynthetic starting materials would be curtailed at just the time when they could be most utilised - when the energy charge is high. This indicates that the glycolytic rate must respond to the need for supplying intermediates, for example acetyl-CoA, as well as the need for generating ATP. The appropriate response of these two types of input is illustrated in Figure 7.6.

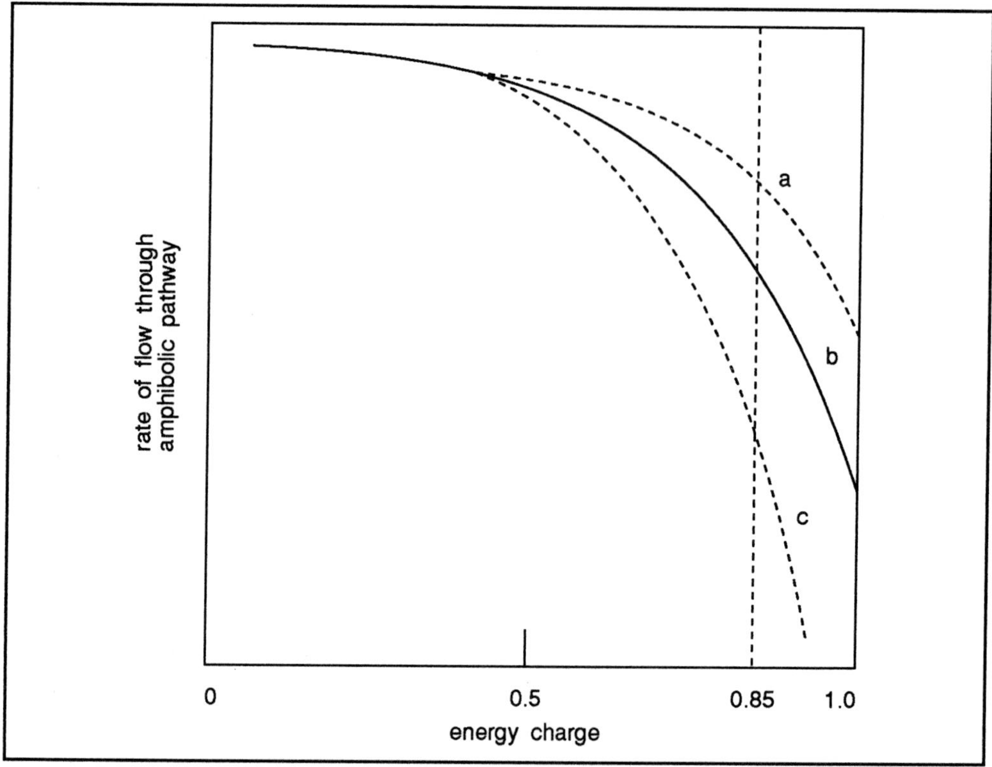

Figure 7.6 Generalised response curves for interaction between energy charge and concentration of pathway intermediates for an amphibolic pathway at: a) low b) normal and c) high concentrations of the intermediates.

Let us consider the lines shown in Figure 7.6 in a little more detail. If the cell has an adequate supply of the products of an amphibolic pathway needed for biosynthesis and the energy charge is high (that is, close to 1), then the pathway is not needed for either metabolite or energy generation and its rate will be low (line c). If on the other hand, the level of metabolites is low, but the energy charge is high, then the pathway will need to operate in order to supply the metabolites. Thus, even though the energy charge is high, the pathway will still operate (line a in Figure 7.6). Note that when the energy charge is low, the pathway will need to operate at a high rate.

We can see from the generalised response curves shown in Figure 7.6 that both energy charge and metabolite circulation may regulate the rate at which a pathway may operate. The regulation of amphibolic pathways jointly by energy charge and metabolite concentration is not however always the case.

There are many examples of regulatory enzymes in amphibolic pathways that are modulated by the concentration of a metabolite rather than the concentration of the adenine nucleotides.

Other regulatory inputs, such as the $NADH/NAD^+$ ratio and $NADPH/NADP^+$ ratio, are thought to interact with the adenylate energy charge in a similar manner.

overlapping control in biosynthesis

Rates of biosynthesis pathways (ATP-utilising sequences) are similarly influenced by factors other than energy. The type of interaction to be expected between energy charge and the concentration of the end-product of a biosynthetic pathway is shown in Figure 7.7. The cell will expend ATP and intermediates to synthesise a product only if the level of the product is low and the energy charge is high. We will consider this further in the next chapter.

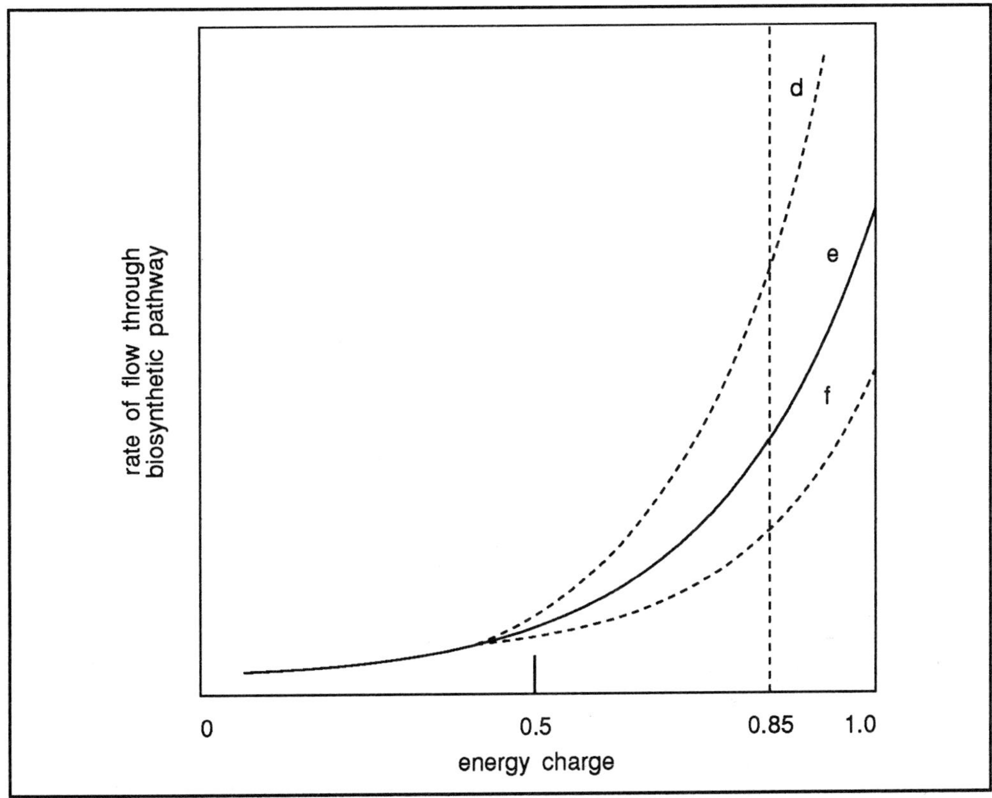

Figure 7.7 Generalised response curves for interaction between energy charge and concentration of end-product for control of a regulatory enzyme of a biosynthetic sequence: d) low e) normal and f) high concentration of the end-product.

7.3.3 Interaction between ATP-utilising and ATP-generating sequences

Let us apply these principles to a specific case. The tricarboxylic acid cycle acts as an amphibolic pathway. It is concerned with the production of cell energy and reducing power and it supplies essential intermediates for the biosynthesis of amino acids and porphyrins. We have provided a simplified version of this in the following figure. We have also represented the metabolic relationships between the TCA cycle and carbohydrate and fatty acid metabolism. In this figure we have indicated some key intermediates.

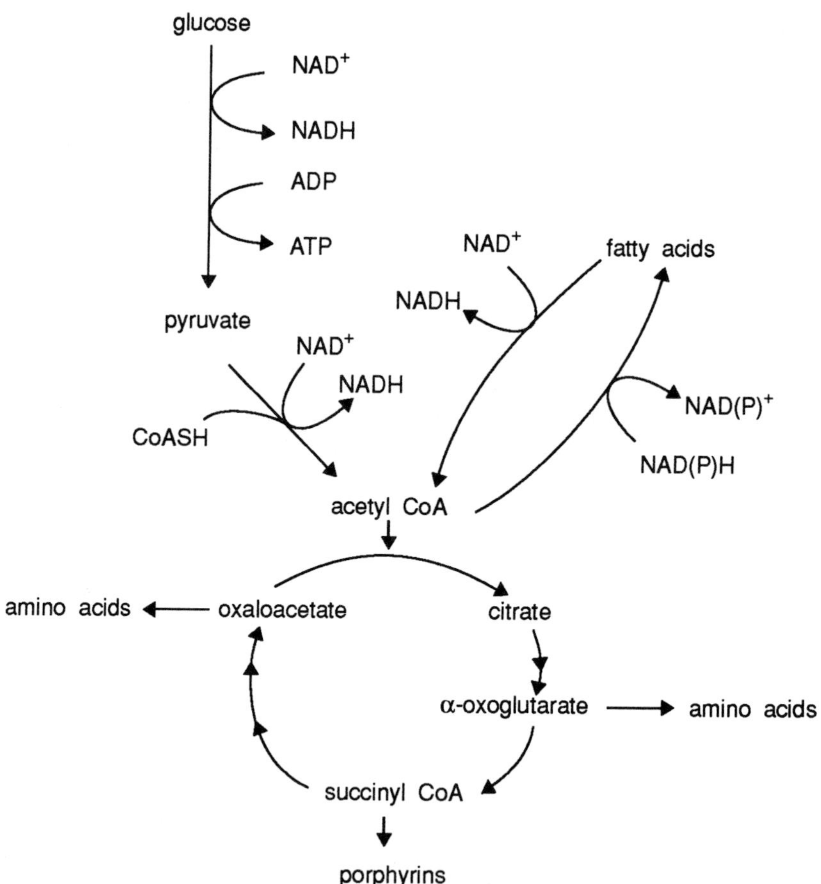

Π When the energy charge of the cell is low (ie ADP/ATP and NAD^+/NADH ratios are high), will glycolycis and fatty acid oxidation proceed quickly?

The answer is yes, both pathways require NAD^+ to proceed. If however all of the NAD^+ had been reduced to NADH, the pathways will naturally stop because of the unavailability of NAD^+. Likewise if the concentration of ADP is high, then oxidative and substrate level phosphorylation may proceed quickly, thus NADH will be rapidly oxidised. Thus under these circumstances these catabolic pathways will have a large throughput.

A key enzyme in the regulation of glycolysis, lipid metabolism and the tricarboxylic acid cycle is pyruvate dehydrogenase. We will use this enzyme to show how pathways can be regulated by both energy charge (ie ATP, ADP, NADH) and by metabolites (in this case acetyl CoA and succinyl CoA).

Pyruvate dehydrogenase may be inhibited by acetyl CoA and by NADH. (If the cell has plenty of both of these, it has no need for pyruvate dehydrogenase to make NADH or acetyl CoA). This inhibition can be either directly by these molecules binding to the enzyme or indirectly by activating another enzyme (pyruvate dehydrogenase kinase which inactivates pyruvate dehydrogenase).

Thus:

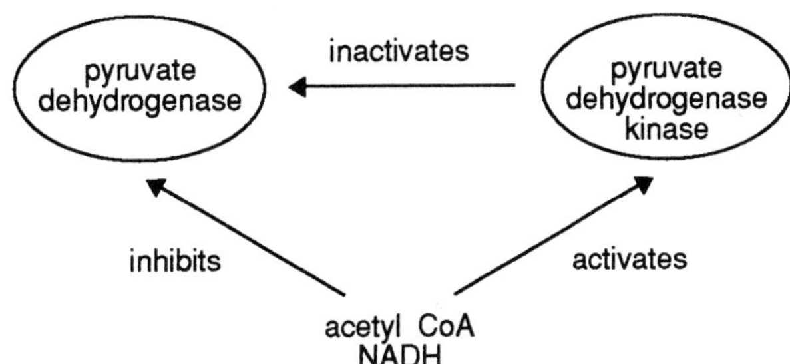

This inactivation of pyruvate dehydrogenase by the kinase enzyme requires ATP. We can stylise this inactivation diagrammatically. Thus:

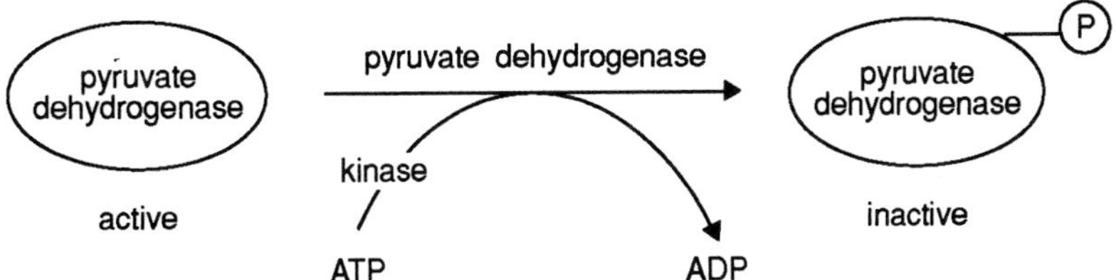

Thus, if the energy level of the cell is high because of carbohydrate catabolism and there is a plentiful supply of reducing power and acetyl CoA, acetyl CoA can be diverted into fatty acid synthesis. Altenatively, if acetyl CoA concentrations and the levels of ATP and NADH are all high because fatty acids are being catabolised, then the same enzyme is inhibited and pyruvate can be converted into carbohydrates.

If, however, the levels of acetyl CoA, ATP and NADH fall because they are being used for biosynthesis, then pyruvate dehydrogenase is active. Thus the rate of carbon entering the amphibolic TCA cycle is governed both by the energy charge and the availability of metabolites.

The rate-limiting step of the TCA cycle is catalysed by the enzyme isocitrate dehydrogenase.

Π Have a guess what compound might stimulate (activate) this enzyme - there are several candidates.

In most systems it is activated by ADP. This makes sense because if the cell is low in energy (ie ADP is high), then the cycle is needed to generate more energy. ATP and NADH both inhibit this enzyme.

The regulation of the cycle by metabolites is mainly confined to the enzyme α-oxoglutarate dehydrogenase. Succinyl CoA and NADH both inhibit the enzyme.

The desciption given here is rather simplified, but it does show how the regulation of metabolic flux through amphibolic pathways can be regulated by energy charge and metabolites. Citrate, for example, will inhibit glycolysis by interacting with phosphofructokinase.

SAQ 7.7

The graph shows generalised response curves for ATP-utilising and ATP-generating pathways.

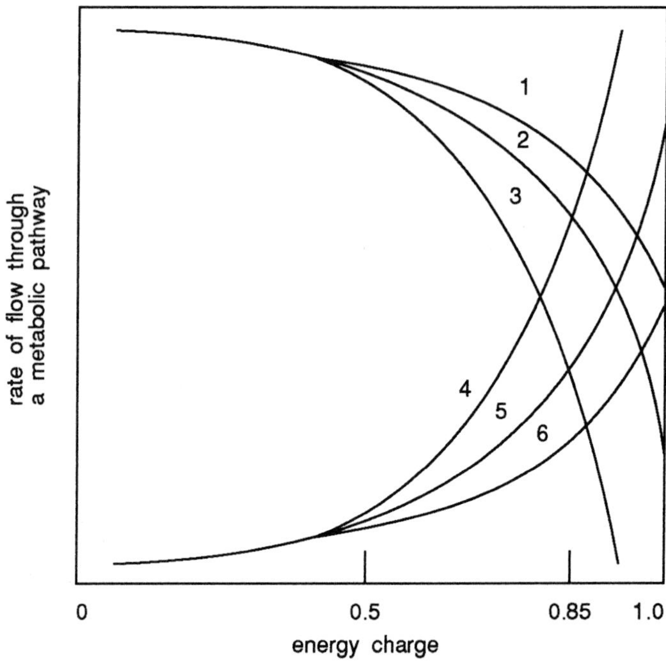

1) Select the response curve appropriate for glycolysis when the intracellular concentration of citrate is high.

2) Select the response curve appropriate for a biosynthetic pathway when the intracellular concentration of the pathway end-product is low.

3) The energy charge of a bacterial culture was found to be 0.5. If the requirement for aromatic amino acids was high, would the synthesis of aromatic amino acids be low or high? Give a reason for your response.

4) The end-product of a biosynthetic pathway is being used unusually rapidly and the end-product then becomes available from the environment. How would you expect the response curve to energy charge for the biosynthetic pathway to change?

SAQ 7.8

Some of the key crossroads of carbon metabolism are shown below.

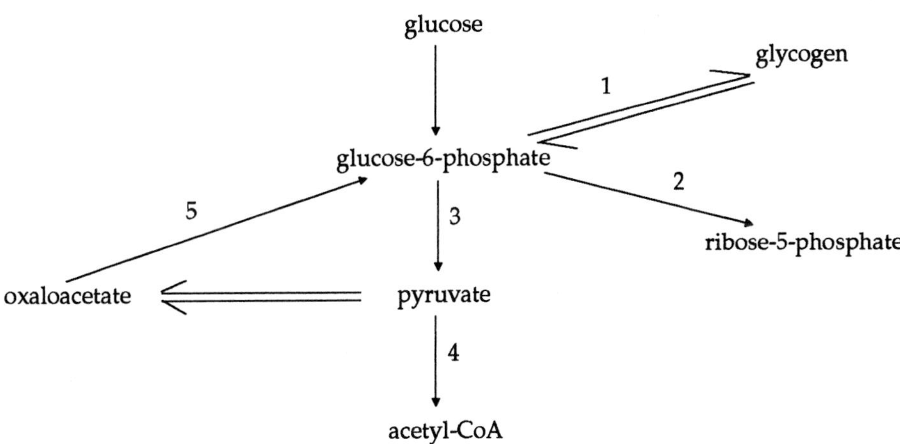

For each of the numbered routes, select *two* conditions from the list provided most likely to *stimulate* the rate of carbon flow through the route.

Low energy charge, high energy charge, nitrogen limitation (cell unable to grow), low NADPH, high NADPH, low amino acids, high amino acids, low lipid content, high lipid content, low ribose-5-phosphate, high ribose-5-phosphate; high pyruvate.

Summary and objectives

We have seen that the second-to-second needs of the cell are maintained through sophisticated integration and control of metabolism. Such regulation prevents the operation of futile cycles and enables the cell to achieve maximum economy and efficiency. Metabolic pathways are stoichiometrically coupled through carrier molecules and activated metabolites. Many of these coupling agents are modulators of allosteric enzymes that control the flow of metabolites through the key crossroads of carbon metabolism. Because of the importance of ATP in coupling catabolic and anabolic pathways, other coupling agents may be assigned ATP-equivalents generated or used in specific metabolic conversions. The unique role of ATP in stoichiometric coupling is accompanied by the ubiquitous participation of the adenylate energy charge in the kinetic regulation of metabolism. At high energy values ATP-generating pathways are inhibited whereas ATP-utilising pathways are stimulated; a decrease in energy charge reverses both effects. The response of regulatory enzymes to energy charge may be influenced by the concentration of relevant metabolites, such as the end-product of a biosynthetic pathway, giving rise to overlapping controls. Stabilisation of energy charge is linked to the maintenance of metabolic homeostasis.

Now that you have completed this chapter you should be able to:

- demonstrate the need for both integration and control of metabolism;
- give named examples of carrier molecules and activated metabolites that participate in coupling;
- predict how the concentration of certain metabolites might influence the flow of carbon through key crossroads of carbon metabolism;
- determine the energy charge from a knowledge of relative concentrations of adenine nucleotides;
- draw generalised response curves for ATP-generating and ATP-utilising pathways to energy charge;
- predict how energy charge and other coupling agents interact to effect control of ATP-generating and ATP-utilising pathways;
- predict the effect of positive and negative modulators of allosteric enzymes on the levels of metabolites;
- describe the operation of a futile cycle;
- name intermediates of amphibolic pathways that are precursors of certain biosynthetic pathways.

Control of metabolic pathway flux

Introduction	192
8.1 Control through $[S]_{active}$	192
8.2 Control through $[E]_{active}$	196
8.3 Patterns of regulation of metabolic pathways	210
8.4 Identification of the rate-limiting step	216
8.5 Optimisation of the flux through a pathway	220
Summary and objectives	223

Control of metabolic pathway flux

Introduction

In the preceding chapter we considered the role of coupling agents in the integration and control of metabolism. We saw that many of these agents influenced the rates of metabolic pathways (metabolic pathway flux) by altering the activity of allosteric (regulatory) enzymes. There are many other ways by which the rate of a metabolic pathway can be controlled. In this chapter we will consider: 1) the factors that influence the rates of metabolic pathways; 2) mechanisms and patterns of pathway flux control; 3) some of the experimental approaches used in studies of metabolic regulation.

All enzyme-catalysed reactions may be described by appropriate modifications of the simple reaction scheme:

$$[E] + [S] \leftrightarrow [ES] \rightarrow [E] + [P].$$

The rate of the reaction is a function of the concentration of 'active' substrate, $[S]_{active}$ and of 'active' enzyme, $[E]_{active}$. $[S]_{active}$ is the proportion of total substrate which is in an appropriate state for direct binding to the enzyme. $[E]_{active}$ is the proportion of enzyme molecules which is able to react directly with substrate.

Changes in reaction rate are effected only through changes in $[S]_{active}$, $[E]_{active}$, or both. An important point to appreciate is that $[S]_{active}$ and $[E]_{active}$ are not necessarily directly related to the total concentrations of substrate or enzyme; we shall see later in this chapter how this can be so. We will first examine the effects of control through $[S]_{active}$.

8.1 Control through $[S]_{active}$

In this section we will consider how substrate availability, and the factors governing substrate availability, can influence enzyme-catalysed reactions.

8.1.1 Rate limitation by total substrate availability

Any metabolic pathway could, in theory at least, be regulated very simply by the availability of total substrate. A reduction in the concentration of the substrate will decrease the reaction rate (provided it is not saturated with substrate) and this could result in a decreased flux through the pathway. Similarly an increase in substrate concentration could stimulate the pathway. This type of control is illustrated by the Michaelis Menten curve shown in Figure 8.1.

For many enzymes the intracellular concentration of their respective substrates is of the order of magnitude of the Michaelis constant. We can see from Figure 8.1 that for these enzymes small changes in substrate concentration cause large changes in the rate of the reaction. This itself brings about a homeostatic effect that stabilises the intracellular substrate concentration.

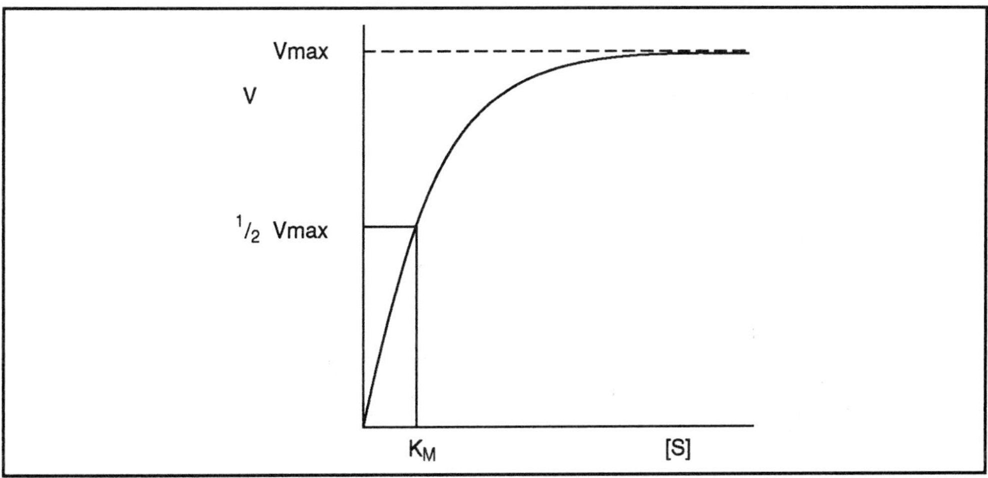

Figure 8.1 The Michaelis Menten curve for an enzyme-catalysed reaction. K_M is the Michaelis Menten constant ie the substrate concentration that will support enzyme velocity at half its maximal rate.

Π An intracellular substrate concentration is normally maintained at a steady-state concentration by the balance between its formation and its conversion to other products. Explain what would happen if the concentration of this substrate rises for any reason.

The rate of the conversion reaction rises also, tending to depress the substrate concentration again. This, in turn, would reduce reaction rate. The overall effect is a tendency to revert to a steady-state concentration.

Π Would you expect this to occur if the intracellular substrate concentration was much higher than the K_M value? Give a reason for your answer.

No. Because the velocity of the reaction will be close to the maximum and the substrate concentration will have little effect on it (see Figure 8.1).

It is clear that substrate availability will play some part in maintaining the constancy of the environment in the cell. However, it is still reasonable to ask, is substrate availability a regulatory mechanism that controls pathway flux? The answer to this is yes, substrate availability should be regarded as a potential regulatory mechanism for almost any pathway, even if it is regulated by other mechanisms. For example, stimulation of a metabolic pathway rate by increased activity of an enzyme may reduce the concentration of the pathway substrate which, in turn, could minimise the extent to which stimulation can occur. However, for some pathways the complex organisation of metabolic regulation can ensure that an increase in substrate supply is provided either simultaneously or immediately prior to the increase in flux through a pathway. For these pathways, substrate concentration is not a limiting factor despite a very high rate of substrate utilisation.

When considering the effect of substrate concentration on the rate of a metabolic pathway, it is important to realise that the substrate may not be evenly distributed within the cell. This may be caused by the physical makeup of the cell or the chemical nature of the substrate.

8.1.2 Physical unavailability of substrate

We know that in the eukaryotic cell there are numerous distinct membrane-bound compartments (eg mitochondria, cytoplasm, nuclei) and that many common metabolites occur in more than one compartment of the cell. For example, nucleotides are involved both in transcription in the nucleus, and in translation in the cytosol, as well as serving as precursors and energy sources throughout the cell. It is not unreasonable, therefore, to expect that these metabolites may not be evenly distributed throughout the cell, and that they may be present at different concentrations in the organelles involved. This is indeed the case.

The concentration of metabolites within intracellular compartments is often related to the activities of membrane proteins which catalyse selective transport; such as those found in the inner mitochondrial membrane. It follows that regulation of the activity of these proteins provides a means of coordinating the rates of metabolic pathways located in different cell compartments. For example, suppose two pathways in different cell compartments require NAD^+ for activity. The distribution of NAD^+ between the two compartments could then determine the relative activity of these competing pathways, and the pathway with access to the most NAD^+ would be favoured.

Even in prokaryotes (eg bacteria), where the general absence of membrane-bound organelles make it more difficult to envisage simple compartmentalisation, 'pools' of metabolites can exist which give an uneven distribution of metabolites in the cytosol. This can also occur within compartments in eukaryotic cells. Such concentration gradients are formed because intracellular compartments are generally composed of dense material which prevent rapid diffusion. Metabolite gradients will therefore build up in the vicinity of localised enzymes, such as those attached to membranes.

Π Complete the following statement which describes the formation of a metabolic gradient generated by a localised enzyme.

Substrate concentration will be [ ] near the enzyme and will increase with [ ] distance from it.

We anticipate that after some thought you would insert the words 'low' and 'increasing'.

Such concentration gradients can directly affect the activity of enzymes if substrate concentration is within the rising portion of the hyperbolic substrate saturation curve (Figure 8.1).

Π Two localised enzymes in different pathways compete directly for the same metabolites. If the pathways are located in the same subcellular compartment would you expect the enzyme winning this competition to: 1) have a lower or higher K_M for the metabolites than the other? 2) generate a shallower or steeper gradient for substrate than the other?

1) Lower. A lower K_M means that the winning enzyme has a higher affinity and therefore operates closer to full capacity. 2) Steeper. Since the winning enzyme has a higher affinity it binds more substrate. This reduces the concentration of substrate in the vicinity of the enzyme to the greatest extent.

As well as binding to their specific enzymes, some substrates can interact with other cellular components and, in so doing, may reduce their 'active' concentration. An example of this is the interconversion of citrate and isocitrate catalysed by the enzyme aconitase (a TCA cycle enzyme). For this enzyme, $[S]_{active}$ is related to the intracellular concentration of Mg^{2+}; Mg^{2+} binds more tightly to citrate than to isocitrate and the enzyme is unable to bind to either metal-substrate complex. The overall reaction is shown below.

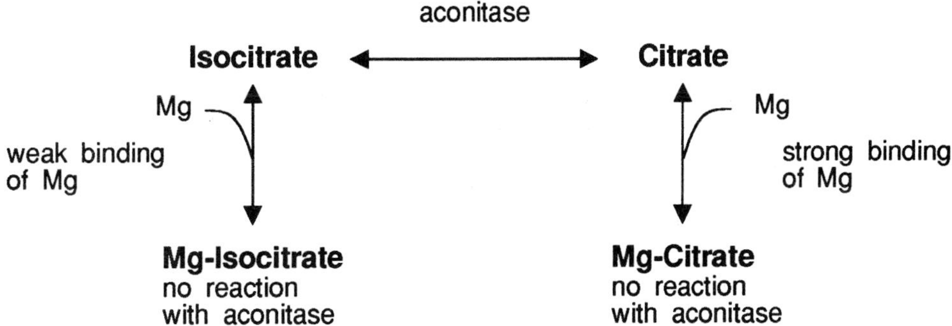

Π Increasing amounts of Mg^{2+} are added to an equilibrium mixture of isocitrate and citrate in the presence of aconitase. How does this affect: 1) the proportion of citrate in the mixture and 2) $[S]_{active}$ relative to $[S]_{total}$?

1) The proportion of citrate in the mixture is increased because Mg^{2+} binds tighter to citrate than to isocitrate. 2) $[S]_{active}$ is reduced relative to $[S]_{total}$ because enzyme is unable to bind the metal-substrate complex.

It is clear to see from this example that $[S]_{active}$ could be only a small proportion of $[S]_{total}$. For some enzymes which are present at a concentration greater than their substrate, $[S]_{active}$ may again be very much less than $[S]_{total}$ simply because a very high proportion of substrate is already bound to the enzyme forming [ES].

8.1.3 Chemical unavailability of substrate

Some compounds show purely chemical interactions which modify the concentration of $[S]_{active}$ for a particular enzyme. For example glyceraldehyde-3-phosphate (G3P) undergoes conversion to the hydrated form:

CH_2OⓅ		CH_2OⓅ
HO – CH	⟵⟶ (H_2O)	HO – CH
$CH(OH)_2$		CHO
(95.7% at 37°C)		(4.3% at 37°C)

and at 37°C the equilibrium is very far to the left. G3P dehydrogenase will act only on the free aldehyde form. Therefore, the concentration of the active form of G3P at 37°C is very low. You can see that chemical unavailability of substrate depends upon the occurrence of chemical interconversions and the high specificity of enzymes.

Π The K_M value for an enzyme acting on G3P calculated from the total G3P concentration of 37°C was 100μmol l^{-1}. What is the corrected K_M value calculated for $[S]_{active}$.? (Use the data on the previous figure).

Taking into account the proportion of $[S]_{total}$ that is in $[S]_{active}$ (aldehyde) form the corrected K_M value for the enzyme = 4.3/100 x 100 = 4.3μmol l^{-1}.

A different type of substrate interconversion is displayed by sugars which can exist in both α and β forms as well as in furanose, pyranose and straight-chain configurations. Different enzymes exhibit specificities for different forms and the rate of interconvertibility of these forms may create steep gradients of $[S]_{active}$, generated from pools of chemically unavailable substrate.

It should be clear from the examples provided that the physical and chemical unavailability of substrate can influence the rate of an enzyme-catalysed reaction and should be taken into account both in the design of experiments and interpretation of data. We have also seen that the availability of substrate can play a role in co-ordinating metabolic pathway flux in different intracellular compartments. However, you should note that metabolic pathway flux is determined mainly by control through $[E]_{active}$, not by control through $[S]_{active}$, all pathways have an $[E]_{active}$ control mechanism but may also be controlled through $[S]_{active}$.

8.2 Control through $[E]_{active}$

There are two levels at which the rates of metabolic pathways may be varied in response to changing environmental conditions: 1) there are rapid mechanisms (operating within seconds or minutes) for the regulation of enzyme activity - these depend upon changes in the catalytic activity of individual enzyme molecules; 2) there are slower mechanisms (operating within hours or days) that are dependent upon an increase or decrease in the number of enzyme molecules through modification of their rate of synthesis or degradation.

In addition to these mechanisms, coordination and regulation of pathways can be influenced by the intracellular location of enzymes.

8.2.1 Physical compartmentation

membrane transport proteins

This, of course, overlaps with substrate compartmentation, since the effect is physical separation of enzymes from substrate. We know that, in eukaryotic cells, pathways are located in different membrane-bound compartments. This makes possible the simultaneous, but separate, operation and regulation of similar pathways. Since the operation of these pathways relies upon the interplay of reactions that occur in more than one compartment, membrane transport proteins may determine pathway flux and their activity is often regulated.

Compartmentation of some pathways offers unique possibilities for pathway flux control. For example, long-chain fatty acids are transported into the mitochondrial matrix as esters of carnitine, a carrier that enables these molecules to traverse the inner mitochondrial membrane (Figure 8.2).

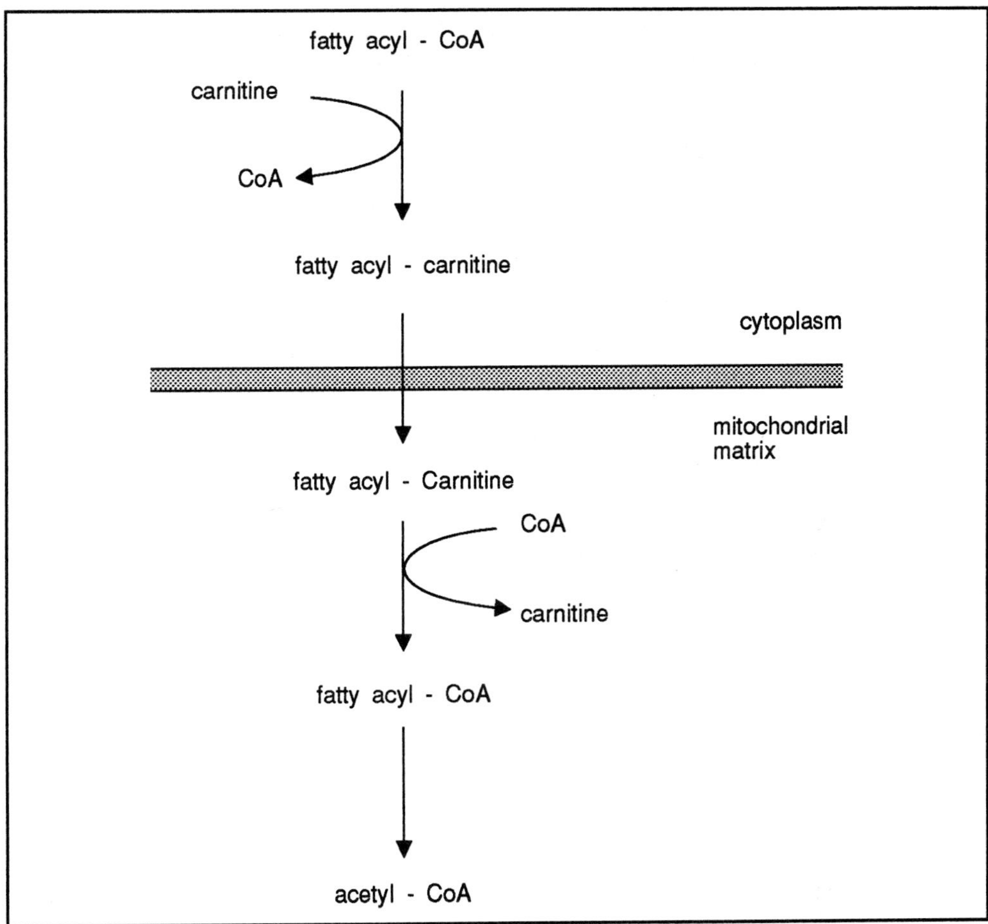

Figure 8.2 The role of carnitine in fatty acid oxidation.

It is possible that variation in the concentration of carnitine could regulate the rate of fatty acid oxidation. Since carnitine is not a cofactor for other enzymes, regulation of fatty acid oxidation through its concentration could occur without affecting other metabolic processes.

Π Write down a reason why variation in the level of carnitine is not a particularly sensitive mechanism of pathway flux control.

We can see from Figure 8.2 that the free carrier carnitine is regenerated and re-used. This means that carnitine would have to be reduced to very low levels before significant inhibition of fatty acid oxidation occurred.

Π Can you think why variation in the levels of most cofactors, other than carnitine, is not suitable as a mechanism of pathway flux control?

You should have realised that most cofactors are involved in more than one pathway and so the control mechanism would lack pathway specificity.

8.2.2 Chemical compartmentation

As with substrates, enzymes may exist in one or more forms of different activity within the same subcellular compartment. Regulatory enzymes can be switched on and off by reversible covalent modification. Usually, this occurs through the addition of a particular group, such as phosphate or adenylic acid; one form of the enzyme being more active than the other. For example, glycogen phosphorylase can exist in phosphorylated ('a' form) and dephosphorylated ('b' form) forms. Phosphorylase 'a' is active whereas phosphorylase 'b' is inactive, except in the presence of high levels of AMP which are not usually found in cells. Phosphorylation and dephosphorylation of the protein are catalysed by separate enzymes, which are also regulated.

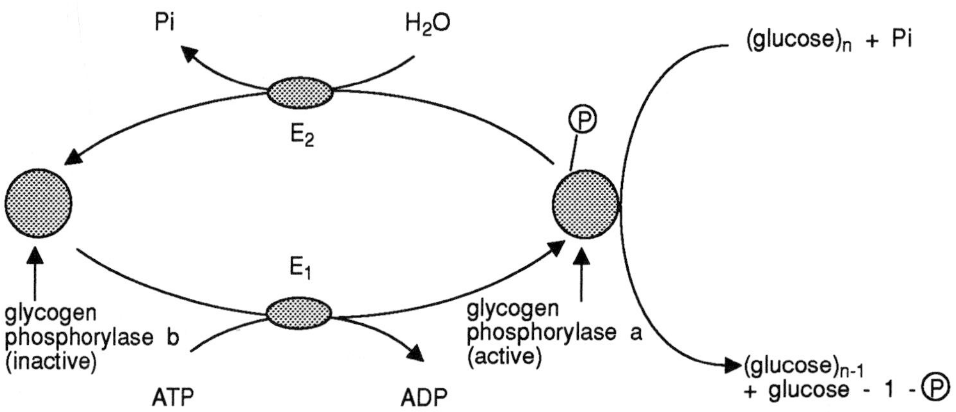

Π Can you think why the involvement of so many enzymes in the control of glycogen phosphorylation (shown in the illustration above) is advantageous to the cell?

This results in a large number of regulatory sites which enable the fine tuning of the reaction to be carried out. This confers the ability to respond to many different inputs. We will not expand on the regulation of this at this stage, but a fuller account is available in the Biotol text, 'Functional Physiology'.

8.2.3 Enzyme degradation

Variation in $[E]_{active}$ could also be due to changes in the rate of protein degradation. Indeed, the rates at which cell proteins turn over are very variable and, in general, regulatory enzymes are turned over more rapidly than structural proteins and non-regulatory enzymes. For most enzymes, however, variation in the rate of protein degradation is not an important regulatory mechanism.

For some enzymes a limited degradation can actually increase $[E]_{active}$. These are the enzymes which are synthesised as inactive precursors, known as zymogens or pre-enzymes. They are subsequently activated by irreversible attack by proteolytic enzymes which cleave a specific peptide bond thus removing a small peptide sequence. For example, many digestive enzymes that themselves hydrolyse proteins are synthesised as zymogens in the stomach and pancreas in mammals. This has the obvious advantage that the cells which produce the zymogen are not themselves digested by it.

8.2.4 Allosteric enzymes and metabolism regulation

Effects of substrate

Whilst many enzymes display a hyperbolic relationship when initial velocity is plotted against substrate concentration and thus obey Michaelis Menten kinetics, allosteric enzymes do not. Instead a plot of initial velocity against [S] gives a sigmoidal curve. Such enzymes usually consist of several subunits, although possession of multiple subunits is not a pre-requisite for allostery. What must occur, however, is communication between binding sites. This is termed a co-operative effect or co-operativity. What co-operativity means is that the binding of one substrate molecule facilitates or enhances the binding of subsequent substrate molecules.

co-operative effect or co-operativity

How is co-operativity achieved? Let us assume several subunits (often 4) occur, each is believed to be able to adopt two alternative conformations. These were originally called R (for 'relaxed') and T (for 'tight'). These alternative conformations differ in their kinetic properties, particularly in terms of their affinity for substrate(s). The T form has a low affinity for substrate whereas the R form has a high affinity for substrate. Now consider our tetrameric enzyme which has an active site on each enzyme monomer. In the absence of substrate, the T form predominates. On addition of substrate, one of the subunits binds the substrate and this binding is coupled to conversion of the other subunits into the R form. Diagrammatically we can represent this as:

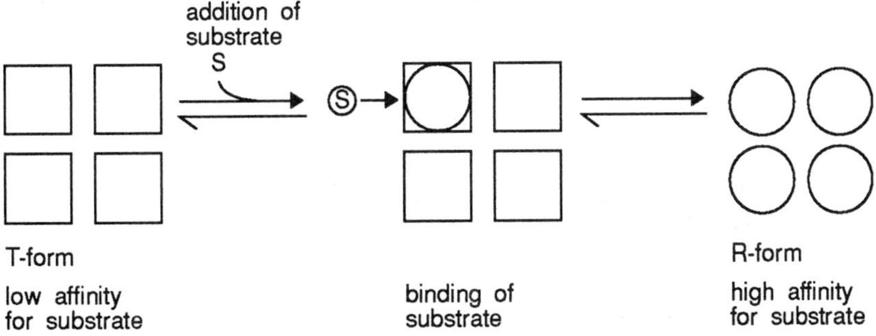

Effectively what is happening is that the conversion of one subunit into the R form by the bound substrate forces adjacent subunits to also adopt the high-affinity R form. This results in a sigmoidal velocity against substrate concentration plot. The actual degree of sigmoidicity depends on the relative affinities of the two forms for substrate and the equilibrium between T and R forms. Essential, when substrate concentrations are high, the enzyme will be predominantly in the R form, while at low substrate concentrations it will be in the T form.

What is the significance of sigmoidal kinetics? Let us examine the degree of saturation of an enzyme at various substrate concentrations.

Π Examine Figure 8.3 carefully. It shows both Michaelis Menten and sigmoidal kinetics. What is the relative increase in substrate concentration needed to increase the velocity from 20% to 80% of V_{max}?

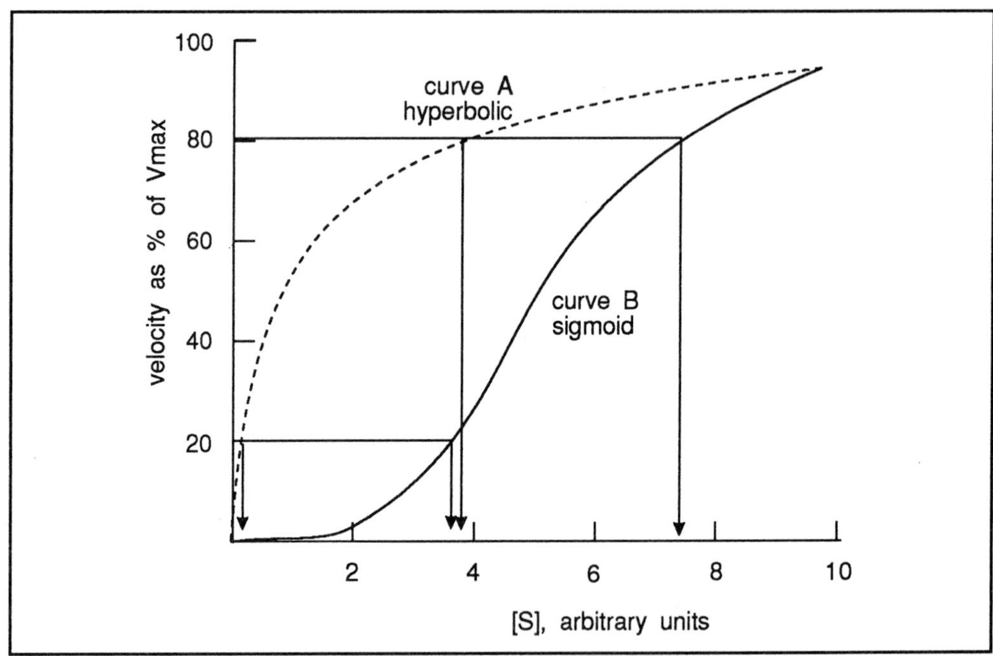

Figure 8.3 Influence of sigmoid and hyperbolic V against [S] relationships on the responsiveness to changes in [S]. (See text for details).

Inspection of Figure 8.3 should have shown you that to increase the rate from 20% to 80% of V_{max} required a concentration increase of substrate of 16 times. (You can verify this for yourself using the Michaelis Menten equation.) With sigmoid kinetics, the same relative activity increase requires only a 2-fold increase in substrate concentration. This means that a modest increase in [S] over a particular concentration range causes more marked rate increases with allosteric enzymes. In other words allosteric enzymes are thus more responsive to changes in [S] over particular substrate concentration range.

Influence of effectors

A second aspect of allosteric enzymes plays a vitally important role in the regulation of metabolic flux. Many allosteric enzymes bind molecules other than substrates. They bind these molecules at sites other than the substrate-binding sites. The binding of these compounds can alter the activity of the enzyme and such molecules are often called 'effectors' (or 'modulators' or 'signal molecules'). The effector molecule may either activate or inhibit the enzyme (ie positive or negative effectors). Their effect is to shift the velocity against [S] plots as shown in Figure 8.4).

Since the V against [S] plot can be changed by inhibitors and activators and the curve is not hyperbolic, Michaelis Menten kinetics do not apply.

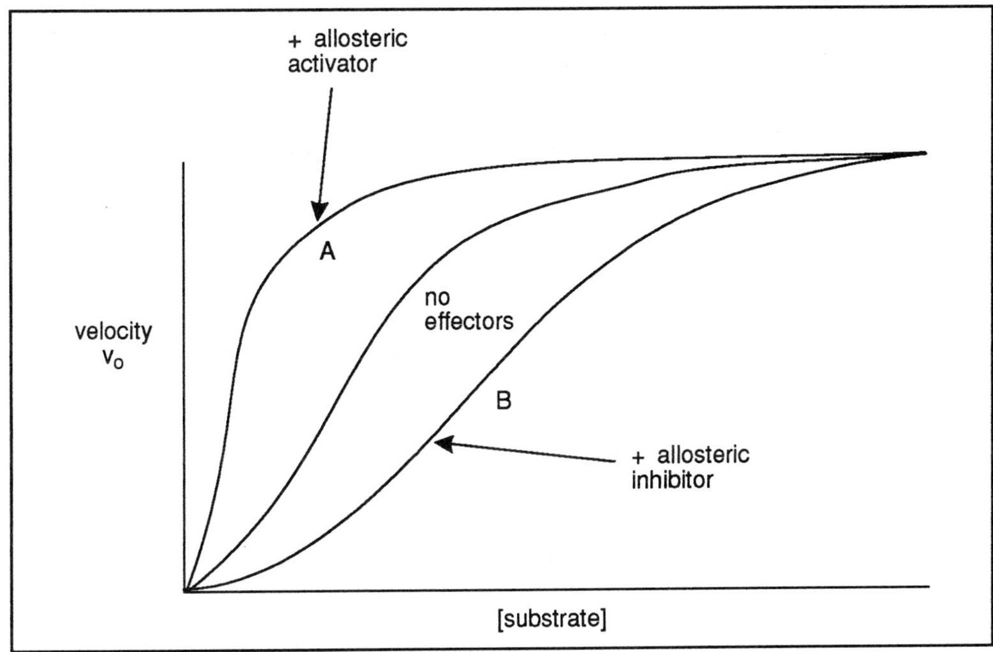

Figure 8.4 Effect of activator a) and inhibitor b) on the activity of an allosteric enzyme. An allosteric activator gives a higher rate at a given [S]; an allosteric inhibitor lowers the activity at a given [S].

The benefit of allosteric regulation is that compounds other than substrate or product can regulate the activity of a crucial enzyme. Typical allosteric inhibitors are the products of a metabolic pathway acting to inhibit the first step. When adequate levels of the end-product of the pathway are present, it switches off or lessens further synthesis.

Consider the pathway:

$$A \xrightarrow{1} B \xrightarrow{2} C \xrightarrow{3} D \xrightarrow{4} E$$

(where A.....E are intermediates, 1....4 are enzymes), enzyme 1 is often an allosteric enzyme. It will bind E which results in the enzyme being held in an inactive form (ie it is inhibited). Thus E inhibits its own synthesis. Such a process is usually referred to as feedback inhibition. In this case E (the effector) binds to a different site to A on enzyme 1. Perhaps this is not surprising since if the pathway is long the end-product (E) may bear little resemblance to the substrate (A) of the first enzyme.

As we have pointed out the first enzyme of a pathway is usually the enzyme subjected to regulation. This is not always the case.There may be cases where the end product of one pathway is needed to react with the product of a second pathway (Figure 8.5). What if there is an imbalance in the levels of these two compounds? One resolution of this is if the more abundant compound activates the synthesis of the compound which is deficient. This can be accomplished by an allosteric activator, acting on the first (and rate-limiting) enzyme of the less active pathway (Figure 8.5). The effect of this is to raise the rate at a given [S], as shown in Figure 8.4. This is achieved by binding to the enzyme and 'locking' it into the high-affinity R form.

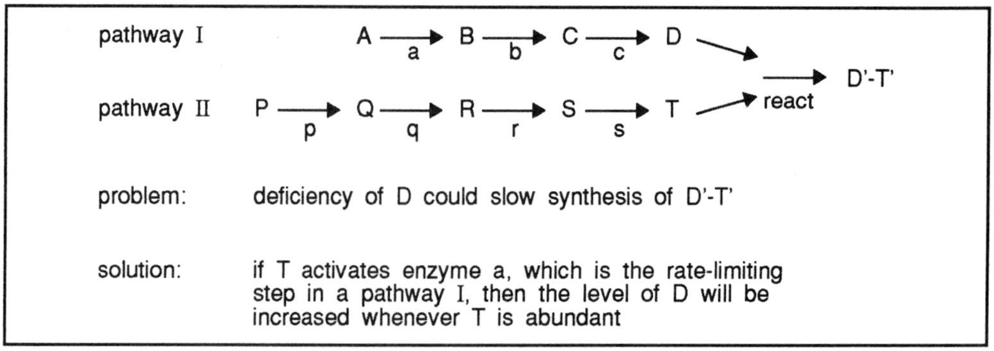

pathway I	A →(a) B →(b) C →(c) D ↘
	→ D'-T'
pathway II	P →(p) Q →(q) R →(r) S →(s) T ↗ react
problem:	deficiency of D could slow synthesis of D'-T'
solution:	if T activates enzyme a, which is the rate-limiting step in a pathway I, then the level of D will be increased whenever T is abundant

Figure 8.5 The activity of two pathways may need to be co-ordinated. Deficiency of the end-product of one pathway can be rectified by activation of the first enzyme of this pathway by the product of the second pathway.

These activation and inhibitory effects are exemplified by the enzyme aspartate transcarbamylase (ATCase). This enzyme catalyses the condensation of carbamyl phosphate and aspartate to form carbamyl aspartate, which is the first step in the biosynthesis of pyrimidines (Figure 8.6). subsequent reactions result in the formation of cytidine triphosphate (CTP). CTP is a highly effective allosteric inhibitor of ATCase, thereby shutting off entry of material into the pathway when CTP is abundant.

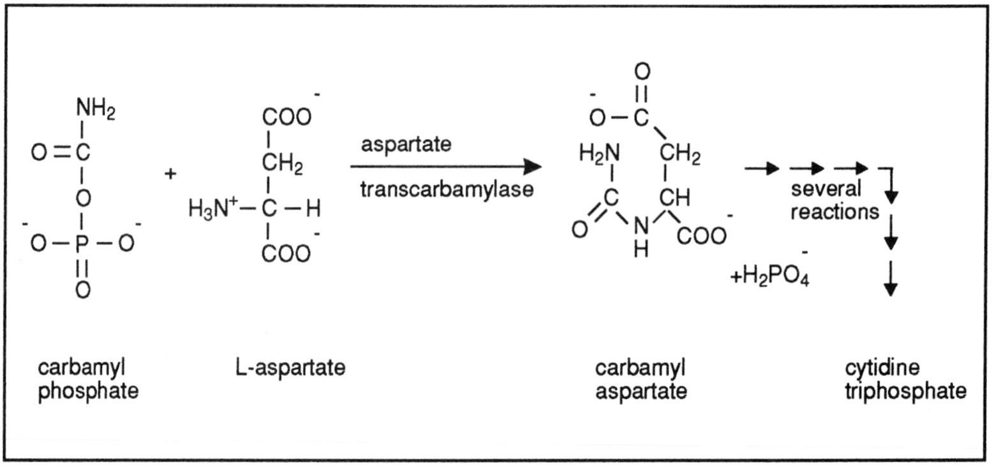

carbamyl phosphate + L-aspartate → (aspartate transcarbamylase) → carbamyl aspartate $+H_2PO_4^-$ → → → (several reactions) → cytidine triphosphate

Figure 8.6 Reaction catalysed by aspartate transcarbamylase (ATCase) L-aspartate and carbamyl phosphate condense to form carbamyl aspartate. Further reactions result in synthesis of cytidine triphosphate.

Binding of CTP to ATCase causes a decrease of affinity for the substrates, without affecting Vmax. Inhibition (resulting from the decrease in affinity) can be as high as 90%. This is much more efficient than waiting for product inhibition of each step to feed back down the pathway (high [CTP] inhibits the enzyme which made it; the substrate of this enzyme builds up and then inhibits the preceding enzyme, etc).

ATP as activator

ATP also acts as an activator of ATCase. It increases the affinity of the enzyme for its substrates, whilst having no effect on Vmax. ATP can be thought of as shifting the equilibrium of the enzyme in favour of the high-affinity R form. Both ATP and CTP bind at the same site. They therefore compete with each other, and high levels of ATP prevent CTP inhibiting the enzyme. We have already seen that CTP inhibition will occur when [CTP] is high. ATP activation (which will offset or override the effect of CTP) occurs when [ATP] is high and therefore energy is available for growth. Growth will require the synthesis of nucleic acids. These can only be synthesised if pyrimidine nucleotides are available. Hence the desirability of positive control by ATP, to ensure that purine and pyrimidine nucleotides are simultaneously available. It would be sensible to draw yourself out a scheme representing the ATCase system.

loss of regulating features

The most remarkable aspect of ATCase emerged from structural studies on the enzyme. Treatment with mercury-containing reagents, (such as p-hydroxymercuribenzoate) which react with sulphydryl groups, resulted in loss of regulation by ATP or CTP. In addition, the relationship between rate and [S] was now hyperbolic (ie the co-operativity had been lost), although the modified enzyme was still fully active. This loss of regulation by effectors is called desensitisation. Analysis by ultracentrifugation revealed that treatment with the mercurial reagent had caused dissociation of the enzyme into smaller particles. These were found to be either trimers of a subunit of 34 000 daltons (ie 3 x 34 000) or dimers of 17 000 dalton subunits (ie 2 x 17 000; Figure 8.7). The 34 000 dalton subunits are catalytically active but do not bind ATP or CTP. They are called C subunits. The smaller subunits bind ATP and CTP (competitively) but do not interact with the substrates.

desensitisation

C and R subunits

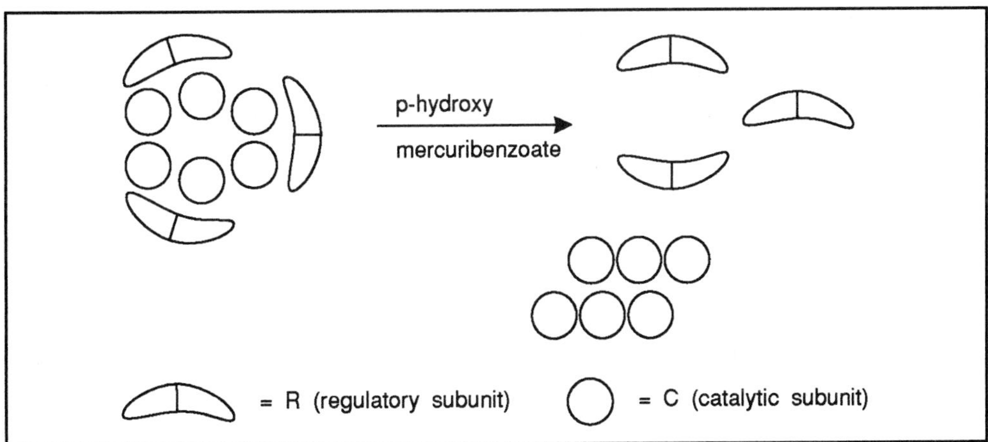

Figure 8.7 Structure of aspartate transcarbamylase. Treatment with mercurials produced dissimilar subunits. C-type subunits are catalytically active but do not bind ATP or CTP. R-type subunits bind ATP and CTP (competitively) but do not bind substrates.

effectors

These are known as R subunits. Thus in this case there are 2 distinct types of subunits. The sigmoid binding relationship must be caused by substrate binding to one C-type subunit causing conformation changes in other C subunits. Superimposed on this, binding of effectors (ATP and CTP) to the regulatory (R) subunits also influences the conformation of the catalytic subunits. A complex system indeed! If the mercurial is removed (by treatment with 2-mercaptoethanol) then reassociation of C and R subunits occurs to form native enzyme:

$$2C_3 + 3R_2 \rightarrow C_6R_6$$

The reformed complex displays similar allosteric properties to those of untreated enzyme.

Allosteric feedback inhibition and activation is a strategy widely used in metabolism. This allows a pathway to be regulated without having excessive concentrations of intermediates. This, of course, only works if the regulated enzyme is the rate-limiting step. One simple way in which this can be visualised is via a pipe analogy. If you think of water flowing through a pipe, which has regions of varying diameters, it is the region of narrowest diameter which determines the flow rate (Figure 8.8).

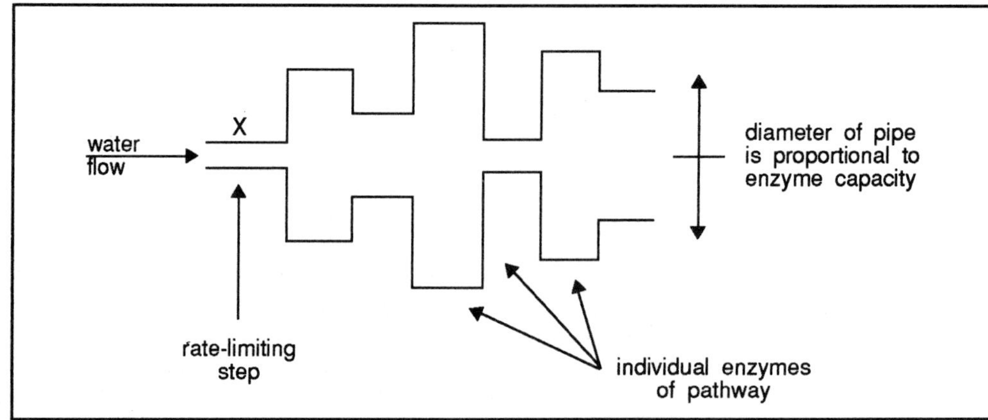

Figure 8.8 The pipe analogy. Pipe diameter represents enzyme capacity: the step with the smallest capacity will be rate-limiting; affecting this step (either activation or inhibition) will have the most effect on the overall throughput of the pathway.

enzyme capacity

rate-limiting step

If the water pressure remains the same, then varying the diameter anywhere other than point X will have minimal effect. If we look at a metabolic pathway, thinking of the enzyme capacity at each stage as the diameter of a pipe, then step X in Figure 8.8 is clearly the 'rate-limiting' step. Activation of this enzyme (widening of the pipe) will allow faster throughput, whereas activating any other step will not affect the overall flux through the pathway. We can thus summarise this area by stating that:

- in a metabolic pathway, the enzyme present with least activity will be the rate-limiting step. The activity of an enzyme will result from the level of gene expression and the efficacy of the enzyme as a catalyst;

- if one wishes to regulate the flux of material through the pathway, it is this step which would be expected to be regulated. We should, however, note that this model is a simplification. Actual pathways are often found to have more complicated and subtle control mechanisms than this;

- allosteric regulation (whether inhibition or activation) is a widely used method for achieving 'second-by-second' adjustments in enzyme activity; this is known as 'fine control' or 'fine tuning' and is distinct from altering the rate of synthesis of the enzyme, which is called 'coarse control'.

8.2.5 Regulation of transcription

The control of metabolism by allosteric regulation is often termed 'fine' control: it acts rapidly to respond to the cells' second-to-second needs. Cells are also able to control the expression of their genome (transcription) which requires longer and is termed 'coarse' or 'long-term' control.

induction, repression and constitutive enzymes

Regulation of gene expression complements the regulation of enzyme activity. It serves to conserve energy and raw material by maintaining a balance between the amounts of the various cell proteins. For example, the chromosome of *Escherichia coli* can code for around 4,000 peptide chains, although only around 800 enzymes are present in *E. coli* growing with glucose as a carbon and energy source. Such regulation of gene expression is achieved mainly through induction and repression of enzyme synthesis. It is useful to introduce the term constitutive here. A constitutive enzyme is synthesised continuously and is not subject to induction or repression.

Induction is when the concentration of an enzyme rises because of the effect of a small molecule called an inducer.

Repression is when the concentration of an enzyme decreases because of the action of a small molecule. Such molecules are often end products of a metabolic pathway in which the repressible enzymes take part. These molecules are called co-repressors, not repressors, the reason for which will be explained below.

Inducible enzymes are synthesised only when the substrate is available (ie they are absent in the absence of the inducer). Conversely, repressible enzymes are always present unless the end-products of their pathways become available (eg an amino acid provided in the growth medium of a bacterial culture). Although induction and repression involve different mechanisms, they both ensure that enzymes are present in the cell only if they are needed. Many genes coding for catabolic enzymes are regulated by induction whereas many genes coding for biosynthetic enzymes are regulated by repression.

The regulation of gene expression is a complex subject and a detailed consideration is beyond the scope of this chapter. It is dealt with more fully in the Biotol texts, 'Genome Management in Prokaryotes,' and, 'Genome Management in Eukaryotes'. However, to fully appreciate how metabolism is regulated, it is essential to be familiar with the principal mechanisms of induction and repression. The key points concerning the mechanisms of both induction and repression are listed below. These are illustrated in Figure 8.9.

- Induction and repression arise principally from changes in the rate of transcription (mRNA synthesis).

- The rate of mRNA synthesis is controlled by repressor proteins synthesised under the direction of regulatory genes.

- The repressor protein binds to a specific site on DNA called the operator.

- The enzyme that catalyses mRNA synthesis (RNA polymerase) binds to a promoter which is located adjacent to the operator. If the repressor is bound to the operator, RNA polymerase binding to the promoter is prevented. Thus transcription is inhibited in the presence of bound repressor.

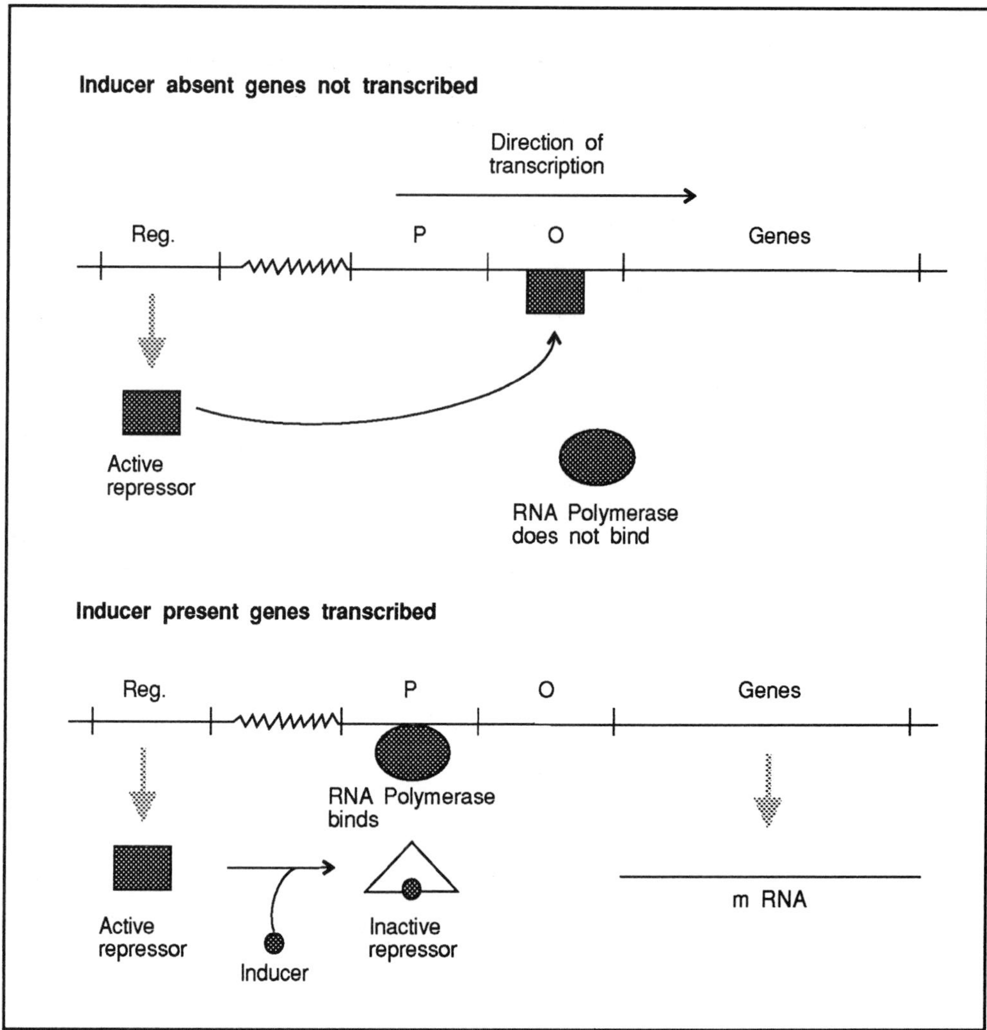

Figure 8.9 The mechanism of gene induction. Key: regulatory gene (Reg.); operator (O) and promoter (P).

Let us now examine the control of synthesis of a repressible enzyme.

Repression exists in both active and inactive modes; this forms the basis of the distinction between induction and repression from a mechanistic point of view. In inducible systems, the inducer stimulates transcription by reversibly binding to the repressor and causing it to change to an inactive shape: this prevents binding to the operator. In repressible systems, the repressor protein is initially inactive and only becomes an active shape (able to bind to the operator) when the co-repressor is bound to it. You can see now why the small molecule which causes repression is called a co-repressor.

In prokaryotes, the synthesis of several proteins can often be regulated by a single repressor. Here, the structural genes coding for polypeptides are lined up together on the DNA, and a single mRNA molecule carries all the messages. The sequence of bases coding for one or more polypeptides, together with the operator controlling its expression, is called an operon.

Π Now complete the illustration of the mechanism of repression by the addition of an active repressor (■), inactive repressor (Δ), co-repressor (•) and RNA polymerase (●) molecules.

Genes transcribed

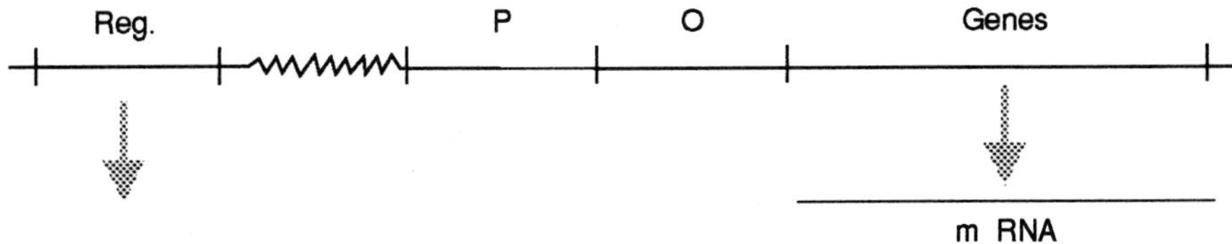

Genes not transcribed

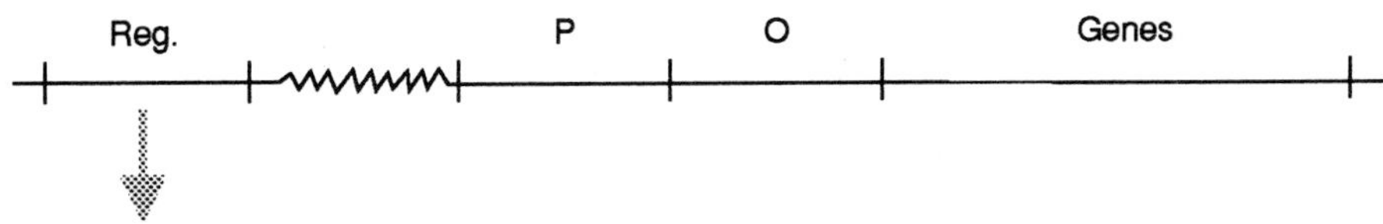

The mechanism of repression can be illustrated as follows:

Genes transcribed

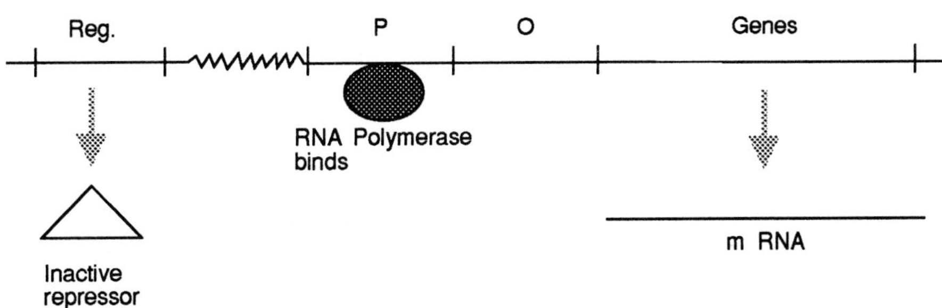

Genes not transcribed

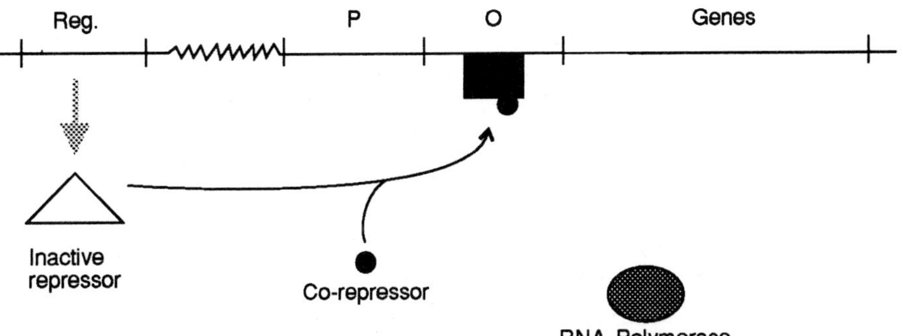

SAQ 8.1

For the induction and repression mechanisms already described, match each of the statements (a-e) with one of the following: 1) Induction; 2) Repression; 3) Both induction and repression; 4) Neither induction nor repression of mRNA synthesis.

Statements:

a) Binding of RNA polymerase to the operator.

b) The presence of an external molecule activates the repressor molecule.

c) Repressor binding to the operator prevents gene transcription.

d) Binding of co-repressor prevents the synthesis of inactive repressor.

e) The presence of an external molecule inactivates the repressor molecule.

The induction and repression mechanisms described so far are forms of negative control because transcription is prevented by the controlling factor, the repressor. Some operons are under positive operon control; they are stimulated by the controlling factor. The best studied operon under positive control is the lactose operon (lac operon) of *E. coli*. The operon codes for the enzymes responsible for lactose breakdown, among others, β-galactosidase. It shows both negative and positive control. The negative control is expressed as shown in Figure 8.9 with lactose filling the role of inducer. The lac operon function is further regulated by: 1) the catabolic activator protein (CAP) and 2) the small cyclic nucleotide 3′, 5′ -cyclic adenosine monophosphate (cAMP). Positive regulation of the lac operon is illustrated in Figure 8.10.

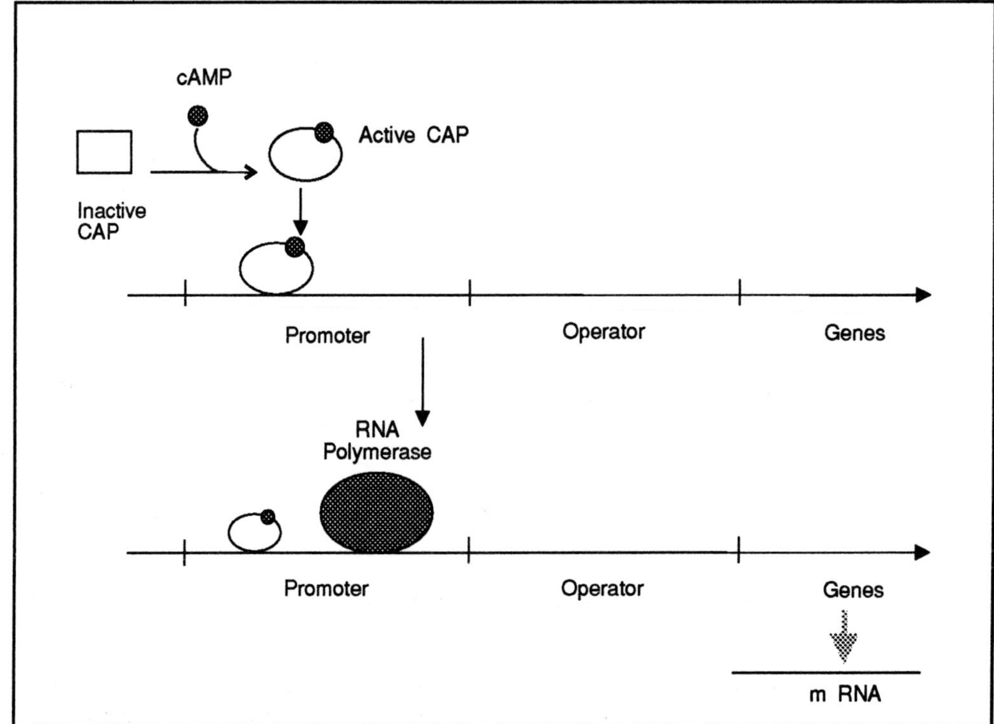

Figure 8.10 Positive control on the lac operon.

The lac promoter is different from those referred to earlier in that it contains a site to which CAP must bind before RNA polymerase can attach to the promoter and begin transcription. CAP is able to bind to the CAP site only when it is complexed with cAMP. The lac operon is, therefore, controlled by two control points rather than one; these being the CAP control and the lactose induction control. Thus expression of the lac operon occurs only when both of these control points are 'switched on'. We will see below how cAMP levels are controlled.

catabolic repression

A very interesting variety of positive operon control is called catabolite repression and regulates the synthesis of many catabolic enzymes. This involves the inhibition of utilisation of a catabolic substrate in the presence of another utilisable substrate. Catabolic repression is displayed, for example, by the lac operon of *Escherichia coli* and can be demonstrated simply by growing the bacterium in a medium containing both glucose and lactose. Glucose is used preferentially until it is exhausted, then, after a short lag, growth resumes with lactose as the carbon source. The biphasic pattern of growth obtained is called diauxic growth (Figure 8.11).

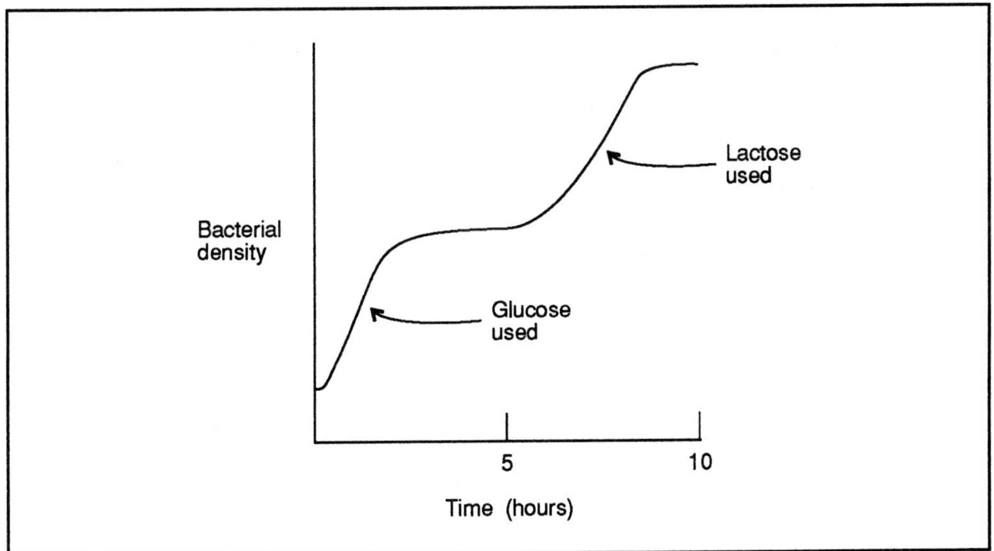

Figure 8.11 The diauxic growth curve obtained if *E. coli* is grown in a mixture of glucose and lactose. Analysis of glucose and lactose levels shows that no lactose is used until all the glucose has been consumed.

The enzymes for glucose catabolism are constitutive (active all the time) and unaffected by CAP activity. When the organisms are supplied with glucose, the cAMP level drops, CAP is thus no longer able to bind to the lac promoter and positive expression of the lac operon does not occur. We might regard cAMP as a 'hunger' signal. High levels of cAMP indicate that the cell is hungry and alternative catabolic pathways are switched on.

Π Can you think why control by catabolite repression is advantageous to *E. coli*?

It enables the bacterium to be more efficient ie it will use the most easily catabolised sugar first rather than unnecessarily synthesising the enzymes required for another carbon and energy source.

SAQ 8.2

For each of the three types of control mechanisms listed, select controlling factors from the other list provided that would i) enhance gene transcription and ii) prevent gene transcription.

Control mechanisms

1) Negative control by induction.

2) Negative control by repression.

3) Positive control by the lac operon.

Controlling factor

a) High levels of cAMP.

b) Inactive repressor.

c) RNA polymerase binding to the promoter.

d) Repressor binding to the operator.

e) Inactive CAP.

f) Inducer binding to repressor.

We have now covered the fundamental elements of control and are ready to examine how these are used to control the operation of metabolic pathways.

8.3 Patterns of regulation of metabolic pathways

Certain generalisations can be made about the regulation of metabolic pathways:

Biosynthetic pathways

- The first enzyme of the pathway is often an allosteric enzyme and is regulated by end-product inhibition ie inhibition of enzyme activity caused by the final product in the pathway.

- The synthesis of all enzymes of the pathway is often controlled by end-product repression ie repression of synthesis of enzymes caused by the final product in the pathway.

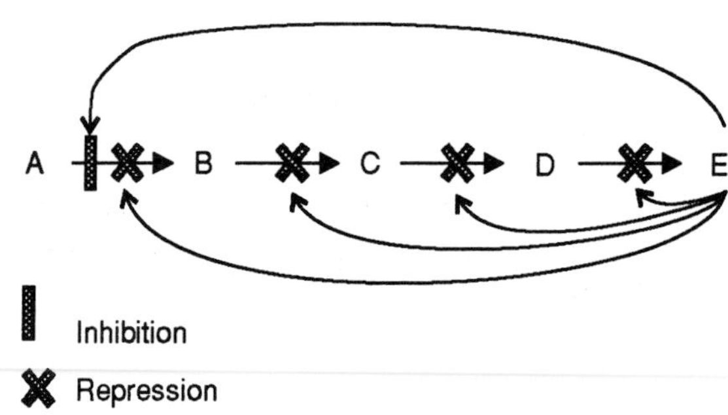

Catabolic pathways

- Pathways responsible for the degradation of substrates commonly encountered by the cell are catalysed by enzymes which are synthesised constitutively. They are regulated mainly by modifying enzyme activity.

- Pathways responsible for the metabolism of substrates that occur rarely in the environment of the cell are regulated primarily by enzyme induction and by catabolic repression, rather than by change of enzyme activity. We will now examine each of these in more detail.

8.3.1 Regulation of biosynthetic pathways

Highly efficient regulation of biosynthetic pathways is achieved through a combination of end-product repression and end-product inhibition the latter also being known as feedback inhibition.

Although the effect of end-product repression is immediate in regulating the rate of enzyme synthesis, were it to act alone, the pathway would continue to function until the concentration of the already existing enzymes was reduced to low levels. This would occur through protein turnover and by dilution of the enzyme through growth. End-product inhibition, however, brings about an immediate cessation of the operation of the pathway. Thus, if the end-product becomes too concentrated, it allosterically inhibits the control enzyme which slows down the end-product synthesis. As the end-product concentration decreases, pathway activity once again increases and more product is formed. In this way, end-product inhibition automatically matches supply and demand of the end-product.

Π Can you think of a situation in a growing bacterium in which the concentration of an end-product might be reduced?

Numerous answers could be suggested, one of which is the incorporation of free amino acid into proteins. Amino acids are the end-products of their biosynthetic pathways but their synthesis occurs only when protein synthesis lowers the free amino acid concentration. You probably cited the case of aspartate transcarboxylase and pyrimidine biosynthesis described earlier.

Frequently, a biosynthetic pathway is branched and a single intermediate is involved in the synthesis of two or more end-products. In these systems the synthesis of pathway end-products must be precisely co-ordinated; it would not be acceptable to have one end-product present in excess while another is lacking. The patterns of end-product inhibition that enable co-ordination of branched biosynthetic pathways will now be considered.

- End-product inhibition control of the first enzyme after the last common intermediate.

If an end-product is present in excess, it often inhibits the branch-point enzyme leading to its formation, in this way regulating its own formation without affecting the synthesis of other products.

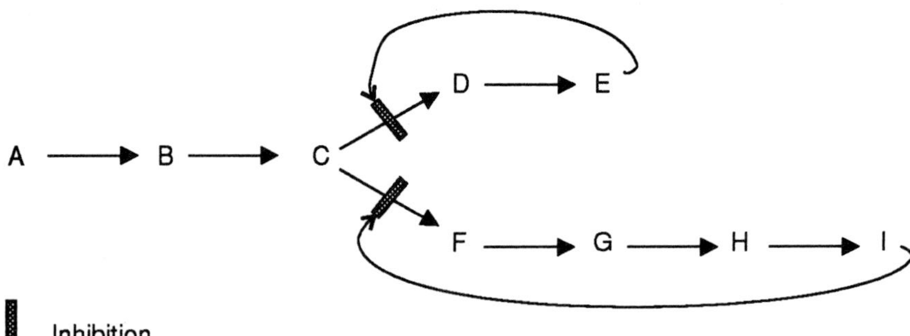

- Partial end-product inhibition of the first enzyme in the overall pathway by each end-product.

In this case an excess of one product slows the flow of carbon into the whole pathway at the same time as it inhibits the appropriate branch-point enzyme.

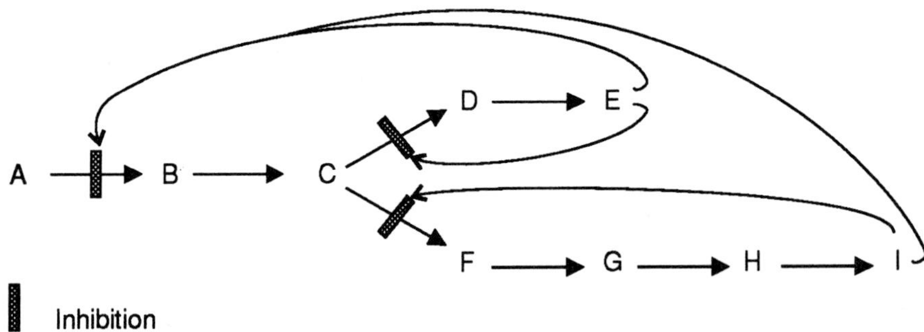

Because less carbon is required when one branch of a pathway is not functioning, partial feedback inhibition of the initial pacemaker enzyme helps match the supply and demand in branched pathways. In some cases, high levels of both end-products are needed to inhibit the first enzyme.

Alternatively each end-product can partially inhibit the first enzyme independently of the other, but maximum inhibition is only achieved by both products being in excess. This is termed cumulative end-product inhibition.

Π Look at the regulatory scheme above and see if you can draw two more ways that the reaction step A $\rightarrow$B may be controlled by feedback inhibition.

In multiple branched pathways the initial pacemaker step may be catalysed by several isoenzymes. These are different enzymes catalysing the same reaction and each may be under separate and independent control ie each isoenzyme may be inhibited by one particular end-product. In these situations, an excess of a single end-product reduces pathway activity, but does not completely block it because some isoenzymes are still active.

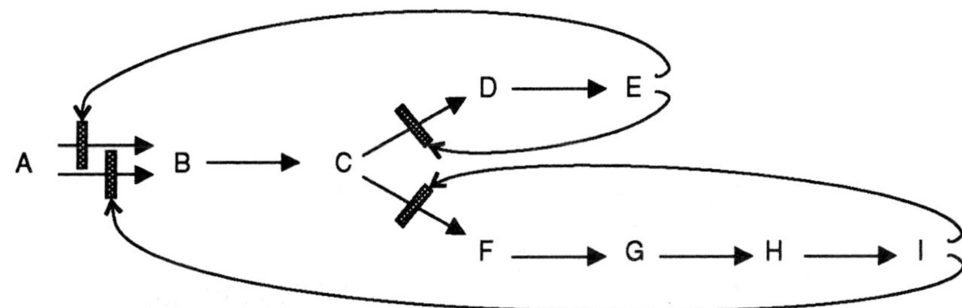

Sequential feedback inhibition.

In this type of control the first enzyme of a branched pathway is not controlled by the end-product of the pathway but by the metabolic intermediate preceding a branch point. End-product inhibition of branch point enzymes leads to an accumulation of this intermediate, which then inhibits the first enzyme in the pathway.

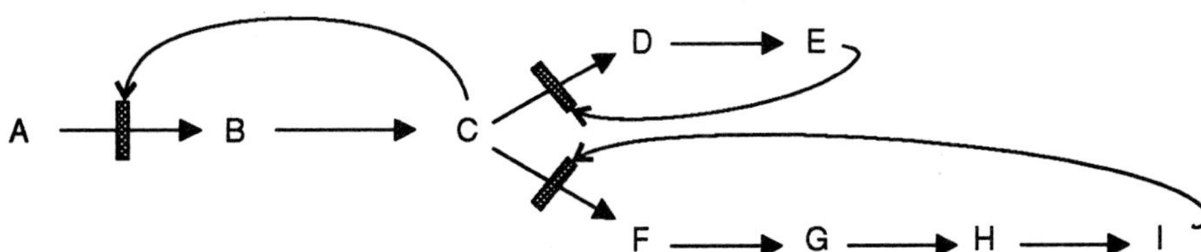

A good example of this multiplicity of feedback inhibition is provided by the biosynthesis of aromatic amino acids.

We can simplify this pathway in the form:

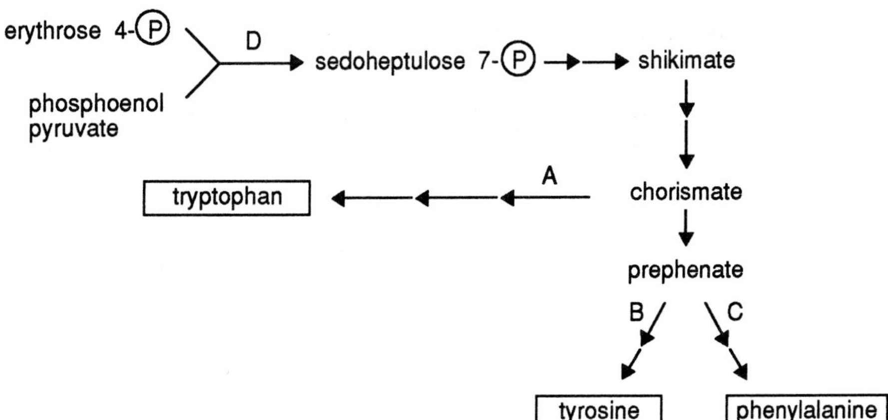

In many cases typtophan will inhibit enzyme A, tyrosine will inhibit enzyme B and phenylalanine inhibit enzyme C.

Enzyme D often exists as isoenzymes, one inhibited by tryptophan, one by tyrosine and one by phenylalanine.

Thus if all of the three amino acids are present, the pathway will be completely switched off. If on the other hand both tyrosine and phenylalanine are present, the pathway is diverted solely to tryptophan production.

It should however, be pointed out that some variation in the control of this pathway exists in different organisms. In some cases, the chorismate → prephenate step is also catalysed by isoenzymes, one being inhibitor (and/or repressed) by tyrosine, the other by phenylalanine. Chorismate may also regulate step D.

SAQ 8.3

A pathway is regulated as follows:

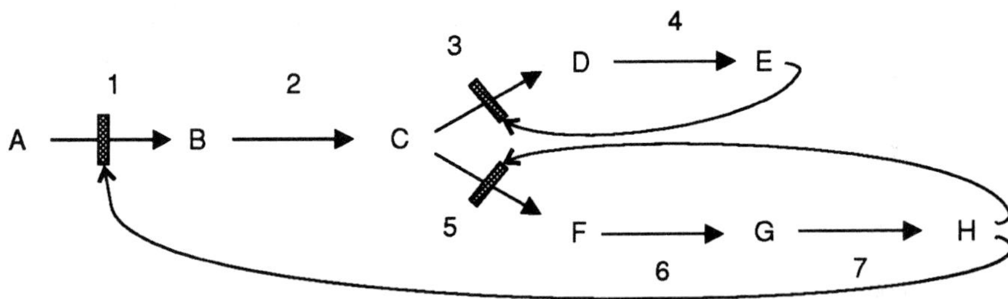

1) Explain how addition of excess E could reduce the flow of metabolite A into the pathway.

2) What happens to the synthesis of each of the metabolites (A-H) on addition of excess C?

3) Modify the pattern of regulation shown to illustrate the type of sequential and cumulative inhibition you might expect for this sequence of reactions.

8.3.2 Regulation of catabolic pathways

sequential induction

Many bacteria can use a very wide range of organic compounds as carbon and energy sources. The pathway for conversion of many of these substrates merge, and some substrates are intermediates in the metabolism of others. Such pathways are controlled in sections, rather than as a single unit. This enables bacteria to grow on any one of a number of substrates without synthesising enzymes unnecessarily. This is achieved by a process of sequential induction, where the product of one set of induced enzymes acts as an inducer for the next group of enzymes in the pathway. Figure 8.12 shows the pattern of sequential induction for catabolic enzymes during growth of a bacterium on aromatic substrates.

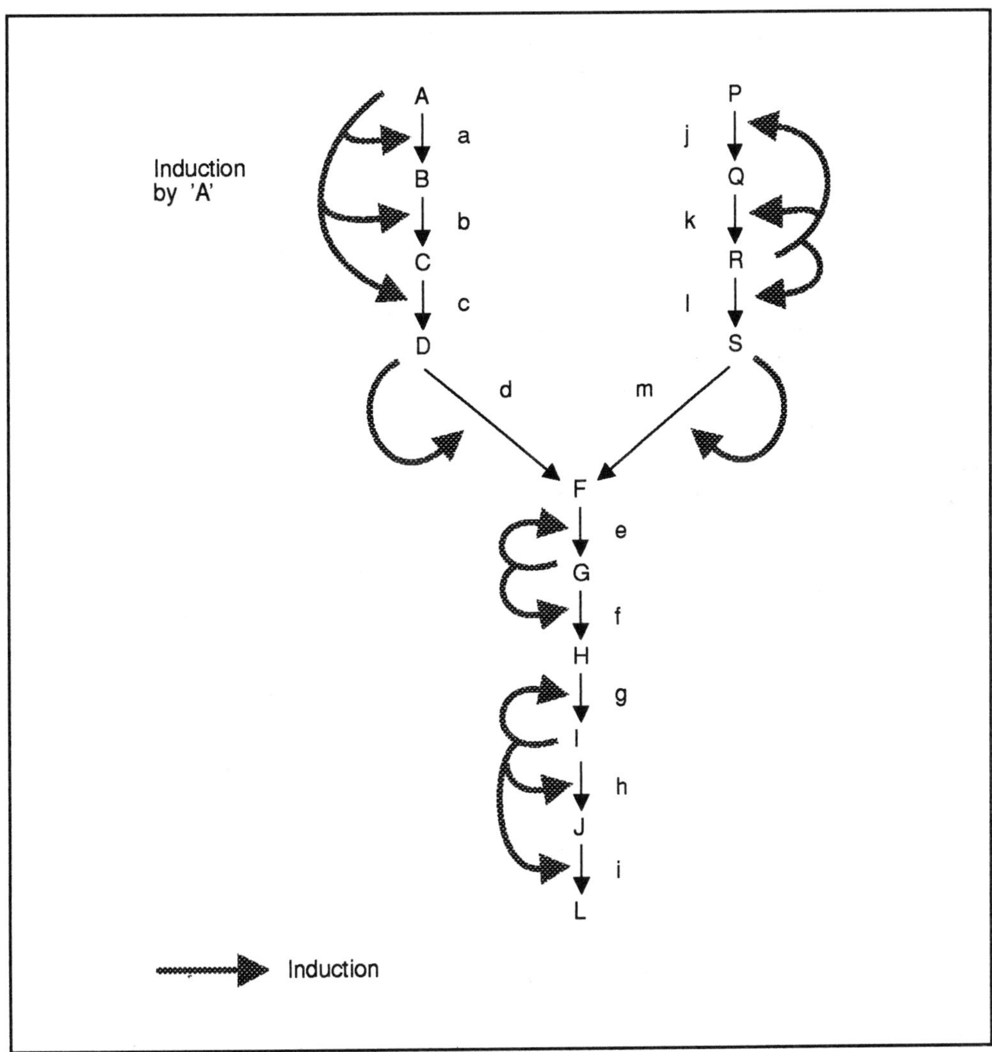

Figure 8.12 Pattern of sequential induction of catabolic enzymes.
Note the groupings of regulated parts of the pathway and that some compounds actually induce the enzyme which catalyses their own synthesis.

Π We can see from Figure 8.12 that metabolite R induces the synthesis of enzymes leading to its own formation. Suppose that the bacterium was provided with P as growth substrate and was shown to produce L. Write down how the bacterium was able to do this?

Low level of enzymes j,k,e and g must already have been present in the cell. These are required to form the inducer metabolites R,S,G and I which, in turn, boost the level of all the enzymes in the pathway.

8.4 Identification of the rate-limiting step

pacemaker enzyme

In the previous section we noted that only certain enzymes in a pathway are allosterically controlled. One of these enzymes determines metabolic flux through a pathway and is therefore termed the pacemaker enzyme. A pacemaker enzyme catalyses the rate-limiting reaction in the pathway. Because other reactions proceed more rapidly than the pacemaker reaction, changes in the activity of this enzyme directly alter the speed with which a pathway operates. We have already seen that the first step in a biosynthetic pathway is usually catalysed by an allosteric enzyme; as you might expect this is often the pacemaker enzyme for the whole pathway.

A full understanding of the regulation of a pathway requires the identification of the rate-limiting step and the factors responsible for changes in activity of the enzyme catalysing this step.

Identification of the rate-limiting step relies on the fact that it will always be a non-equilibrium reaction and the rate of the forward reaction is greater than the rate of the backward reaction. Other reactions in the pathway will be close to equilibrium ie the rates of the forward and reverse reactions will be almost identical and there will be little net flux in either direction. The ratio of the forward and the backward rates of reaction depends only on the ratio [products] : [substrates]. In a reversible reaction which reaches equilibrium, where the rates of the forward and reverse reactions are identical, this ratio is the apparent equilibrium constant (K_{eq}).

Π For a reaction at equilibrium, addition of an enzyme speeds up the rate of the forward reaction by a factor of 10^4. What happens to the speed of the backward reaction?

The backward reaction also speeds up by a factor of 10^4 and the position of equilibrium is unaffected.

Π If the reaction was not at equilibrium, would you expect the addition of enzyme to change the concentration ratio of reactants?

Yes. In this case the addition of enzyme would change the relative concentrations of reactants until the ratio was as close as possible to k_{eq} ie the enzyme speeds the move to equilibrium.

In order to explain why rate-limiting steps do not reach equilibrium, we will now consider a pathway in a steady-state condition ie when the concentration of the intermediates of the pathway remains constant. This requires the continual supply of substrate and the continual removal of products. Under steady-state conditions the net rate through each enzymatic step is the same and is the amount by which the 'forward' exceeds the 'back' rate. Where enzymes are not rate-limiting, both 'forward' and 'back' rates will be substantially greater than the net rate and the reaction will be essentially at equilibrium. In contrast, the rate-limiting enzyme will have little or no back rate and thus a forward rate close to the net rate. This is a consequence of the low activity of the pacemaker enzyme which is insufficient to bring the reaction to equilibrium ie the rate-limiting step is a non-equilibrium reaction.

To identify the rate-limiting step in a pathway, the first stage is to classify the intermediary reactions as near equilibrium (equilibrium) or far removed from equilibrium (non-equilibrium). The main experimental approaches that have been used to identify non-equilibrium reactions will now be examined.

8.4.1 Comparison of equilibrium constants and mass-action ratios

mass-action ratios

The mass-action ratio (MAR) is defined as the ratio of the concentration of the product(s) to those of the substrate(s) under given conditions. For the reaction:

$$A + B \rightleftarrows C + D$$

$$MAR = \frac{[C]\ [D]}{[A]\ [B]}$$

If an enzyme catalyses an equilibrium (or near equilibrium) reaction in the cell, the ratio [Products] : [Substrates] should approximate to the apparent equilibrium constant (K_{eq}) for the reaction. If the calculated value for MAR is similar to K_{eq} then it can be concluded that the reaction is near-equilibrium in the cell. However, if MAR is markedly smaller than the K_{eq} then this suggests that the reaction is displaced far from equilibrium and the reaction is a non-equilibrium reaction.

In general enzymes that are rate-limiting exhibit MAR values 2-4 orders of magnitude smaller than K_{eq}. The discrepancy between MAR and K_{eq} for near-equilibrium reactions is usually far less and is considered to be due to experimental error. Remember that to calculate MAR values you must measure the concentration of the substrate and the products. There will always be some error in carrying this out.

SAQ 8.4

Concentrations of substrates and products for two enzymes are given, along with K_{eq} values. Use the data to determine mass-action ratios for the enzymes. Decide whether each enzyme is in equilibrium *in vivo*. Which of the two enzymes do you think is the more likely to catalyse a rate-limiting step in a pathway?

Nucleoside diphosphate kinase

ATP + GDP $\rightleftarrows$ ADP + GTP; $K'_{eq} = 1$

Conc. in liver (mol l^{-1})

ATP 2.36×10^{-3}

ADP 0.97×10^{-3}

GTP 0.63×10^{-3}

GDP 0.21×10^{-3}

Pyruvate kinase

Phosphoenol pyruvate + ADP $\rightleftarrows$ Pyruvate + ATP; $K'_{eq} = 0.6 \times 10^{4}$

Conc. in liver (mol l^{-1})

Phosphoenol pyruvate 7×10^{-5}

Pyruvate 3×10^{-5}

ATP 3.75×10^{-3}

ADP 1.88×10^{-3}

8.4.2 The measurement of reaction rates

We know that a non-equilibrium reaction in a metabolic pathway arises because the enzyme which catalyses this reaction is not sufficiently active to allow the forward and backward reaction to take place sufficiently quickly to bring the substrates and products to equilibrium. It follows that enzymes catalysing the reactions leading to and from the rate-limiting step in the pathway must possess higher catalytic capacities for the formation of the substrate and removal of the product than the non-equilibrium enzyme has for its substrate and product. This provides a second method of identifying non-equilibrium reactions in a metabolic pathway. The maximum activities of all the enzymes in the pathway are measured *in vitro* under optimal conditions and on the basis of these results the enzymes are classed as low or high-activity enzymes. The enzymes in the low-activity group are considered to catalyse non-equilibrium reactions in the cell. In general it has been found that the difference in activities of these two groups of enzymes is 10- to 100-fold. However, if an enzyme is classified in the high activity group it will not necessary catalyse an equilibrium reaction in the cell; it may be inhibited to a very great extent in the cell. Consequently data on maximal enzyme activities are not considered in isolation but are usually supported by mass-action ratio measurements.

8.4.3 Isotopic labelling

Where measurements of maximal enzyme activities or metabolite concentrations are difficult to obtain or interpret it is not possible to decide whether a reaction is in non-equilibrium. In these cases the identification of non-equilibrium reactions can sometimes be achieved by using isotopically-labelled substrates.

isotopic method for identifying rate-limiting step

Consider the reaction sequence A $\rightarrow$ B $\rightarrow$ C $\rightarrow$ D each step of which is reversible and imagine that ^{14}C-labelled A is added. Each of the intermediates B to D will be labelled in turn with ^{14}C. If the rate of the reaction A $\rightarrow$ B is rapid in comparison with the net flux along the pathway, the amount of labelled B will rise rapidly compared with that of A. Similar arguments apply to all other near-equilibrium reactions in the pathway. In the case of a non-equilibrium reaction, the 'back' reaction will be very low and the labelled product will be relatively slow in reaching equilibrium with the corresponding labelled substrate. In practice, therefore, a study of the change in the amount of radioactivity associated with each metabolic intermediate as a function of time will identify the controlling step in the pathway.

8.4.4 Addition of intermediates

If, in a sequence A $\rightarrow$ B $\rightarrow$ C $\rightarrow$ D $\rightarrow$ E $\rightarrow$ F, the reaction C $\rightarrow$ D is rate-limiting, then addition of D or E will lead to an increase in the rate of appearance of F. Since A,B or C have to pass the rate-limiting step at C $\rightarrow$ D, addition of these compounds will not greatly alter the rate of appearance of F. By addition of suitable intermediates, therefore, it should be possible to locate the rate-limiting step.

Two problems often arise with this approach: 1) membrane barriers may prevent intermediates from reaching the site where the reaction is occurring and 2) the amount added may be such that the regulatory properties of the systems are altered.

SAQ 8.5

In the reaction sequence A $\rightarrow$ B $\rightarrow$ C $\rightarrow$ D $\rightarrow$ E, which of the following statements are consistent with C $\rightarrow$ D being the rate-limiting step.

1) Addition of ^{14}C-labelled A leads to the rapid appearance of ^{14}C-labelled B and C.

2) Addition of B leads to an increase in the rate of appearance E.

3) Addition of ^{14}C-labelled E leads to the rapid appearance of ^{14}C-labelled D and C.

4) The steady-state concentration of A, B and C are high relative to those of D and E.

8.4.5 The addition of regulatory metabolites inhibitors

crossover theorem

The use of allosteric modulators which inhibit enzyme - catalysed reactions to identify the rate-limiting step of a pathway relies on the crossover theorem. This states that when a pathway is inhibited at a specific reaction, the substrate concentration will increase and the product concentration will decrease.

Consider the pathway:

$$A \xrightarrow{E_1} B \xrightarrow{E_2} C \xrightarrow{E_3} D \xrightarrow{E_4} E \xrightarrow{E_5} F$$

for which compound X inhibits pathway flux. Let us assume that the analysis of metabolites shows that there is an accumulation of the equilibrium mixture of A, B and C and a reduction in D, E and F compared with the levels before addition of inhibitors. This is illustrated in Figure 8.13 with the minus inhibitor values expressed as 100%.

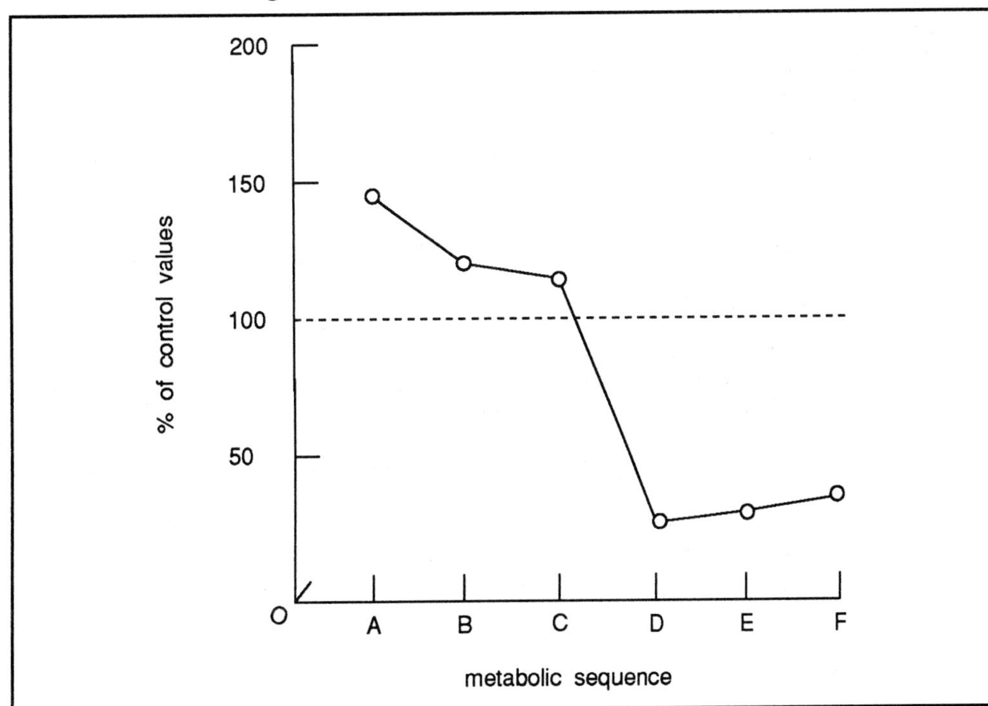

Figure 8.13 Crossover plot. Control values (100%) are the steady state concentrations of the metabolite in the pathway before addition of an inhibitory regulator.

crossover plot

Crossovers are obtained when the plus inhibitor lines cross the 100% line (minus inhibitor line). We can see from Figure 8.13 that C to D is the rate-limiting step for the pathway. Not surprisingly Figure 8.13 is known as a crossover plot.It is also possible to obtain a crossover plot by adding a positive modulator. In this case the result would be the inverse of that shown in Figure 8.13. Thus the levels of A, B and C would decrease, D, E and F would increase.

SAQ 8.6

Addition of an allosteric regulator was found to inhibit the reaction sequence A $\rightarrow$ B$\rightarrow$ C$\rightarrow$D. When the concentrations of the metabolites were determined in the presence and absence of the regulator, the following data were obtained:

	- inhibitor	+ inhibitor
A	0.197	0.551
B	0.006	0.022
C	0.239	0.037
D	0.059	0.029

Determine the rate-limiting step from a crossover plot of the data provided.

SAQ 8.7

Identify each of the following statements as True or False giving a reason for your response.

1) A rate-limiting enzyme will have a mass-action ratio far higher than its K_{eq}.

2) A rate-limiting reaction will have a forward rate close to the backward rate.

3) A pathway in steady-state has no net rate.

4) An enzyme with high activity *in vivo* is likely to catalyse an equilibrium reaction.

5) Measurement of reaction rates under optimum conditions *in vitro* is likely to overestimate the *in vivo* rate of equilibrium reactions to a greater extent than it does for rate-limiting reactions.

6) All irreversible reactions in a pathway are non-equilibrium reactions and are thus all rate-limiting for the pathway.

8.5 Optimisation of the flux through a pathway

Efforts are made in biotechnology to change cells (bacteria in particular) by means of genetic manipulation so that they carry out certain processes more rapidly. An obvious way to do this is to identify the rate-limiting step in the pathway of interest and then genetically manipulate the organism so that more of the pacemaker enzyme is expressed. However, before starting such experiments it is important to consider what might be expected from such an approach.

We know that the velocity of a non-equilibrium enzyme-catalysed reaction is directly proportional to the concentration of that enzyme. This conclusion can be expressed mathematically in the so-called flux-control coefficient:

flux-control coefficient

$$\text{Flux–control coefficient} = \frac{dv\ /v}{d\,[E]\ /[E]}$$

ie the relative change in velocity divided by the relative change in enzyme concentration.

Π If a 1% increase in the concentration of the enzyme leads to an increase in velocity of 1%, what is the flux-control coefficient?

The flux-control coefficient is unity (one).

If we have a flux control coefficient of 0.5, this means that for each 1% rise in enzyme activity, we increase the flux (velocity) of the reaction by 0.5%.

A flux-control coefficient of one is usually obtained when dealing with a single enzyme-catalysed reaction. However, if we consider a biosynthetic pathway, an increase in concentration of the first enzyme may lead to an increase in velocity of that reaction but the flux-control coefficient is usually not one. This indicates that the control of metabolic flux through the pathway is usually distributed among more than one enzyme in the pathway. Consider Figure 8.14 which shows three hypothetical outcomes of an experiment in which the flux through a pathway is plotted against the amount of the first enzyme of the pathway.

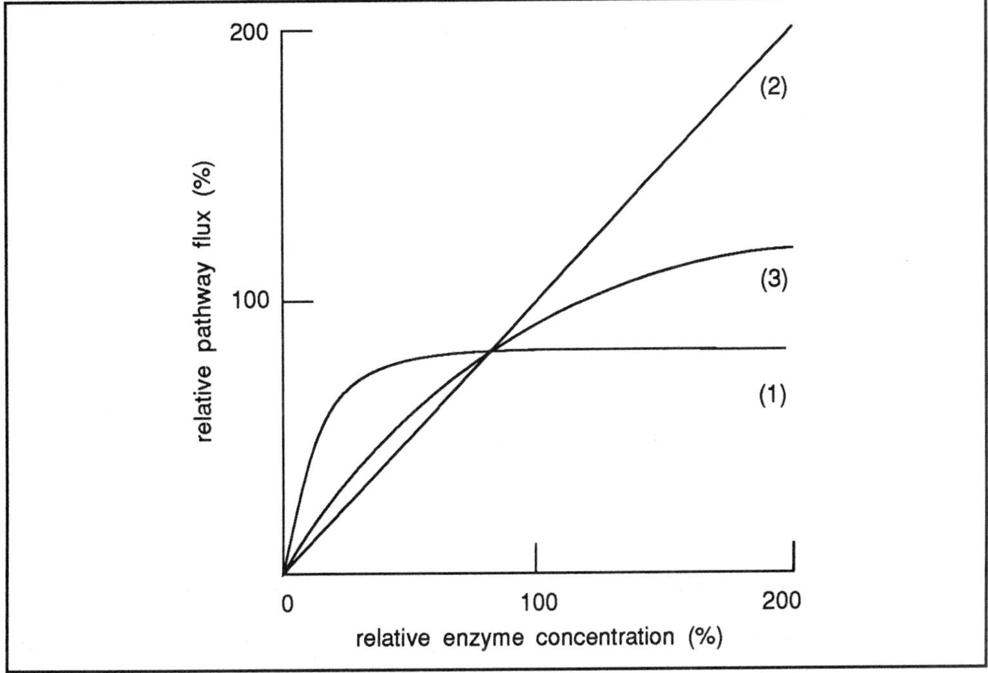

Figure 8.14 Possible effects on the flux through a pathway by altering the concentration of the first enzyme in the pathway. 1, 2 and 3 are hypothetical examples of the relationship between enzyme concentrations and flux through a pathway. Changes in pathway flux are plotted relative to the unmodified (wild-type) organism ie 100% = pathway flux in the unmodified organism. The normal amount of enzyme is indicated by 100% (see text for detailed discussion).

In case 1), halving or doubling the enzyme concentration hardly affects the pathway flux; this indicates that the enzyme is not catalysing the rate-limiting step. In case 2), halving or doubling the concentration of the enzyme changes the pathway flux in direct proportion; the enzyme clearly catalyses the rate-limiting step. However, in case 3), the enzyme is not as rate-limiting as in case 2) yet it is more rate-limiting than in case 1). Clearly a two-valued logic system (rate-limiting versus non-rate-limiting) is not always appropriate.

The experimental approaches described previously (section 8.4) actually identify the reaction that is most rate-limiting for the pathway. If the amount of this enzyme is increased pathway flux is also increased but it is obvious that this cannot continue for ever. At some point control of flux through the pathway would be taken over by another enzyme ie another enzyme would become rate-limiting. In practice scientists involved in genetic engineering are facing such questions as to what extent will 'improvement' of one of the enzymatic steps in a complete metabolic pathway accelerate the complete pathway? It follows that it would be useful to be able to quantify the degree to which an enzyme under study is rate-limiting for the pathway. The flux-control summation theorem allows us to do this.

flux-control summation theorem

The flux-control summation theorem shows that the sum of the flux-control coefficients for all the enzymes in a pathway equals unity. So, what use is the flux-control summation theorem when devising a strategy for genetic manipulation of an organism? Well, if we were to find that an enzyme in a pathway had a flux-control coefficient of 0.2, it would be sensible to investigate other enzymes in the pathway to establish where the rest (1.0-0.2=0.8) of the flux-control operates. If it was found to be a single enzyme with this flux-control coefficient, then that enzyme would be a much better candidate for genetic manipulation: the same percentage increase in its concentration would produce a four-fold (0.8/0.2) greater increase.

SAQ 8.8

For the reaction sequence:

$$A \xrightarrow{E_1} B \xrightarrow{E_2} C \xrightarrow{E_3} D$$

a 2% increase in enzyme concentrations increases pathway flux by 1.4% for E_1 and by 0.2% for E_2.

1) What is the flux-control coefficient of E_3?

2) Which enzyme is most rate-limiting for the pathway?

3) Quantify the extent to which E_1 is a better candidate for genetic manipulation than other enzymes in the pathway.

4) If an enzyme in another pathway competed for B would you expect this enzyme to have a negative or positive flux-control coefficient with respect to the pathway $A \rightarrow B \rightarrow C \rightarrow D$. Give a reason for your response.

Summary and objectives

Limitation of total substrate is a potential regulatory mechanism for almost any pathway. In addition, the rates of some pathways are influenced by physical or chemical compartmentation of substrate within cells. Metabolic pathway flux is also controlled by the availability of enzyme in a form able to bind to substrate. Factors that influence this include: compartmentation; the interconversion of active and inactive forms of an enzyme; protein degradation;the activation of inactive enzyme precursors.

'Fine control' of metabolic pathway flux is achieved mainly by allosteric regulation and common patterns of fine control by end-product inhibition have been identified for biosynthetic pathways. 'Coarse control' of metabolic pathway flux is achieved by the regulation of gene expression; mechanisms involved include negative control by induction or repression and positive control by catabolite repression. Rate-limiting steps in a pathway can be identified by a variety of experimental approaches that rely on the fact that the rate-limiting reaction is far from equilibrium. The degree to which an enzyme is rate-limiting for a pathway may be determined according to the flux-control summation theorem. This can be used to optimise the metabolic flux through a pathway.

Now that you have completed this chapter you should be able to:

- give examples to illustrate how physical and chemical compartmentation can influence the rate of enzyme catalysed reactions.
- describe mechanisms of negative and positive control of gene transcription.
- predict how different control patterns would influence the operation of pathways.
- interpret data from experiments designed to identify the rate-limiting step in a pathway.
- use the flux control summation theorem to predict the most rate-limiting step in a pathway.

Responses to SAQs

Responses to Chapter 2 SAQs

2.1

1) Glucose is likely to act as a source of carbon and energy for the cells. Methionine is likely to be a growth factor for the synthesis of proteins.

2) Nitrogen enters the cell in the form of ammonium ions (NH_4^+) derived from NH_4Cl. Sulphur enters the cell in the form of sulphate ions (SO_4^{2-}) derived from $MgSO_4.7H_20$.

3) Methionine is a sulphur containing amino acid and so could serve as a combined source of carbon, nitrogen and sulphur for the cells. The low level of methionine in the medium would, in fact, support very little growth. The glucose, NH_4Cl and $MgSO_4.7H_2O$ are readily utilisable sources of C, N and S and would be preferentially used.

4) A more diverse population of organisms would grow because the yeast extract would satisfy the growth factor requirement of many organisms.

2.2

Permeability will decrease in the following order: H_2, O_2, H_2O, glycerol, glucose, Na^+, glucose-6-phosphate.

The uncharged, non-polar molecules are the ones that diffuse most readily across a membrane. Since H_2 is smaller than O_2 it will pass through the membrane more readily. As to the polar uncharged molecules; on the basis of their molecular size, we can establish the following order of permeability: H_2O, glycerol, glucose. At pH 7 glucose-6-phosphate is charged (the phosphate group is ionised) and will therefore be less able than Na^+ to pass through the membrane.

2.3

1) False. Passive diffusion can take place through protein channels. Although the transported molecule is not bound to the protein, as in carrier-mediated transport, proteins are nevertheless involved in this type of passive diffusion.

2) True. All carrier-mediated transport processes (passive and active) show saturation kinetics because they involve the binding of transported substrate to the carrier.

3) True. Unidirectional transport occurs if energy is used (active) and bidirectional transport if energy is not coupled to the carrier (passive).

2.4

a) 3), 4), 5) and 6);

b) 2) and 7);

c) 2);

d) 1),2) and 7).

2.5 Since HPr is a common phosphotransferase in the transport of sugars, a non-functional HPr will disturb the transport of several sugars. With a defective enzyme II the transport of only one kind of sugar will be disturbed.

2.6

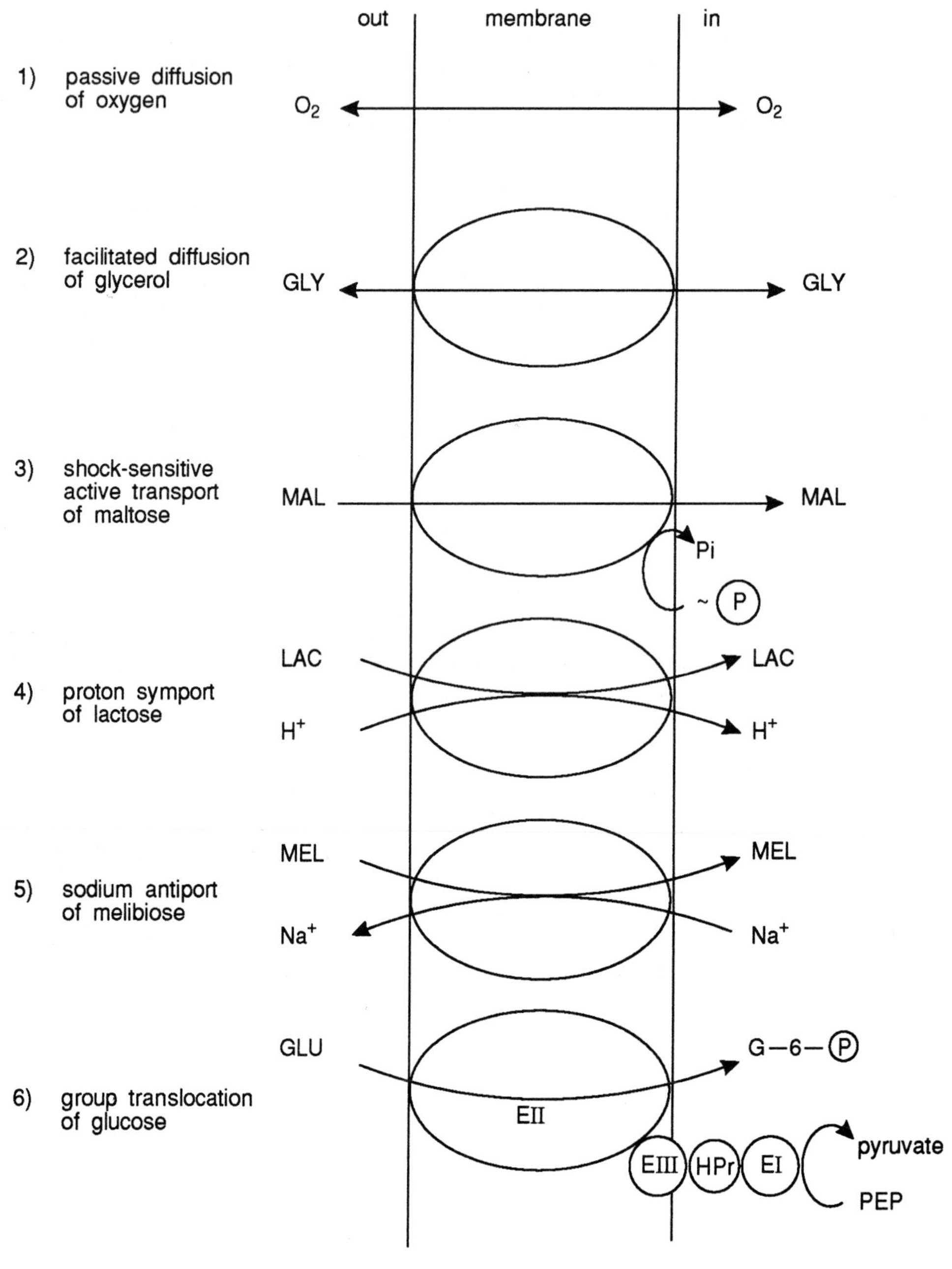

2.7 1) The membrane will have a glucose carrier but no citrate carrier. Glucose genetically represses the formation of the citrate carrier.

2) The membrane contains both glucose and citrate carriers. The glucose carrier is constitutively formed ie it is produced even in the absence of glucose. The citrate carrier is induced by the citrate in the medium.

3) The membrane will not contain a functional citrate carrier protein. The membrane will either not have a citrate carrier protein or will have a non-functional protein coded by the mutated citrate carrier gene. This will depend on the type of mutagenesis used to obtain the transport negative mutant. The glucose carrier will be present since it is constitutively formed.

4) The membrane will not contain a functional glucose carrier. The citrate carrier will be absent because there is no citrate to induce its synthesis.

2.8 The results could suggest that the transport system by which the first class of solute enters the cell is dependent upon phosphate bond energy. Transport of the second class of substrates was independent of phosphate bond energy; transport of these substrates is presumably driven directly by the proton-motive force.

2.9 Transport rate in the presence of a diffusional carrier depends on the diffusion across the membrane of the whole carrier. This diffusion will take place more easily in a liquid crystalline bilayer than in a bilayer in which the lipids are in the gel phase. If transport rate is determined as a function of temperature, we will see a sudden increase in transport rate at the transition temperature of the liquid phase.

A channel-former migrates laterally in the membrane. It is open when the two halves come together. It closes again when they move apart. The rate of migration will of course increase at the transition temperature, but this will not increase the proportion of time in which the channel is open so the transport rate will gradually increase with increasing temperature (temperature dependence of diffusion).

2.10 Conductivity will increase if a channel is created by means of the association of two gramimycin molecules in both monolayers of the membrane. After approximately one second the complex dissociates and the 'link' is broken: conductivity decreases. For each open channel one particular electrical surge (ion flux) takes place, which means a discrete increase in conductivity. When more channels are open at the same time a multiple of the increase in conductivity found with one channel will be observed. Therefore the conductivity changes in a stepwise manner.

Responses to chapter 3 SAQs

3.1 2) and 5)

3.2

1) True. This is likely because sulphite and hydrogen sulphide are intermediates in the route of sulphate assimilation. However, this assumes that sulphite and hydrogen sulphide can enter the cell and indeed in most cases they can.

2) False. The sulphur in hydrogen sulphide is at the same oxidation state as the sulphur in cysteine so reduction does not occur.

3) False. Serine is not a sulphur-containing amino acid and so cannot serve as a source of sulphur. Only cysteine or methionine could serve as combined sources of carbon, nitrogen and sulphur.

4) True. Assimilation of sulphate requires two ATP, four NADPH and the synthesis of four enzymes that are not necessary for the assimilation of cysteine.

5) True. Although only two ATP are used three, high energy bonds are cleaved: two in the formation of adenosine 5′-phosphosulphate and one in the formation of phosphoadenosinephosphosulphate (adenosine 3′-phosphate 5′-phosphosulphate).

3.3

Enzyme 1) matches with statement j) and g).

Enzyme 2) matches with statements b), d), f) and j).

Enzyme 3) matches with statement b).

Enzyme 4) matches with statements a) and d).

Enzyme 5) matches with statements a), d), e), h) and i).

Enzyme 6) matches with statement a).

Enzyme 7) matches with statement e).

3.4 Pyruvate will support nitrogen fixation in anaerobes but not aerobes. Therefore, ATP must come from substrate level phosphorylations.

3.5

1) True. Nitrogen fixation is an anaerobic process and facultative anaerobes do not have any special mechanisms of O_2 protection.

2) True. NADH and NADPH have been shown to function as electron donors for nitrogen fixation in aerobic organisms.

3) False. Glutamate dehydrogenase has a low affinity for NH_3 and will only function when ammonia concentrations are high. Because nitrogen fixation is an energy demanding process, nitrogenase activity is regulated and ammonia concentrations

are always low. Ammonia assimilation following nitrogen fixation thus proceeds via GOGAT and glutamine synthetase (the low ammonia route).

4) False. Symbiotic nitrogen fixers rely on the production of leghaemoglobin by the plant.

5) False. Binding of nitrogen does occur at the MoFe-co site but this is located on the nitrogenase component, not on the nitrogenase reductase component.

6) True. *Azotobacter* species are chemotrophic (heterotrophic) nitrogen fixers and rely on the carbon source for the generation of reductants and ATP for nitrogen fixation.

7) True. Only two electrons must be transferred to acetylene to form ethylene, whereas six must be transferred to nitrogen to form ammonia.

8) True. Molybdenum has two oxidation states, Mo^{5+} and Mo^{6+}. It can thus accept an electron becoming Mo^{5+} and donate an electron becoming Mo^{6+}. You will recall that molybdenum also functions in this way during nitrate assimilation.

3.6 For *Rhizobium* responses 6), b) and iii) are appropriate.

For *Azotobacter* responses 2), b) and iv) are appropriate.

For *Frankia* responses 6), b) and iii) are appropriate.

For *Klebsiella* responses 4), a) and i) are appropriate. Note *klebsiella* is a facultative anaerobe. It only fixes nitrogen under anaerobic conditions.

For purple bacteria responses 3), a) and i) are appropriate.

For Cyanobacteria the following combinations are appropriate:

1), b) and ii); 3), a) and i); 5), b) and iii.

3.7 Conditions 1): the organism assimilates nitrate via nitrate reductase c) and nitrite reductase e). These will produce ammonia which will then be assimilated by GOGAT and glutamine synthetase d).

Conditions 2): the organism assimilates ammonia via the high ammonia route - glutamate dehydrogenase a). As the concentration of available ammonia drops in the medium the low ammonia route will become operative - GOGAT and glutamine synthetase d).

Conditions 3): assimilation of nitrogen cannot proceed because nitrogenase is inactivated under the aerobic conditions and the organism is unable to grow. You will recall that there is no special mechanism of O_2 protection in facultative anaerobes.

Condition 4): the organism assimilates nitrogen via nitrogenase b) and the low ammonia route, GOGAT and glutamine synthetase d).

Condition 5): NH_3 initially suppresses nitrogenase synthesis and the organism assimilates ammonia via the high ammonia route, glutamate dehydrogenase a). As the concentration of available ammonia drops in the medium the low ammonia route will become operative; GOGAT and glutamine synthetase d). As the medium becomes depleted of ammonia the organism will assimilate dinitrogen via nitrogenase b), GOGAT and glutamine synthetase d).

3.8

1) *Rhizobium*. The limiting step in the nitrogen cycle is nitrogen fixation and *Rhizobium* is a symbiotic nitrogen fixer.

2) All organisms contribute to the nitrogen cycle because they assimilate NH_3.

3) *Azotobacter*. This organism is a free-living aerobe and chemotrophic. It will utilise organic material in this dark aerobic environment.

4) *Klebsiella*. Swamp mud is often an anaerobic environment with high levels of utilisable carbon. This organism is a free-living anaerobe and chemotrophic and is thus well-suited to this environment.

5) *Anabaena*. Shallow fresh water often has a low mineral and organic carbon content and chemotrophs are not suited to these conditions. However, cyanobacteria like *Anabaena* are phototrophic and can use light as an energy source and CO_2 as carbon source.

Responses to Chapter 4 SAQs

4.1

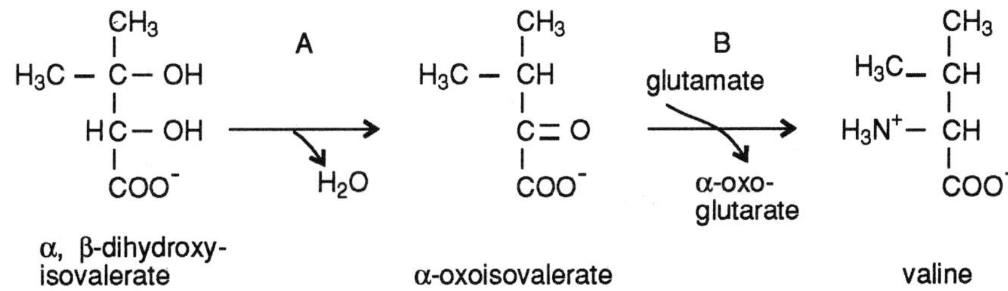

Enzyme A dihydroxyacid dehydratase
Enzyme B aminotransferase

4.2

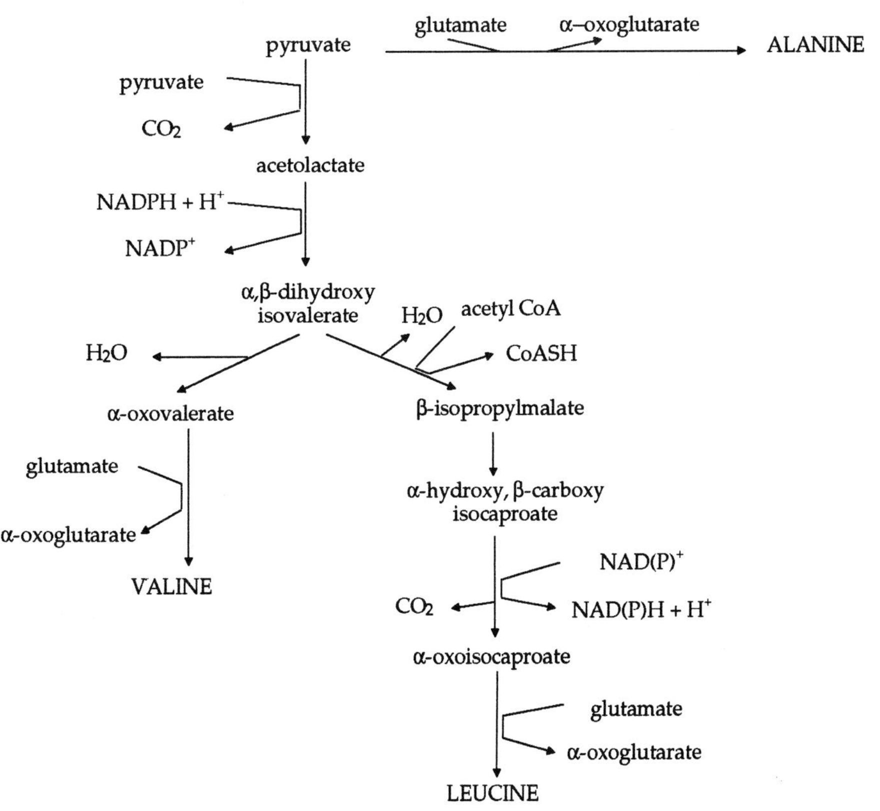

a) pyruvate + glutamate → alanine + α-oxoglutarate;
b) 2 pyruvate + NADPH + H^+ + glutamate → valine + CO_2 + $NADP^+$ + H_2O + α-oxoglutarate;
c) 2 pyruvate + acetyl CoA + glutamate → leucine + $2CO_2$ + H_2O + CoASH + α–oxoglutarate.

4.3

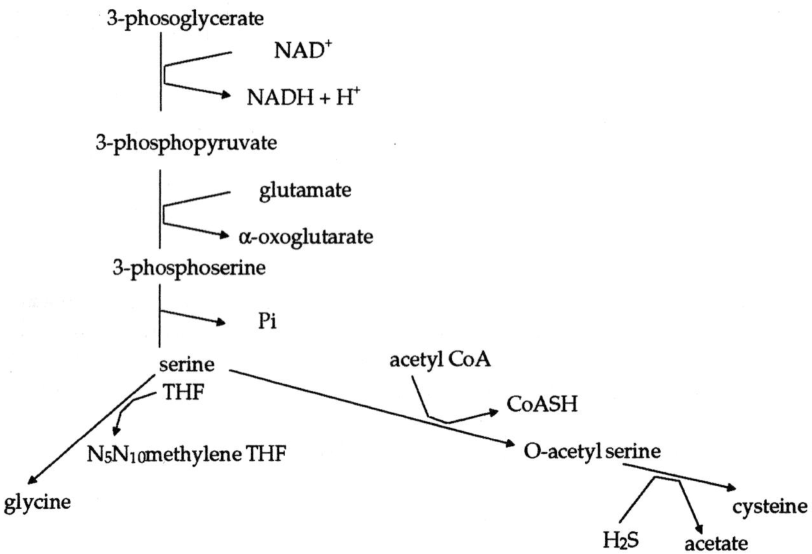

4.4 1) Oxaloacetate is one of the intermediates of the TCA cycle.

2)

$$\begin{array}{l} COO^- \\ | \\ C=O \\ | \\ CH_2 \\ | \\ COO^- \end{array}$$

oxaloacetate, a four-carbon, α-oxo dicarboxylic acid

3)

$$\begin{array}{r} COO^- \\ | \\ H_3N^+-CH \\ | \\ CH_2 \\ | \\ COO^- \end{array}$$

aspartate, a four carbon, α-amino dicarboxylic acid

4) An aminotransferase (transaminase) usually using glutamate as the donor.

5)

$$\begin{array}{r} COO^- \\ | \\ H_3N^+-CH \\ | \\ CH_2 \\ | \\ H_2N-C=O \end{array}$$

asparagine, a four-carbon, α-amino-δ-amido dicarboxylic acid

6) Aspartate + NH_3 + ATP $\rightarrow$ asparagine + AMP + PPi + H_2O.

The important points in 6) are that energy is required in the form of ATP and usually ammonia is incorporated rather than an organic group transaminated.

4.5 This answer atypically is somewhat lengthy but it establishes basic rules of metabolic control so it will pay you to concentrate on it and stay with it to the end: if you can do so it will help you with many such future metabolic systems.

There are so many individual enzymes within cells that control of all of them would in practice be impossible. When trying to deduce where controls occur in complicated systems such as the one in Figure 4.4 there are one of two 'rules of thumb' which you can apply and generally you will be correct. Thus it is better to remember the rules rather than many individual reactions as this avoids unnecessary learning and leaves time to promote understanding.

Firstly there must be a mechanism to switch on and off the whole pathway. This mechanism must occur very early in the system to avoid the cell doing something which is unnecessary and wasteful. The logical place therefore is to block immediately after aspartate in the system. If you guessed or concluded this you are correct, as the first enzyme (aspartokinase) is subject to complex repression or inhibition. (Note: remember repression refers to control of enzyme synthesis whereas inhibition refers to control of activity of enzymes already present.

Many organisms have several aspartokinases, for example *Escherichia coli* has three:

- aspartokinase I is inhibited by threonine;
- aspartokinase II is repressed but not inhibited by methionine;
- aspartokinase III is inhibited and repressed by lysine.

There is no evidence of separate channelling of the substrate aspartate, it is available to all enzymes. The net result of control of aspartokinase(s) is a lowering of production of the first compound after aspartate if one or more amino acids are present in excess. Further control is exercised at the branch points because it makes sense to control pathways at the first reaction which is unique to a given amino acid and this is following a branch. The obvious compound to inhibit or repress will be the product of the branch point and this process is called end-product inhibition or end-product repression. Thus lysine inhibits conversion of aspartate-β-semialdehyde to dihydropicolinic acid; methionine inhibits the conversion of homoserine-O-phosphate to O-succinylhomoserine and isoleucine inhibits conversion of threonine to α-oxobutyrate.

Further control exists in the conversion of aspartate-β-semialdehyde to homoserine where threonine and methionine respectively inhibit the two isoenzymes. This control point avoids over-production of threonine if methionine and isoleucine are in excess.

4.6

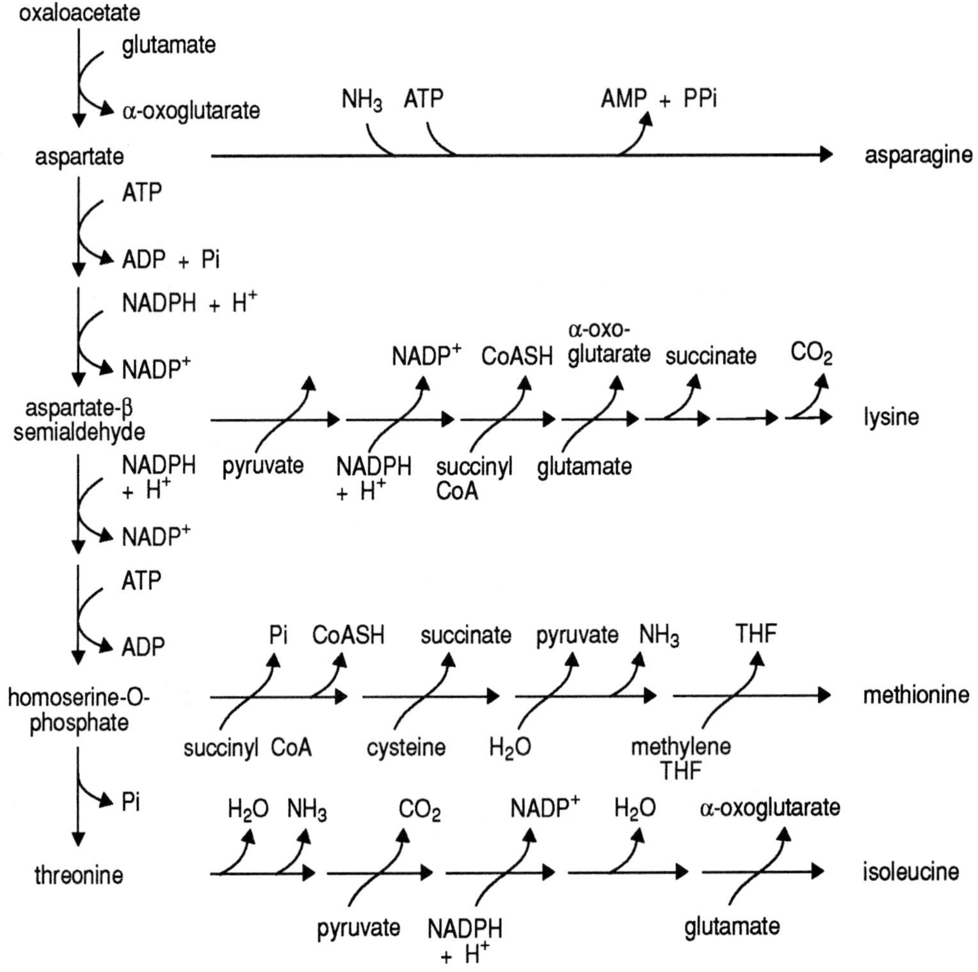

4.7

1) The seven precursors are in italics in our answer to part 2): 3-phosphoglycerate, phosphoenol pyruvate and pyruvate originate in glycolysis. Erythrose 4-phosphate and ribose 5-phosphate occur in the pentose phosphate pathway and oxaloacetate and α-oxoglutarate arise in the TCA cycle.

2)

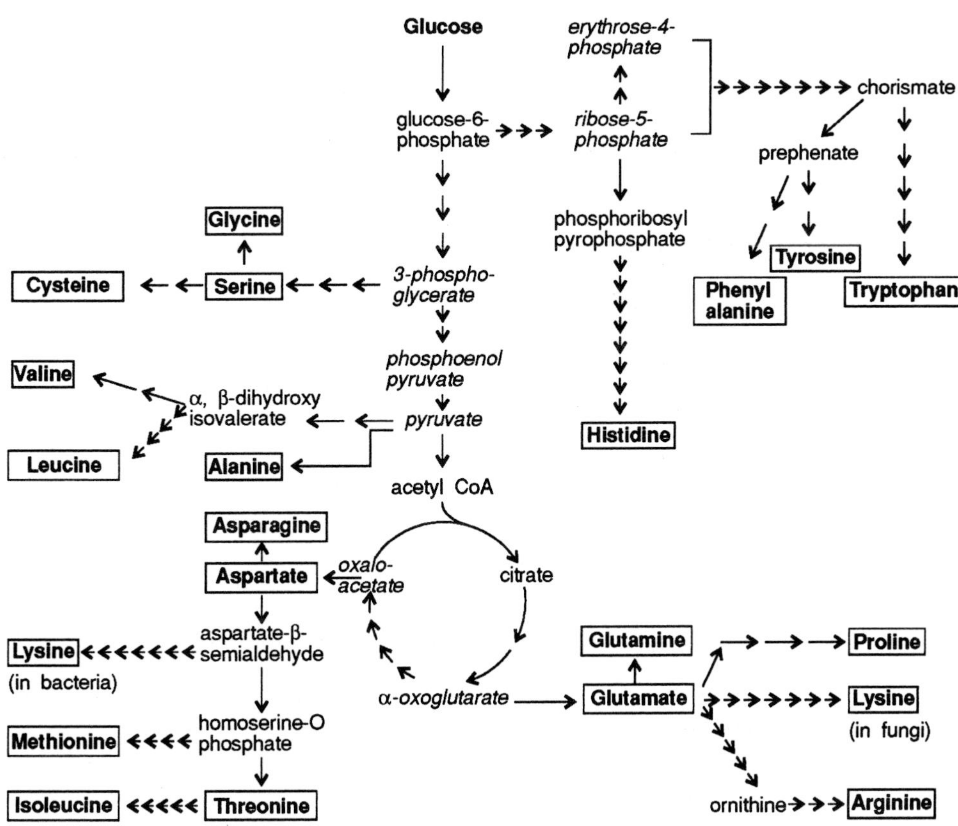

4.8

ribose-5-phosphate + ATP ⟶ PRPP + AMP

ribose-5-phosphate PRPP

The enzyme is PRPP synthetase, sometimes called ribose phosphate pyrophosphokinase. Ribose-5-phosphate is produced in the pentose phosphate pathway.

4.9 1) B, Glycine; 2) A, Formyl THF; 3) C, Glutamine; 4) C, Glutamine; 5) A, Formyl THF; 6) D, Aspartate; 7) E, CO_2

4.10

1) True. One N atom forms atom number 1 in the purine and a second forms the 6-substitution which converts IMP to AMP.
2) False. Two N atoms are donated from glutamine to form IMP but a third one is used to convert IMP to GMP.
3) True. One N atom is incorporated from glutamine to produce UMP and a second to convert this from UTP to CTP. In bacteria the second one comes from ammonium ions.

4.11 ATP is adenosine triphosphate.

PRPP is phosphoribosyl pyrophosphate.

IMP is inosine monophosphate.

THF is tetrahydrofolic acid.

TTP is thymidine triphosphate.

TPP is thiamine pyrophosphate.

dUTP is deoxyuridine triphosphate.

CDP is cytidine diphosphate.

GDP is guanosine diphosphate.

AICAR is aminoimidazole-carboxamide ribonucleotide.

Reponses to Chapter 5 SAQs

5.1

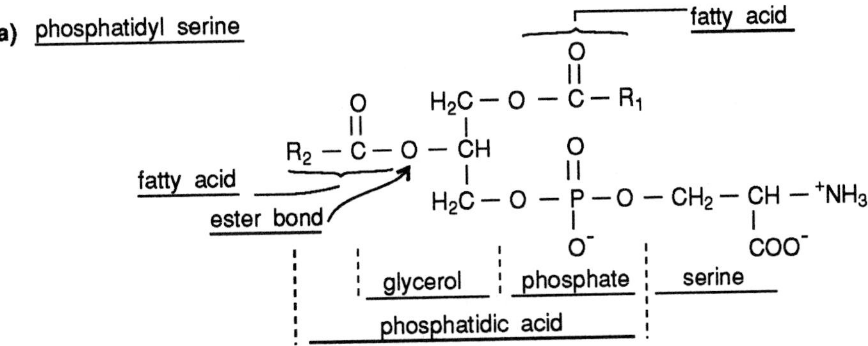

b) an archaebacterial phospholipid

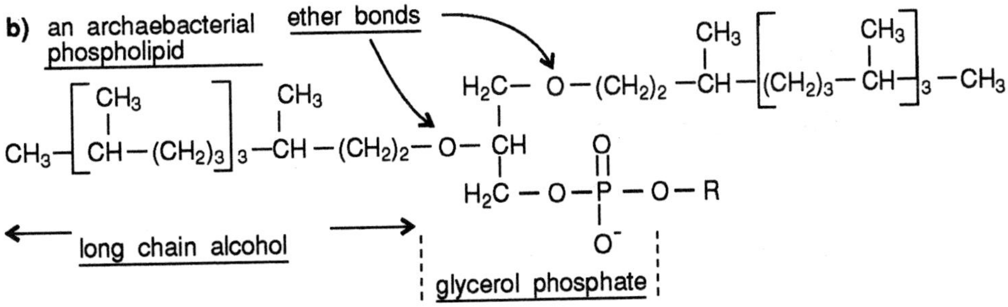

ether bonds

$CH_3-[CH(CH_3)-(CH_2)_3]_3-CH(CH_3)-(CH_2)_2-O-CH$

$H_2C-O-(CH_2)_2-CH(CH_3)-[(CH_2)_3-CH(CH_3)]_3-CH_3$

$H_2C-O-P(=O)(O^-)-O-R$

← long chain alcohol →

glycerol phosphate

c) choline plasmalogen

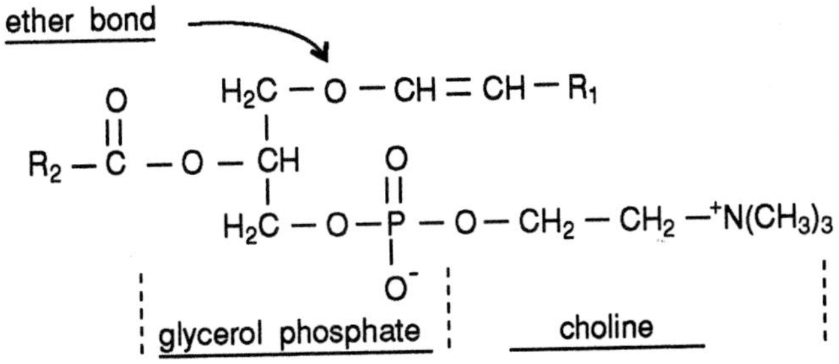

ether bond

$H_2C-O-CH=CH-R_1$

$R_2-C(=O)-O-CH$

$H_2C-O-P(=O)(O^-)-O-CH_2-CH_2-{}^+N(CH_3)_3$

glycerol phosphate

choline

5.2

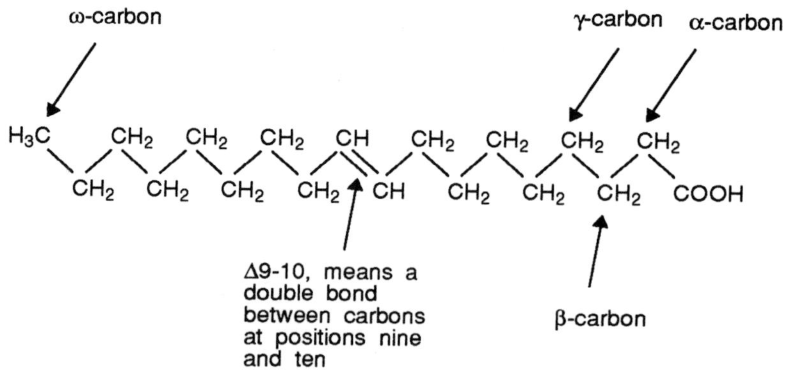

ω-carbon γ-carbon α-carbon

$$H_3C-CH_2-CH_2-CH_2-CH_2-CH_2-CH_2-CH_2-CH=CH-CH_2-CH_2-CH_2-CH_2-CH_2-CH_2-CH_2-COOH$$

Δ9-10, means a double bond between carbons at positions nine and ten

β-carbon

Explanation: The ω carbon is the terminal carbon, that is the carbon at the opposite end to the carboxyl group, irrespective of chain length of the fatty acid. The α carbon is carbon two and not carbon one. This is important and must be remembered. Similarly the β carbon is carbon three, the γ carbon is carbon four and so on. Numbering begins at the carboxyl end. In a molecule such as this convention dictates that numbering starts at the end with a functional group. Finally note that carbons nine and ten have only one hydrogen each.

5.3

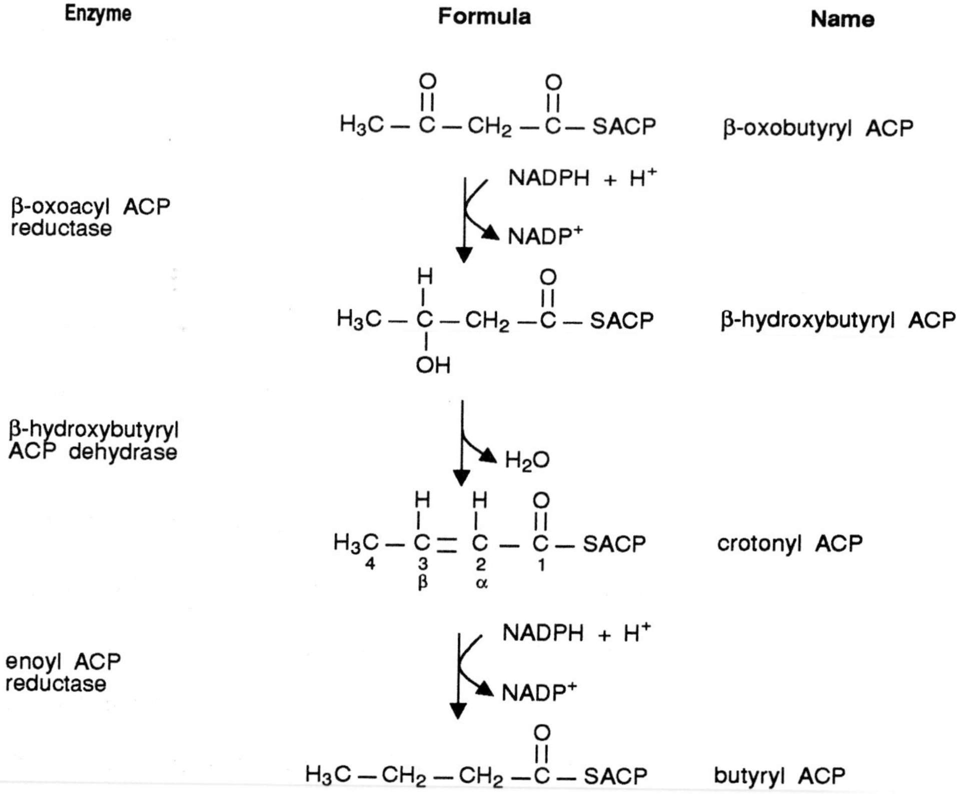

Enzyme	Formula	Name
	$H_3C-C(=O)-CH_2-C(=O)-SACP$	β-oxobutyryl ACP
β-oxoacyl ACP reductase	↓ $NADPH + H^+$ → $NADP^+$	
	$H_3C-CH(OH)-CH_2-C(=O)-SACP$	β-hydroxybutyryl ACP
β-hydroxybutyryl ACP dehydrase	↓ → H_2O	
	$H_3C-CH=CH-C(=O)-SACP$ (carbons 4, 3 (β), 2 (α), 1)	crotonyl ACP
enoyl ACP reductase	↓ $NADPH + H^+$ → $NADP^+$	
	$H_3C-CH_2-CH_2-C(=O)-SACP$	butyryl ACP

The name of the third compound, crotonyl ACP may be new to you. Its chemical name, 2-3-enylbutyryl ACP would be acceptable.

You may have called the first and third enzymes dehydrogenases. The fact that you recognised this is excellent but the reactions here are not reversible, thus only the reduction direction occurs.

5.4

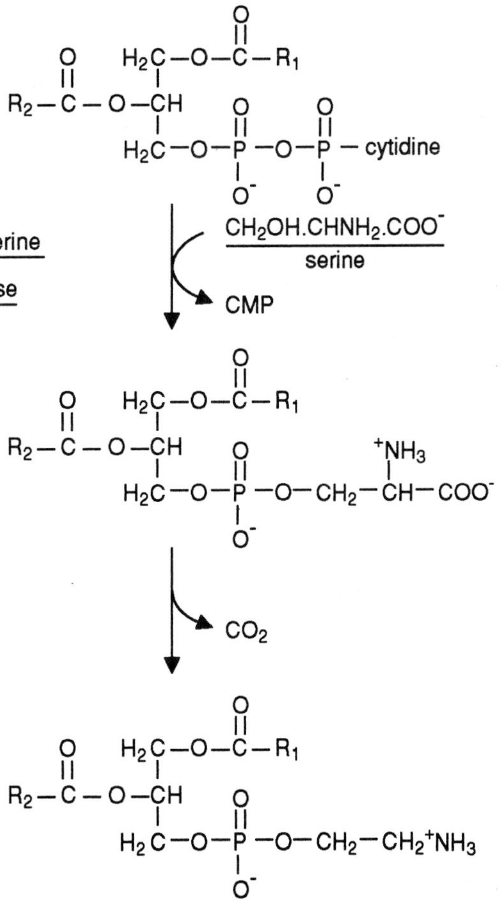

CDP diacyl glycerol:

$$\begin{array}{l} H_2C-O-\overset{O}{\overset{\|}{C}}-R_1 \\ R_2-\overset{O}{\overset{\|}{C}}-O-CH \\ H_2C-O-\overset{O}{\overset{\|}{P}}(O^-)-O-\overset{O}{\overset{\|}{P}}(O^-)-\text{cytidine} \end{array}$$

↓ CDP diacylglycerol: serine phosphatidyl transferase ($CH_2OH.CHNH_2.COO^-$ serine in, CMP out)

phosphatidyl serine:

$$\begin{array}{l} H_2C-O-\overset{O}{\overset{\|}{C}}-R_1 \\ R_2-\overset{O}{\overset{\|}{C}}-O-CH \\ H_2C-O-\overset{O}{\overset{\|}{P}}(O^-)-O-CH_2-CH(^+NH_3)-COO^- \end{array}$$

↓ phosphatidyl serine decarboxylase (CO_2 out)

phosphatidyl ethanolamine:

$$\begin{array}{l} H_2C-O-\overset{O}{\overset{\|}{C}}-R_1 \\ R_2-\overset{O}{\overset{\|}{C}}-O-CH \\ H_2C-O-\overset{O}{\overset{\|}{P}}(O^-)-O-CH_2-CH_2{}^+NH_3 \end{array}$$

5.5

2 acetyl CoA → acetoacetyl CoA + CoASH

acetoacetyl CoA + acetyl CoA → 3-hydroxy-3-methyl glutaryl CoA + CoASH

3-hydroxy-3-methyl glutaryl CoA + 2NADPH + $2H^+$

→ mevalonic acid + 2 $NADP^+$ + CoASH

mevalonate + ATP → 5-phosphomevalonate + ADP

5-phosphomevalonate + ATP → 5 pyrophosphomevalonate + ADP

5-pyrophosphomevalonate + ATP → 3-phospho-5-pyrophosphomevalonate + ADP

3-phospho-5-pyrophosphomevalonate → isopentenyl pyrophosphate + Pi + CO_2

overall reaction:

3 acetyl CoA + 2NADPH + 2H+ + 3 ATP

→ isopentenyl pyrophosphate + CO_2 + 3 CoASH + 2 NADP+ + 3 ADP

In terms of potential energy loss, we derived a value for each acetyl CoA of 12 ATP produced by the TCA cycle and oxidation of NADPH + H^+ could yield 3 ATP via the electron transport chain and oxidative phosphorylation.

Therefore:

3 acetyls	36ATP
2NADPH + 2H+	6ATP
3ATP used directly	3ATP
	45ATP

Thus one isopentenyl pyrophosphate could be said to be equivalent to 45 ATP and that it is a fairly expensive compound to make.

Responses to Chapter 6 SAQs

6.1

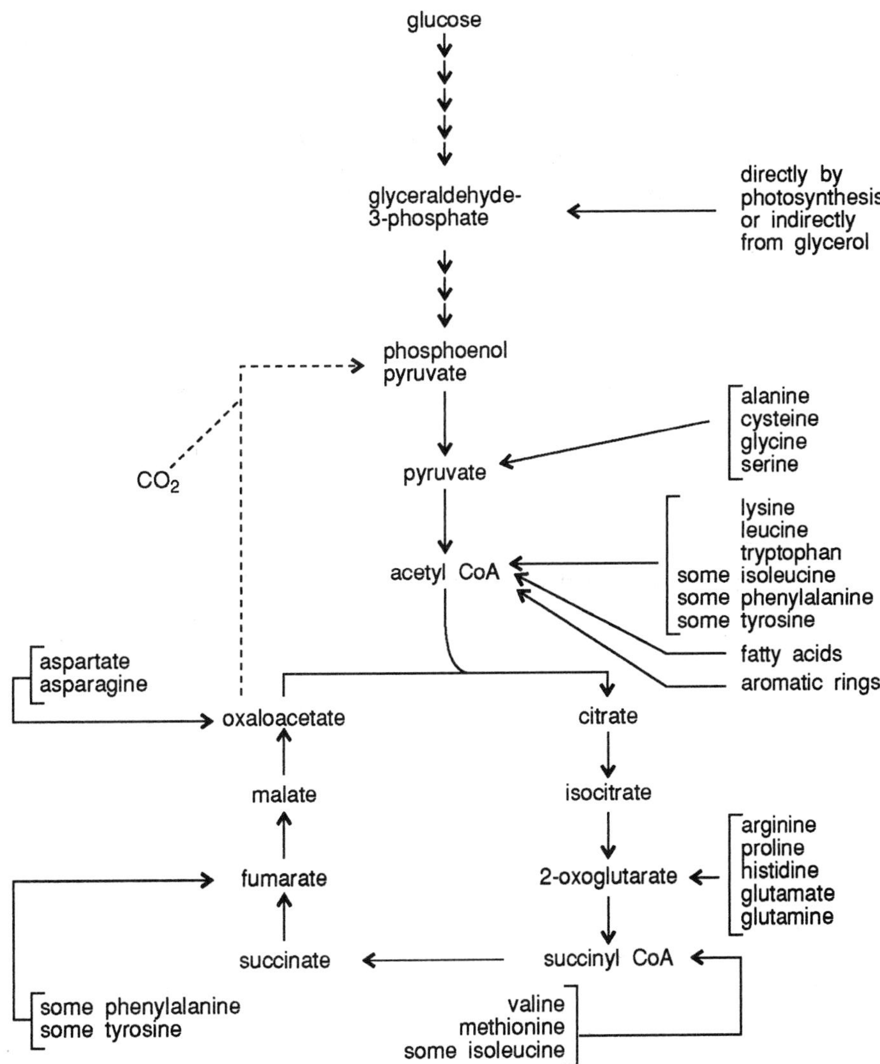

6.2 Any compound converted directly to acetyl CoA cannot be used: all others can.

Explanation: The nature of the actual compound, for example whether it is an amino acid or whether it is a fatty acid, is not the key factor. What is important is whether the compound can be converted back to pyruvate. Although we have not yet studied it in detail we have stated that cells can carry out the process of gluconeogenesis, thus glycerol and amino acids which are converted to pyruvate are useable. The diagram shows that oxaloacetate can be converted to pyruvate via phosphoenol pyruvate:

therefore any compound which is catabolised ultimately to a TCA cycle intermediate is a potential source of glucose. The only remaining compounds are those which are converted to acetyl CoA. Higher animals can use acetyl CoA for energy production but cannot convert it to pyruvate.

6.3

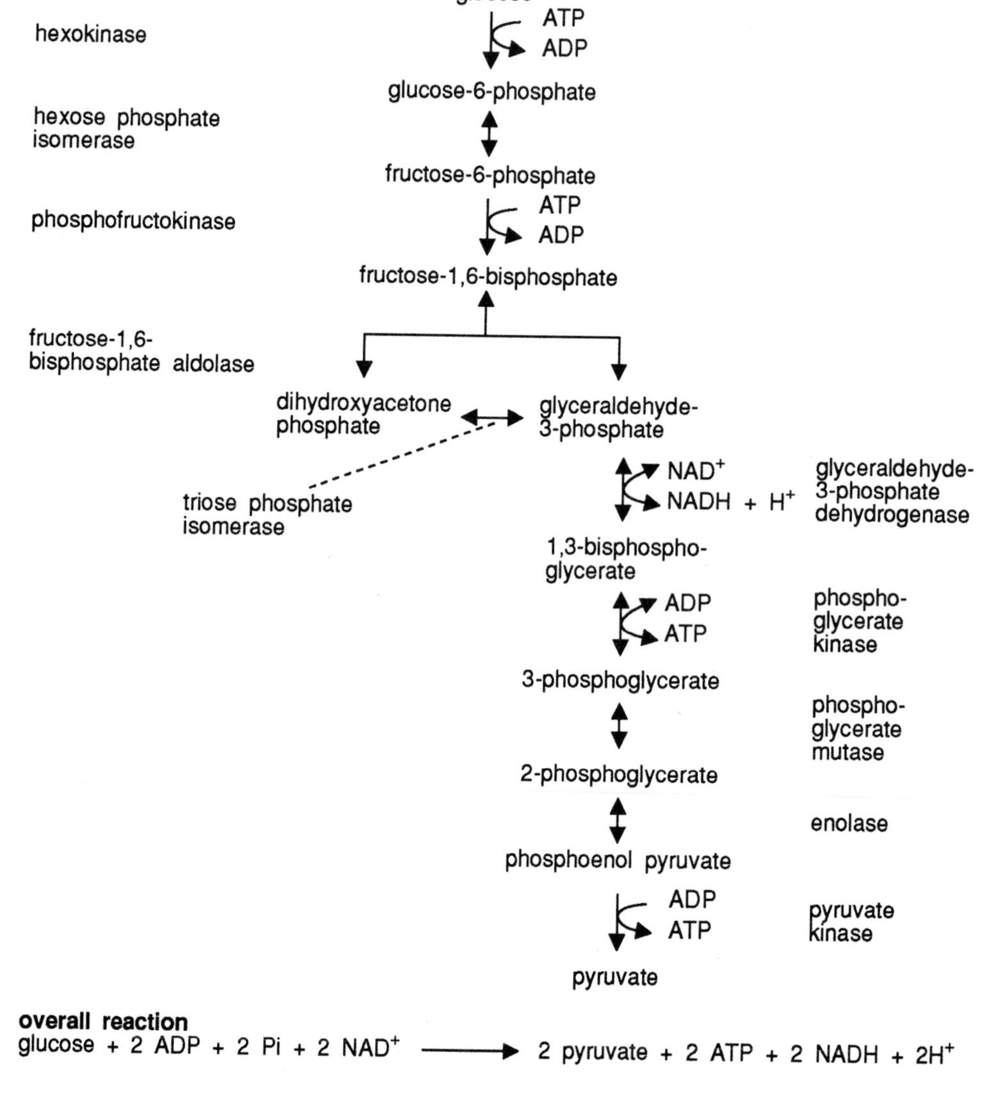

6.4 The Embden Meyerhof pathway would be favoured by 2), 4), and 5). Gluconeogenesis would tend to occur when conditions are as in 1) and 3).

6.5 The correct sequence is: 3), 7), 5), 6), 2), 1), 8) and 4).

6.6

1) True.

2) False; the roles are opposite to that indicated.

3) True

4) False; this enzymes features in the Embden Meyerhof pathway (glycolysis) not in that of gluconeogenesis.

5) True.

Responses to chapter 7 SAQs

7.1

1) True.

2) True.

3) False. The rate of catabolism is controlled by the cell's second-to-second needs for energy in the form of ATP.

4) False. Bacteria carry sufficient ATP for only a few seconds work. However, some bacteria do store energy in the form of fats or carbohydrates and these can be utilised when nutrients are not available externally.

7.2

5-adenosylmethionine	methyl.
guanosine triphosphate	nucleotide
biotin	CO_2
NADPH	electrons
coenzyme-A	acyl
tetrahydrofolate	one-carbon unit
pyruvate does not carry a group in activated form	none

7.3

1) Dihydroxyacetone phosphate formed during glycolysis gives rise to the glycerol backbone of phosphoglycerides.

2) Phosphoenol pyruvate, a glycolytic intermediate, provides part of the carbon skeleton of the aromatic amino acids.

3) Acetyl-CoA, the common intermediate in the breakdown of most growth substrates, carries a two-carbon unit for use in a wide variety of biosyntheses.

4) Succinyl-CoA, formed in the TCA cycle, is one of the precursors of porphyrins.

5) Ribose-5-phosphate, which is formed by the pentose phosphate pathway, is the source of the sugar unit of nucleotides.

6) Tetrahydrofolate is a source of one-carbon units for biosynthesis.

7.4

1) a) High citrate inhibits PFK so fructose-6-phosphate will be high and fructose-1, 6-diphosphate will be low.

b) High AMP stimulates PFK so fructose-6-phosphate will be low and fructose-1, 6-diphosphate will be high.

7.5

1) 0.667 since energy charge $= \dfrac{\text{ATP} + 0.5\,[\text{ADP}]}{\text{ATP} + \text{ADP} + \text{AMP}}$

$$= \frac{3 + 0.5 \times 2}{3 + 2 + 1}$$

$$= \frac{4}{6} = 0.667$$

2) 0.833 since energy charge $= \dfrac{4 + 0.5 \times 2}{6}$

7.6

1) True.

2) False. The energy charge is related to the relative concentrations of ATP, ADP and AMP, not the total size of the pool.

3) False. See comment made for 2).

4) True.

5) True.

6) True.

7) False. During periods of carbon substrate starvation catabolic pathways are unable to regenerate ATP. This would lead to a reduction in energy charge.

7.7

1) Response curve 3 is appropriate. Citrate inhibits phosphofructokinase, a key regulatory enzyme of glycolysis.

2) Response curve 4 is appropriate. The rate of the biosynthetic pathway is maximal when both the energy charge and the requirement for end-product are high.

3) Low. We can see from the graphs that those curves (4,5,6) representing biosynthetic pathways are all low at an energy charge of 0.5.

4) Under the first set of circumstances, the response curve would be like that shown as curve 4. When the end-product became available in the environment, the pathway would not need to operate as quickly to supply the product, so the response curve would become like that shown in curve 6.

7.8

Route 1) will be stimulated by high energy charge and nitrogen limitation, (also by high pyruvate).

Route 2) will be stimulated by low ribose-5-phosphate and low NADPH.

Route 3) will be stimulated by low amino acids and low energy charge. (Pyruvate is a precursor of many amino acids).

Route 4) will be stimulated by low lipid content and low energy charge. (We could also argue that high amino acids might increase this, especially if their catabolism produced pyruvate).

Route 5) will be stimulated by high pyruvate and high energy charge. (We could argue that high amino acid content would stimulate this as the oxaloacetate would not be needed to produce amino acids).

Responses to chapter 8 SAQs

8.1 a) matches with 4): because RNA polymerase binds to the promoter, not the operator.

b) matches with 2).

c) matches with 3).

d) matches with 4): because the co-repressor inhibits transcription by activating the inactive repressor protein, not by inhibiting its synthesis.

e) matches with 1): induction.

8.2 1) Negative control by induction.

i) Gene transcription enhanced: b), c) and f).

ii) Gene transcription prevented: d).

2) Negative control by repression.

i) Gene transcription enhanced: b) and c).

ii) Gene transcription prevented: d).

3) Positive control of the lac operon.

i) Gene transcription enhanced: a) and c).

ii) Gene transcription prevented: e).

8.3 1) Excess E will inhibit step 3. However, all of C will now go to produce H. Consequently, excess H will build up and inhibit Steps 5 and 1. Inhibition of Step 1 reduces the flow of A into the pathway.

2) C will give rise to D, E, F, G and H. Therefore a large amount of C will lead to an excess of the end-products E and H. These will inhibit steps 1,3 and 5. Therefore synthesis of all compounds from A will be reduced.

3)

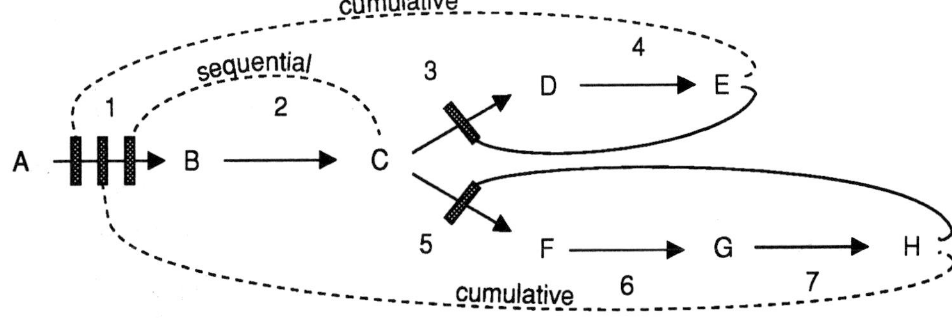

8.4 **Nucleoside diphosphate kinase**

$$\text{MAR} = \frac{[\text{ADP}]\,[\text{GTP}]}{[\text{ATP}]\,[\text{GDP}]} = \frac{(0.97 \times 10^{-3})\ (0.63 \times 10^{-3})}{(2.36 \times 10^{-3})\ (0.21 \times 10^{-3})} = 1.23$$

Since the MAR and K'_{eq} values are so close to each other, we can say that the enzyme is in equilibrium *in vivo.*

Pyruvate kinase

$$\text{MAR} = \frac{[\text{Pyruvate}]\,[\text{ATP}]}{[\text{Phosphoenolpyruvate}]\,[\text{ADP}]} = \frac{(3 \times 10^{-5})\ (3.75 \times 10^{-3})}{(7 \times 10^{-5})\ (1.88 \times 10^{-3})} = 0.85$$

MAR is far lower than K'_{eq}. It follows that this enzyme is not in equilibrium *in vivo* and could constitute a rate-limiting step in a pathway.

8.5 Statements 1) and 4) are consistent with C $\rightarrow$ D being the rate limiting step.

Statement 2) is not consistent because the rate-limiting step (C $\rightarrow$ D) is located between metabolites B and E.

Statement 3) is not consistent because C is located on the other side of the rate-limiting step to E. Addition of ^{14}C-labelled E would therefore lead to the rapid appearance of D but not of C.

8.6

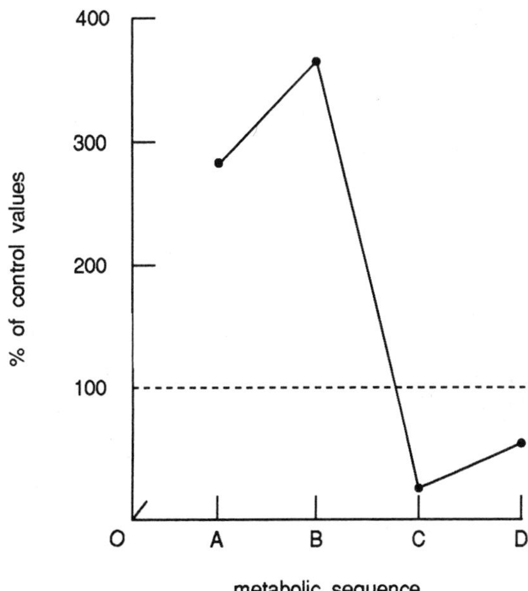

Example of calculation: metabolite A, % control value = (0.551/0.197) x 100 = 279.

The graph shows the reaction B $\rightarrow$ C is the rate-limiting step because the line joining the % control values crosses the 100% line.

8.7

1) False. For a rate-limiting enzyme the MAR is likely to be far lower than the K'_{eq}.

2) False. A rate-limiting reaction is not at equilibrium and the backward rate is lower than the forward rate.

3) False. A pathway in steady-state has a constant net rate with constant levels of metabolites.

4) True.

5) False. The rates of rate-limiting reactions are likely to be overestimated to the greatest extent. The logic here is that rate-limiting reactions will be regulated (activity lowered from optimum) in the cell, probably through allosteric interactions but this may not happen under *in vitro* conditions.

6) False. It is true to say that all irreversible reactions are non-equilibrium reactions. However, only one of these can be rate-limiting for any given set of conditions.

8.8

1) Flux-control coefficient for E_1 = 1.4/2 = 0.7.

Flux-control coefficient for E_2 = 0.2/2 = 0.1.

We can now calculate the flux-control coefficient for E_3 according to the flux-control summation theorem: 1 - (0.7 + 0.1) = 0.2.

2) E_1 is the most rate-limiting because it has the largest flux-control coefficient.

3) 0.7/0.2 = 3.5 ie the same percentage increase in the concentration of E_1 produces at least a 3.5-fold greater increase in pathway flux than that obtained for enzyme E_3. The corresponding value for E_2 is 7.0-fold greater (ie 0.7/0.1 = 7).

4) The competing enzyme would have a negative flux-control coefficient with respect to pathway A $\rightarrow$ B $\rightarrow$ C $\rightarrow$ D ie an increase in its concentration reduces the pathway flux from A through to D.

Appendix 1

Units of measurement

For historical reasons a number of different units of measurement have evolved. The literature reflects these different systems. In the 1960s many international scientific bodies recommended the standardisation of names and symbols and a universally accepted set of units. These units, SI units (Systeme Internationale de Unites) were based on the definition of: metre (m), kilogram (kg); second (s); ampare (A); mole (mol) and candela (cd). Although, in the intervening period, these units have been widely adopted, their adoption has not been universal. This is especially true in the biological sciences.

It is, therefore, necessary to know both the SI units and the older systems and to be able to interconvert between both sets.

The BIOTOL series of texts predominantly uses SI units. However, in areas of activity where their use is not common, other units have been used. Tables 1 and 2 below provides some alternative methods of expressing various physical quantities. Table 3 provides prefixes which are commonly used.

Mass (S1 unit: kg)	Length (S1 unit: m)	Volume (S1 unit: m^3)	Energy (S1 unit: J = kg m^2 s^{-2})
g = 10^{-3} kg	cm = 10^{-2} m	l = dm^3 = 10^{-3} m^3	cal = 4.184 J
mg = 10^{-3} g = 10^{-6} kg	Å = 10^{-10} m	dl = 100 ml = 100 cm^3	erg = 10^{-7} J
μg = 10^{-6} g = 10^{-9} kg	nm = 10^{-9} m = 10Å	ml = cm^3 = 10^{-6} m^3	eV = 1.602 x 10^{-19} J
	pm = 10^{-12} m = 10^{-2} Å	μl = 10^{-3} cm^3	

Table 1 Units for physical quantities

Concentration (SI units: mol m^{-3})

a) M = mol l^{-1} = mol dm^{-3} = 10^3 mol m^{-3}

b) mg1^{-1} = µg cm^{-3} = ppm = 10^{-3} g dm^{-3}

c) µg g^{-1} = ppm = 10^{-6} g g^{-1}

d) ng cm^{-3} = 10^{-6} g dm^{-3}

e) ng dm^{-3} = pg cm^{-3}

f) pg g^{-1} = ppb = 10^{-12} g g^{-1}

g) mg% = 10^{-2} g dm^{-3}

h) µg% = 10^{-5} g dm^{-3}

Table 2 Units for concentration

Fraction	Prefix	Symbol	Multiple	Prefix	Symbol
10^{-1}	deci	d	10	deka	da
10^{-2}	centi	c	10^2	hecto	h
10^{-3}	milli	m	10^3	kilo	k
10^{-6}	micro	µ	10^6	mega	M
10^{-9}	nano	n	10^9	giga	G
10^{-12}	pico	p	10^{12}	tera	T
10^{-15}	femto	f	10^{15}	peta	P
10^{-18}	atto	a	10^{18}	exa	E

Table 3 Prefixes for S1 units

Appendix 2

Chemical Nomenclature

Chemical nomenclature is quite a difficult issue especially in dealing with the complex chemicals of biological systems. To rigidly adhere to a strict systematic naming of compounds such as that of the International Union of Pure and Applied Chemistry (IUPAC) would lead to a cumbersome and overly complex text. BIOTOL has adopted a pragmatic approach by predominantly using the names or acronyms of chemicals most widely used in biologically-based activities. It is recognised however that there remains some potential for confusion amongst readers of different background. For example the simple structure CH_3COOH can be described as ethanoic acid or acetic acid depending on the environment or industry in which the compound is produced or used. To reduce such confusion, the BIOTOL series makes every effort to provide synonyms for compounds when they are first mentioned and to provide chemical structures where clarity and context demand.

Appendix 3

Abbreviations used for the common amino acids

Amino acid	Three-letter abbreviation	One-letter symbol
Alanine	Ala	A
Arginine	Arg	R
Asparagine	Asn	N
Aspartic acid	Asp	D
Asparagine or aspartic acid	Asx	B
Cysteine	Cys	C
Glutamine	Gln	Q
Glutamic acid	Glu	E
Glutamine or glutamic acid	Glx	Z
Glycine	Gly	G
Histidine	His	H
Isoleucine	Ile	I
Leucine	Leu	L
Lsyine	Lys	K
Methionine	Met	M
Phenylalanine	Phe	F
Proline	Pro	P
Serine	Ser	S
Threonine	Thr	T
Tryptophan	Trp	W
Tyrosine	Tyr	Y
Valine	Val	V

Index

A

acetoacetyl ACP, 119
acetohydroxy acid reducto-isomerase, 73
acetohydroxy acid synthetase, 73
acetolactate, 73
acetyl CoA, 118
acetyl CoA carboxylase, 118
acetyl glutamine synthetase, 88
acetyl muramic acid, 158
acetyl transacylase, 119
acetylene
    reduction, 55
acetylglucosamine, 146
aconitase, 195
active transport, 23 , 24 , 27
acyl carrier protein (ACP), 115
adenosine 5′-phosphosulphate, 44
adenylate charge, 154
adenylate energy charge, 178
adenylate kinase, 105 , 181
ADP
    as an allosteric modulator, 181
alanine
    biosynthesis of, 73
    regulation of, 74
    synthesis of, 74
alanine dehydrogenase, 72
Alcaligenes faecalis, 169
alginates, 169
alkaloids, 6
allosteric activation, 179 , 203
allosteric activators, 201
allosteric enzymes, 192 , 199 , 210
    R form, 199
    See also regulation of enzyme activity
    T form, 199
allosteric inhibition, 179
allosteric inhibitors, 202
allosteric interaction, 179
allosteric modulators, 219
allosteric regulation, 204
amide
    formation of, 71
amino acids, 43 , 66
    biosynthetic families of, 70
    essential, 69
    precursors of, 70
    structures of, 67
aminoadipic acid, 89
aminoadipic acid semialdehyde
dehydrogenase, 89
aminoimidazole ribonucleotide, 100
aminoimidazole-4-carboxamide ribonucleotide
    (AICAR), 100
aminotransferase, 76 , 81 , 89
ammonia
    assimilation, 46
ammonification, 62
ammonium salts, 43
AMP
    as an allosteric modulator, 181
amphibolic pathways, 176 , 183
    control of, 184
amylase, 33
amylopectin, 161
amylose, 161
anabolic pathways, 172
anaplerotic reaction, 152
antiport, 25 , 28
antiport transport, 25
arginine
    biosynthesis, 85
aromatic amino acids
    biosynthesis of, 90
Arthrobacter, 141
asparagine, 46
asparagine synthetase, 46
aspartate, 78
aspartate family, 78
    biosynthesis of, 78
aspartate semialdehyde dehydrogenase, 80
aspartate transcarbamylase (ATCase)
    composition of, 203
    regulation of, 202
aspartate-β-semialdehyde, 78 , 80 , 81
aspartokinase, 80
aspartyl phosphate, 80
assimilatory nitrate reduction, 50
ATP
    as an allosteric modulator, 181
ATP-generating sequences, 182 , 185
ATP-utilising sequences, 182 , 185
Auroebasidium pullulan, 169
Azospirillum, 51
Azotobacter, 51
Azotobacter vinelandii, 88 , 169

B

Bacillus, 80 , 127
Bacillus spp., 124
bacitracin, 167
bactoprenol, 139 , 165
binding proteins
    exploration of, 31

biosynthesis, 66
biosynthetic pathways, 5
    regulation of, 210 , 211
biotin, 119
branch-point enzyme, 211
branched pathways
    regulation of, 211
branching enzyme, 161

C

calcium, 12
Calvin cycle, 4
cAMP
    and metabolic regulation, 209
carbamyl phosphate, 103
carbamyl phosphate synthetase
    regulation, 103
carbohydrates, 144
    biosynthesis of, 144
    See also individual carbohydrates
    occurrence of, 144
carbon, 12
cardiolipin, 131
    structural formula, 133
cardiolipin synthetase, 132
carnitine
    as a carrier, 196
carotenoids, 139
carrier molecule, 161
carrier systems, 68
carrier-mediated transport, 18 , 23
    affinity of, 20
    stereospecificity of, 20
carriers, 21
catabolic activator, 208
catabolic pathways, 172
    regulation of, 214
catabolite repression, 209
CDP diacyl glycerol, 130
cell wall
    biosynthesis of, 164
cellulase, 33
cellulose, 162
cephalosporins, 167
cerebrosides, 112
channel-formers, 37
chelating agent, 32
chemoautotrophs, 3 , 62
chitin, 146
cholesterol, 134
choline, 111
citrate, 188
citrate lyase, 117
citrate synthase, 117
Citrobacter, 51
Clostridium, 51 , 80
Clostridium pasteurianum, 58
Clostridium spp., 146
cobalt, 12
compartmentation, 196
concentration gradients, 194
    measurement of, 36
constitutive enzyme, 205
copper, 12
cortisone, 140
Corynebacterium spp., 124
coupling agents in metabolism, 174
crop rotation, 63
crossover plot, 220
crossover theorem, 219
cumulative end-product inhibition, 212
curdlan, 169
cyanide, 43
cyanobacteria, 51
cyclopentane rings, 113
cystathionase, 82
cystathione, 82
cystathione synthetase, 82
cysteine
    biosynthesis of, 77

D

deoxyribonuclease, 33
deoxyribonucleoside
    biosynthesis of, 105
deoxysugars
    biosynthesis of, 158
dextran gels, 169
dextrans, 161 , 168
diagalactosyl diglycerides, 113
diaminopimelate, 72
diauxic growth, 209
dicaylase to L,L-, ε-diaminopimelate, 81
diet
    supply of amino acids, 68
diffusional carriers, 37
diglyceride, 129
dihydropicolinate, 81
dihydropicolinate reductase, 81
dihydropicolinate synthetase, 81
dihydroxyacetone phosphate, 128
dihydroxyisovalerate, 73
dimethylallyl pyrophosphate, 137
diphosphatidyl glycerol, 111
dipicolinate, 72

disaccharides
biosynthesis of, 159
dissimilatory nitrate reduction, 50

E

E. coli, 35 , 123 , 125 , 158
See also *Escherichia coli*
effectors, 200
electrical conductance, 17
elements
physiological functions of, 12
Embden Meyerhof pathway, 77 , 116 , 144 , 145
control of, 155
end-product repression, 210
endocytosis, 33
energy charge, 154 , 181
calculation of, 182
energy coupling, 177
Enterobacteriaceae, 96
enzyme
degradation, 198
enzyme activity
and chemical modification, 198
regulation of, 196
epimerase, 81
equilibrium constants, 217
ergosterol, 141
erythrocyte ghosts, 35
Escherichia coli, 31 , 69 , 129 , 174 , 204 , 209
ester bond, 110
ethanolamine, 111
ethylenediamine-tetraacetic acid, 32
exoenzymes, 33

F

$FADH_2$, 176
farnesyl pyrophosphates, 139
fatty acid oxidation, 197
fatty acid synthetase complex, 123
fatty acids
biosynthesis, 115
biosynthesis of, 119
biosynthesis of cyclopropane, 127
biosynthesis of unsaturated, 126
branched-chain, 128
degradation, 115
essential, 124
odd numbered, 123
precursor, 116
unsaturated, 124
feedback inhibition, 201 , 203 , 213
ferredoxin, 54
flavodoxin, 55
flux-control coefficient, 221
flux-control summation theorem, 222
formamidoimidazole-4-carboxamide, 100
formate dehydrogenase, 53
formyl glycinamide ribonucleotide, 100
formyl glycinamidine ribonucleotide, 100
Frankia, 51
fructose, 157
fructose bisphosphate, 151
fuelling reactions, 2
futile metabolic cycles, 156 , 178

G

galactose, 157
gated channels, 21
gene replication, 8
geological cycling of elements, 42
geranyl pyrophosphate, 137
glucans, 161
See also specific examples
gluconeogenesis, 144 , 150 , 153
control of, 155
location of, 152
glucose
biosynthesis of, 146
from oxaloacetate, 148
from pyruvate, 148
glucose-1-phosphate, 144
glucose-6-phosphatase, 152
glucose-6-phosphate, 144
glutamate, 46
glutamate dehydrogenase, 46 , 71 , 85
glutamate family
biosynthesis of, 85
glutamate kinase, 86
glutamine, 46 , 72
biosynthesis, 86
glutamine synthase, 47
glutamine synthetase, 46 , 47
glutamine-oxoglutarate amino transferase, 47
glutamyl phosphate reductase, 86
glycan, 169
glycerol, 110
glycerol kinase, 128
glycerol phosphate phosphatidyl transferase, 132
glycinamide ribonucleotide, 100
glycine
biosynthesis, 77
glycogen, 145
glycogen phosphorylation, 198
glycolipids, 112

glycolysis, 144
  inhibition of, 188
  regulation of, 186
glycoside bond, 159
GOGAT
  See also glut-oxoglut amino tranferase
Golgi apparatus, 8
gramimycin, 37
green bacteria, 51
group translocation, 28 , 29
growth factors, 14

H

heterocysts, 59
heteropolysaccharides, 144
heterotrophs, 3 , 146
hexose bisphosphate, 151
hexoses
  biosynthesis of, 157
high ammonia route, 47
histidine
  biosynthesis, 91
homeostatic effect, 192
homoaconitase, 89
homocitrate synthase, 89
homocysteine, 82
homocysteine methyl transferase, 82
homoisocitrate dehydrogenase, 89
homopolysaccharides, 144
homoserine kinase, 82
homoserine succinyl transferase, 82
homoserine-O-phosphate, 78 , 82
hopanoid, 139
hormones, 6
hydrocarbons
  aliphatic, 4
  aromatic, 4
hydrogen, 12
hydrogenase, 53
hydroxy-3-methyl glutaryl CoA reductase, 135
hydroxy-3-methylglutaryl CoA, 135
hydroxy-b-carboxyisocaproate, 75
hydroxylamine, 43
hydroxymercuribenzoate, 203
hypoxanthine xanthine phosphoribosyl, 96

I

inducible enzymes, 205
induction, 205
inosine monophosphate (IMP), 100
intracellular compartmentation, 194
ionophores, 37
iron, 12
iron transport, 32
ironophores, 37
isoenzymes, 213
isoleucine
  biosynthesis of, 73 , 83
isopentenyl pyrophosphate, 134 , 135
isoprenoid units, 113
isopropyl malate isomerase, 75
isopropyl malate synthetase, 75
isopropylmalate, 75
isotopic labelling, 218

K

Klebsiella, 51
Klebsiella pneumonide, 58
K_M, 19 , 20
K_M and affinity, 194

L

lac operon, 208
lac promoter, 209
leghaemoglobin, 58
leucine, 75
  biosynthesis of, 73
  regulation of, 74
  synthesis of, 74
Leuconostoc dextranicum, 161
Leuconostoc mesenteroides, 69 , 168
linoleic acid, 124
linolenic acid, 124
lipid vesicles, 16
lipids, 110
  classes, 110
  from isoprene derivatives, 133
  function, 114
  industrially important, 110
  occurence, 114
  See also specific lipids
lipopolysaccharides, 112
liposomes, 16
low ammonia route, 47
lysine, 81
  biosynthesis, 85
  biosynthesis of, 81
lysophosphatidate, 129

M

magnesium, 12
malate dehydrogenase, 117
malonyl CoA, 118

malonyl transacylase, 119
maltose, 160
manganese, 12
mannose, 157
mass-action ratio (MAR), 217
membrane preparations, 35
membrane transport
    inhibitors of, 36
membranes, 8
    permeability barriers, 15
mercaptoethanol, 203
mesodiaminopimelate, 81
mesodiaminopimelate decarboxylase, 81
metabolic homeostasis, 177 , 182
metabolic integration, 172
metabolic links
    amino acids and bases, 106
metabolic pathway flux
    control of, 192
metabolic pathways
    regulation of, 210
metabolic regulation, 172
metal-substrate complex, 195
methionine, 82
mevalonate
    biosynthesis, 134
mevalonic acid, 134
Michaelis Menten constant, 19 , 192
    See also K_M
Michaelis Menten kinetics, 19 , 21 , 192 , 199
mitochondria, 116
mixed function oxidase, 124
modulators, 200
molybdenum, 12
mono-oxygenase, 124
monogolactosyl diglycerides, 113
mRNA synthesis, 205
muscle excitation, 26
mutants
    and transport, 36
Mycobacterium, 51
Mycobacterium spp., 124
mycolipenic acid, 128
Mycoplasmas, 134 , 158

N

NAD^+
    as an allosteric modulator, 181
NADH, 54 , 176
    as an allosteric modulator, 181
$NADP^+$
    as an allosteric modulator, 181
NADPH, 49 , 54 , 176
    as an allosteric modulator, 181
nervous system, 26
neutral lipids, 110
nitrate, 43
    assimilation of, 48
nitrate reductase, 49
nitrite, 43
nitrogen, 12
    assimilation of, 71
    requirement, 42
nitrogen (N_2)
    assimilation of, 50
nitrogen assimilation, 42
nitrogen cycle, 61
nitrogen fixation
    electron carriers, 54
    importance of, 61
    reductants, 53
    requirements of, 52
    unifying concept of, 52
nitrogen-fixing organisms, 51
nitrogenase, 53 , 55
    energy requirement, 56
    measurement of activity, 56
    oxygen protection of, 58
    regulation of, 58
    specificity, 56
nitrogenase reductase, 55
Nocardia restrictus, 141
nucleic acid metabolite
    recycling, 94
nucleoside diphosphate, 105
nucleotide, 66

O

'O' polysaccharides, 113 , 146
oleic acid, 114
oligomycin, 36
oligosaccharides
    biosynthesis of, 159
open channels, 22
operator, 206
operon, 206
optimisation of a pathway, 220
oxaloacetate, 78
oxidation state of an atom, 44
oxidative phosphorylation, 3 , 183
oxoacyl ACP synthase, 119
oxobutyrate, 74
oxobutyryl ACP, 119
oxobutyryl ACP synthase, 119
oxoisocaproate, 75

oxoisovalerate, 75
oxygen, 12
oxygen protection, 58

P

pacemaker enzyme, 216
palmitate, 118
pantothenic acid, 72
para-aminobenzoate, 73
para-hydroxybenzoate, 73
passive diffusion, 18 , 22
passive transport, 23 , 24
passive transport systems, 25
penicillins, 167
pentose phosphate pathway, 116 , 118 , 145
pentoses
    biosynthesis of, 157
peptidase, 33
peptone water, 43
phagocytosis, 34
phenylalanine, 214
    biosynthesis, 90
phosphatidate, 129 , 130
    See also phosphatidic acid
phosphatidate cytidyl transferase, 130
phosphatidic acid, 111 , 129
phosphatidyl choline, 16
phosphatidyl ethanolamine, 130
phosphatidyl glycerophosphatase, 132
phosphatidyl serine, 130
phosphoenol pyruvate carboxykinase, 151
phosphofructokinase, 152 , 188
phosphoglycerate, 76
phosphoglycerate dehydrogenase, 76
phospholipids, 111
    archaebacterial, 111
    biosynthesis of, 129
    examples of, 112
phosphomonoesterase, 96
phosphopyruvate, 76
phosphoribosyl pyrophosphate, 91
phosphoribosyl pyrophosphate (PRPP), 98
phosphoroclastic reaction, 53
phosphorus, 12
phosphorylation
    oxidative, 3
    substrate level, 3
phosphoserine, 76
phosphoserine phosphatase, 76
photoautotrophs, 3
photophosphorylation, 3
pigeon liver, 123
pigments, 6
pinocytosis, 33
piperideine-2,6-dicarboxylate, 81
plasma extender, 169
plasmalogens, 111
polyamines, 72
polysaccharides, 144 , 168
    biosynthesis and functional location, 162
    biosynthesis of, 159
    commercial importance, 168
    importance of linkages, 145
porin, 22
    See also open channels
porphyrins, 72
positive operon control, 209
potassium, 12
presqualene pyrophosphate, 139
proline
    biosynthesis, 85 , 86
promoter, 206
protein (CAP), 208
protein channels, 21 , 22
    See also gated channels
    See also open channels
protein synthesis
    energy cost, 179
proton-motive force, 27 , 31
PRPP synthetase
    inhibition of, 98
Psedomonas aeruginosa, 169
pullulan, 169
purine, 43 , 72
    biosynthesis, 94 , 98
purine nucleoside phosphorylase, 96
purple bacteria, 51
putrescine, 88
pyrimidine, 43
    biosynthesis, 94 , 103
pyrroline-5-carboxylate reductase, 86
pyruvate carboxylase, 151
    control of, 155
pyruvate dehydrogenase
    regulation of, 186
pyruvate orthophosphate dikinase, 151

R

rate-limiting reaction, 216
rate-limiting step, 204 , 216
    identification of, 216
reaction rates
    measurement, 218
receptor-bound endocytosis, 33
reductive amination, 71 , 85
regulatory gene, 206

repression, 205
repressor, 208
respiratory protection, 58
Rhizobium, 51
Rhizopus nigricans, 141
ribonucleoside
  phosphorylation, 105
ribose-5-phosphate, 158
ribosomes, 8
ribulose-5-phosphate, 158
RNA polymerase, 205
root nodules, 58

S

saccharopine dehydrogenase, 89
Sephadex, 169
sequential feedback inhibition, 213
sequential induction, 214
serine, 76 , 111 , 130
  biosynthesis of, 76
  phosphorylated pathway, 77
serine acetyl transferase, 77
serine family, 76
serine hydroxymethyl transferase, 77
serine sulphydrase, 77
signal metabolite, 179
signal molecules, 200
spermidine, 88
sphingomyelins, 111
sphingosine, 111
squalene, 134 , 136
Staphylococcus aureus, 162
Staphylococcus aureus peptioglycan
  structure of, 163
starch, 145
steriods
  chemotherapeutic use, 140
steroid hormones, 140
sterols, 114 , 134 , 139
Streptococcus spp., 161
substrate
  gradients, 196
  unavailability, 194 , 195
substrate availability, 192
substrate concentration
  and metabolic rate, 193
substrate level phosphorylation, 3
substrate level phosphorylations, 53
succinyl homoserine, 82
succinyl-α, ε-diaminopimelate, 81
succinyl-ε-oxo-α-aminopimelate, 81
sugar interconversions, 157
sulphate
  assimilation, 43
sulphur, 12
  assimilation, 42
  requirement, 42
symport, 28
synthetase, 72

T

TCA cycle, 7 , 117 , 145 , 156 , 185
teichoic acids, 112 , 167
tetrahydrofolic acid, 77
tetroses
  biosynthesis of, 157
thermoacidophiles, 111
THF, 77
Thiobacillus, 51
threonine, 78
threonine deaminase, 75
thymidine diphosphate (TDP), 157
transamination, 46
transcription, 8 , 208 , 209
  regulation of, 204
translation, 8
transport
  see specific type of transport, 24
transport antibiotics, 37
transport systems
  inhibition of, 20
  shock-insensitive, 31
  shock-sensitive, 31
triglyceride, 110 , 129
triglycerides
  biosynthesis, 128
tryptophan, 214
  biosynthesis, 90
tyrosine, 214
  biosynthesis, 90

U

uniport, 28
uptake of nutrients, 12
urea cycle, 103
uridine diphosphate (UDP), 157

V

vaccines, 168
valine
  biosynthesis of, 73
  regulation of, 74
  synthesis of, 74

valinomycin, 38
vancomycin, 167

X

xanthan gum, 168
Xanthomonos campestris, 168
xylulose-5-phosphate, 158

Y

yeasts, 123

Z

zinc, 12